OXFORD MEDICAL PUBLICAT

A MANUAL OF OPERATIVE DENTISTRY

A MANUAL OF OPERATIVE DENTISTRY

H. M. PICKARD

F.D.S. R.C.S. (Eng.), M.R.C.S. (Eng.),
L.R.C.P. (London)

Emeritus Professor of Conservative Dentistry, University of London; formerly Director of Department of Restorative Dentistry, Royal Dental Hospital of London School of Dental Surgery; formerly Hon. Consultant Dental Surgeon, Royal Dental Hospital of London

FIFTH EDITION

OXFORD
OXFORD UNIVERSITY PRESS
NEW YORK TORONTO
1983

Oxford University Press, Walton Street, Oxford OX2 6DP
London Glasgow New York Toronto
Delhi Bombay Calcutta Madras Karachi
Kuala Lumpur Singapore Hong Kong Tokyo
Nairobi Dar es Salaam Cape Town
Melbourne Auckland
and associate companies in
Beirut Berlin Ibadan Mexico City Nicosia

Published in the United States
by Oxford University Press, New York

First published 1961
Second edition 1966
Third edition 1970
Fourth edition 1976
Fifth edition 1983

British Library Cataloguing in Publication Data

Pickard, H. M.
A manual of operative dentistry.—
5th ed.—(Oxford medical publications)
1. Dentistry, Operative—Handbooks, manuals, etc.
I. Title
617.6'059 RK501

ISBN 0-19-261327-8

Library of Congress Cataloging in Publication Data

Pickard, H. M. (Huia Masters)
A manual of operative dentistry.

(Oxford medical publications)
Bibliography: p.
Includes index.
1. Dentistry, Operative. I. Title. (DNLM:
1. Dentistry, Operative. WU 300 P594m)
RK501.P64 1982 617.6'7 81-22597
ISBN 0-19-261327-8 (pbk.) AACR2

Set by VAP, Kidlington, Oxford.
Printed in Hong Kong

Preface to the fifth edition

The purpose of this book is to describe the performance of the simpler procedures of operative dentistry and to relate this to underlying principles and their application to daily practice. It is not intended that this manual should supplant more detailed textbooks upon the subject, but rather that it should be used as a supplement.

An attempt has been made to describe basic operative procedures in plain language, and, where several satisfactory methods exist, to choose those most generally acceptable. Where repetition has been employed, it is with the intention of emphasis; dogmatic statements have sometimes been used to simplify for the less experienced operator the issues involved. I hope that the experienced practitioner may also find some clinical and technical interest in these pages. The methods described are based upon standard teaching and I must acknowledge my indebtedness first to those who taught me and also to my former colleagues who have given invaluable advice and criticism. In particular I am grateful to Mr Alan Atkinson, Mr Ralph Grundy, Dr Edwina Kidd, and Mr Arthur Stallwood, in addition to those named below, for useful information and material contained in this edition. My thanks are also due to my publishers for their constant help and advice and to my several artists and the Photographic Department of the Royal Dental Hospital and Mrs Glenda Colquhoun for help with the illustrations.

I am indebted to Messrs Siemens-Reiniger-Werke AG for permission to reproduce Fig. 1.6, and to the Kavo Manufacturing Co. of Great Britain and Messrs Cottrell and Co. for the handpieces shown in Fig. 2.7. To Nesor Equipment Co. for Fig. 1.5, to A D International for Fig. 6.9, to DEGUSSA for Fig. 9.18, and to Fairfield Dental Equipment Ltd and Orthomax Dental Ltd for material used in Figs. 7.37 and 7.38 respectively. My thanks are also due to the following who have allowed me to use illustrative material in their possession: Professor D. N. Allen for Fig. 2.13 (a), (b), and (c); to Professors D. S. Shovelton and E. A. Marsland for Fig. 4.7; to Dr I. E. Barnes for Figs. 2.13 (d), 9.7, and 9.8; to Professor A. H. R. Rowe for Fig. 4.8; to Dr G. Roberts for Fig. 8.14; to Mr D. L. Baker and Professor I. Curson for Fig. 10.10, and to the Editor of the *British Dental Journal* for permission to reproduce them here.

Newnham, Northamptonshire
May 1982 H.M.P.

Contents

1
The operator and his environment

The field of operative dentistry is usually held to cover all forms of treatment aimed at the restoration of natural teeth. The principles and methods on which these procedures have developed are common to some aspects of all restorative work in the mouth. There is a recognizable progression of techniques from the filling of teeth, through crown and bridge work, to partial denture practice.

The operative procedures performed on natural teeth described in this book are those which occupy an increasing portion of the dental surgeon's time. In spite of developments in preventive dentistry it seems likely that this preponderance will continue in the forseeable future. It is therefore essential that he should be knowledgeable in the theory underlying these procedures and proficient in their practice.

In the long run, every operator attains his own particular methods of working, and these are of greater or lesser efficiency according to his innate ability, his basic teaching, and the amount of thought and effort applied to the problems which confront him. He must also be aware, and able to take advantage of, the constant improvements in equipment, materials, and methods.

At the centre of the subject of efficient operating are the relative positions of the chair, the patient, the operator, and his operating stool. The relationship of these positions to one another is of fundamental importance to effective and comfortable operating. The dental surgeon may operate standing or sitting, behind or in front of his patient, according to the position of his field of work, the ease of access, the nature of the operation to be performed, and according to his preference. The design of his working area, the amount and position of his equipment, and the availability of expert chairside help will all have direct effect upon the operator's working habits. In normal practice he spends many hours a week at the chairside in work which needs close concentration. It is his duty to adopt any measure which reduces fatigue. To sit for the majority of operating is particularly worth while.

Patient position

Most patients will seat themselves in the dental chair with little regard for their own comfort, or for that of the operator. Both are matters for attention. The patient should sit well back on the seat, and by suitable adjustment of the chair, the lumbodorsal and cervical spine should be supported in a normal position between flexion and extension (see Figs. 1.1 and 1.2).

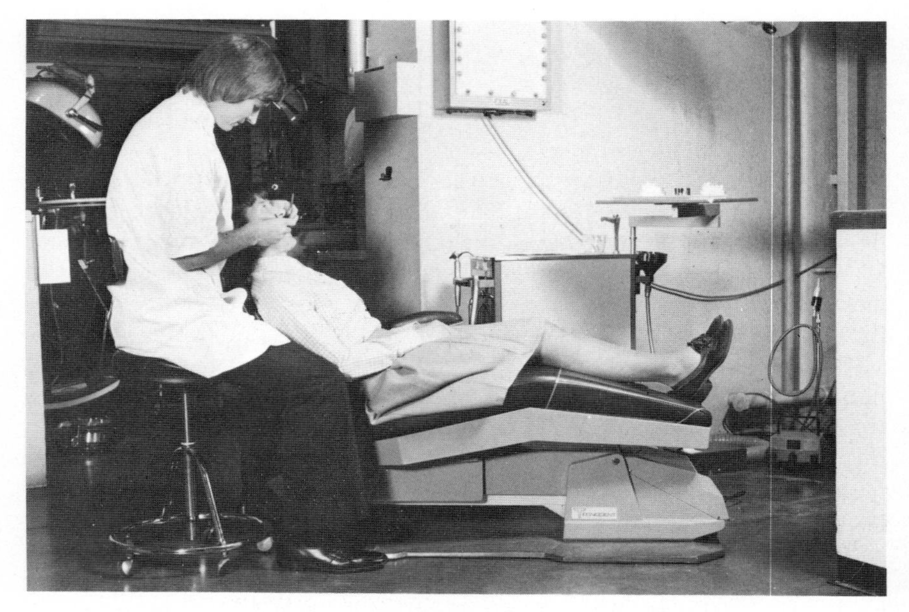

Fig. 1.1. Patient slightly reclined; operator at 12 o'clock position.

If the patient is to be placed in the semi-reclined position, with a warning hand lightly placed on the patient's shoulder the chair should be tipped gently backwards, so that the patient's weight falls into the angle formed by the seat and the back of the chair. In this position most patients experience a comforting sense of support, and relaxation is thereby made easier. By raising the chair, the patient's mouth should be brought to the level of the operator's elbow. Final adjustment of the headrest may be made, extending or flexing the patient's neck to allow adequate access to the field of operation and a clear view for the operator.

It is permissible to ask the patient to assume a position of some discomfort, as of moderate hyperextension of the neck or rotation of the head, *for a short time* while a particular object is achieved. Nevertheless, the positions adopted by both patient and operator for the majority of all operating must allow support and relaxation for the one, and comfort and ease of working for the other.

Operator position

When the operator is standing, it is important that his body be well balanced, with only moderate flexion and rotation of his spine, with his weight evenly distributed on the feet. Similarly, when sitting, his body must adopt a posture which is relatively free from strain. In this case his legs, relieved of the main weight-bearing, assist in balance and mobility.

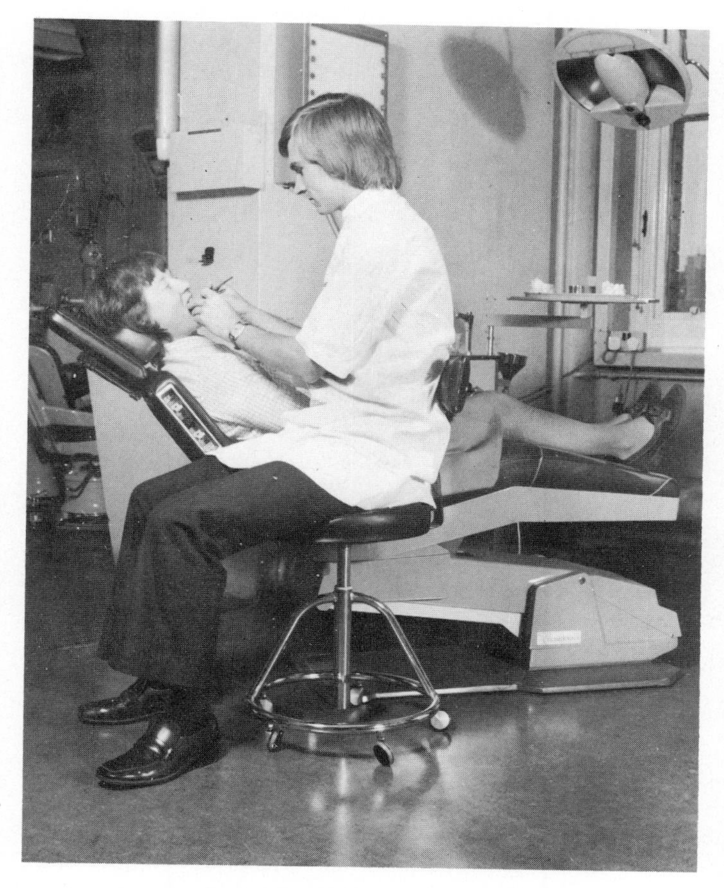

Fig. 1.2. Patient reclined 40 degrees; operator at 9 o'clock position.

It is highly desirable that an operator be equally comfortable when working standing or sitting. The operating stool must be chosen, from many different designs available, to fit in with the design of the chair, with the operator's method, and with the plan of the working area. When a suitable stool is used the patient may be tipped rather further back, looking along a line between 30 and 45 degrees above horizontal. The patient's neck may have to be slightly further extended, or flexed, according to the area of operation, and whether direct or reflected vision is employed.

Figure 1.1 shows a patient in a moderately reclined position, facing about 30 degrees above horizontal, and the operator in the 12 o'clock position, which is used particularly for operations on the maxillary teeth using indirect, that is to say mirror, vision. Figure 1.2 shows the patient further reclined to about 40 degrees, with the operator in the 9 o'clock position. This is a useful position for working by direct vision, particularly on mandibular teeth. There are, of course, circumstances when the operator in this

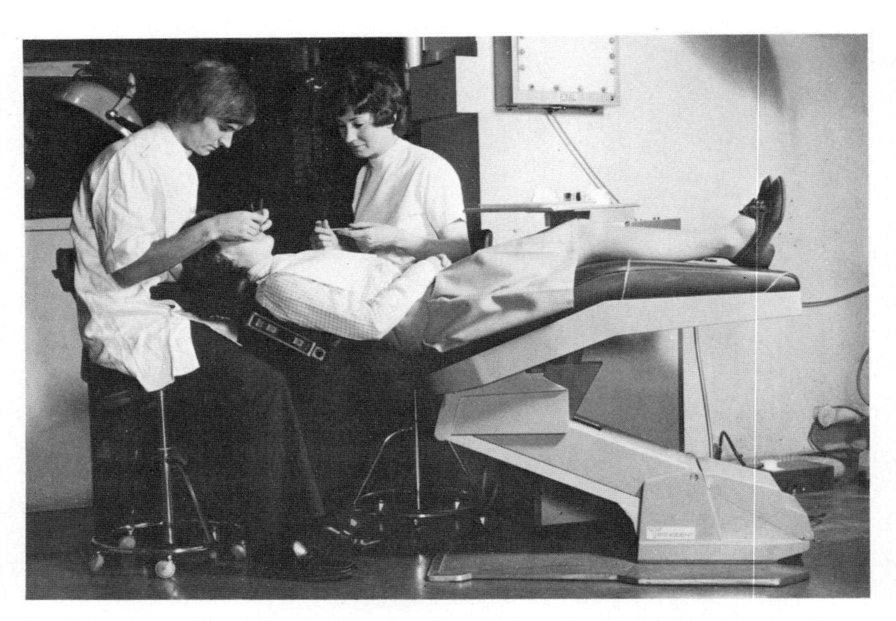

Fig. 1.3. Patient fully reclined; operator at 12 o'clock and assistant at 3 o'clock positions.

position will need his patient practically upright, for example, history-taking, general examination, and recording occlusion.

Figure 1.3 shows the patient fully reclined but still with some flexion at the hips. The operator in the 12 o'clock position can work over a wide area by direct vision and can use the mirror over an equally wide range with little change of body and head position. The assistant at the 3 o'clock position has a good view of the field of operation, which is fully accessible to her.

The plan of the working area, that is to say that area of the surgery immediately around the operator and his patient, is also of primary importance; it is as important as the relative positions of operator and patient, and is closely linked to it. There is a very wide variation in individual solutions to the problems posed by lighting, distribution of equipment and cabinet, and the employment of a surgery assistant. It is, however, possible to bring out the significant points.

Lighting

It is clear that if the operator is to see accurately, the small field in which he operates must be adequately lighted. The level of illumination of the surgery as a whole, whether by day or by artificial light, should not be above that needed for normal purposes. The illumination of the operating field must be higher than that of the surrounding area without being too bright, either to the operator or to the patient. In the northern hemisphere a good north

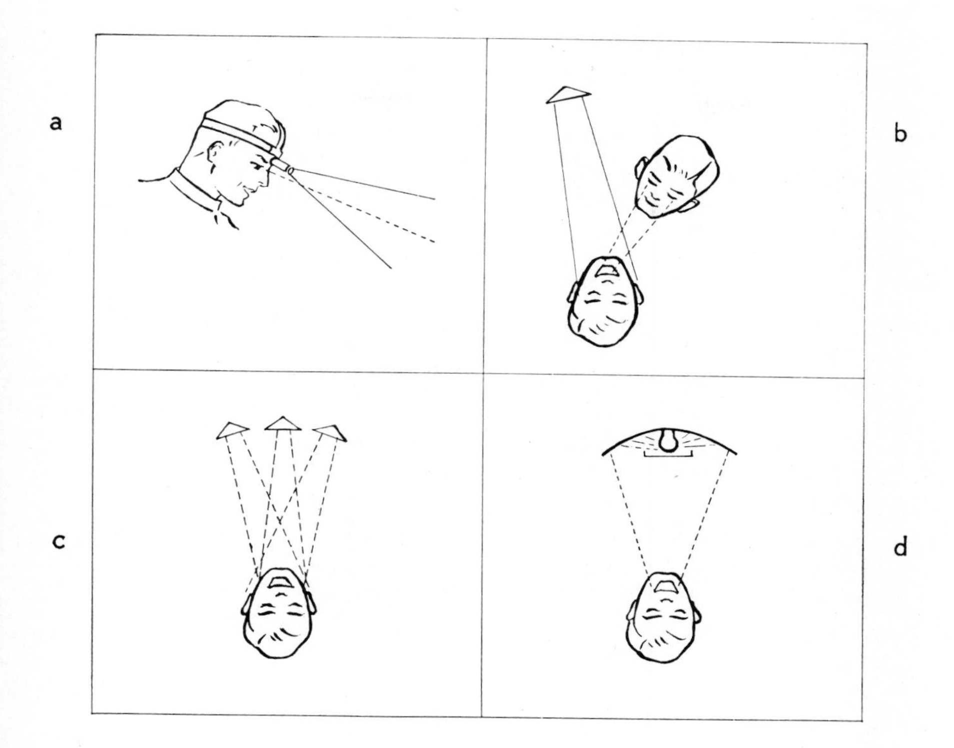

Fig. 1.4. Diagram of (a) light beam co-axial with line of sight; (b) light-beam to left of midline of patient; (c) multiple sources of illumination; (d) principle of the 'shadowless' lamp most common in practice.

window light is a most effective source of daylight, but in many climates and in urban areas much operative work must be done with the assistance, at least, of an operating lamp. The efficiency of the best operating lamps is such that many operators prefer to exclude daylight when working in the mouth.

The lamp must be capable of easy adjustment, and so designed that it illuminates what the operator is looking at and as little else as can conveniently be arranged. This can be achieved either by using a beam which is co-axial with the line of vision, e.g. by using a head-lamp (Fig. 1.4 (a)), or by placing the source so that the beam is not obscured by the operator's head when he is working (Fig. 1.4 (b)).

The head-lamp, which is greatly appreciated by some operators, has never become universally popular because of small practical disadvantages, no one of which is insuperable, but which collectively militate against its continuous use. The fact that the beam is virtually co-axial with the line of vision almost eliminates visible shadows; it also avoids obstruction of the light beam and gives a small field of high illumination which can be conveniently located at the centre of the visual field. It is certainly an adjunct which should be available for the difficult case.

The operating light situated at a distance on the left of the centre line (Fig. 1.4 (b)), is a solution. It is fairly easily accessible for adjustment of position and can be so situated that good illumination may be obtained, though obstruction of the beam may be troublesome.

A further requirement is that the illuminating beam must not be significantly obscured by the operator's hands. This criterion can be approached by the provision of multiple sources of illumination (Fig. 1.4 (c)), or by a single source reflected from circumferentially-placed mirrors, the so-called shadowless lamp (Fig. 1.4 (d)). This arrangement is the one most commonly used in practice.

The quartz-halogen lamp has provided a small but powerful source of light which can be used with a dichroic reflector. This is a metallized glass surface which *transmits* a high proportion of the incident heat but *reflects* a high proportion of the incident light. Thus, a relatively cool beam, providing a very effective operating lamp, results from a high-temperature light source. The area illuminated must be large enough to allow a small range of movement to the patient's head, without necessity for readjustment of the light, but it must not be so large as to dazzle the patient. This means that the lighted area should have a sharp upper margin so that the upper teeth are lighted and the eyes are not.

The illuminated mouth mirror is another example of an instrument which can be invaluable in supplying light of good intensity at the right place, but again it has, like the head-lamp, small disadvantages which make it unsuitable for continuous and exclusive use.

The principle of the fibre-optic has been adapted to give high-level illumination close to the field of work. Outlets of 1.0–2.0 mm diameter are used on handpieces (Fig. 1.5) and mouth mirrors but for reasons of technical design have yet to gain wide acceptance.

Chairside assistance

Given adequate conditions of space and equipment, the careful organization of the work of a surgery assistant and its integration with the operator's procedure is based upon the following considerations. Like the operator, she must be able to maintain a stress-free working position on a mobile stool. She must have ready access to instruments and materials currently in use and to a working top within comfortable reach. She must be able to assist in all intra-oral operations, one of her chief functions being to maintain visibility of the field of operation, for which she needs a three-in-one syringe, an aspirator, and the operating light within reach. In addition, but further away, she needs access to washing and sterilizing facilities.

For these general purposes it is desirable, but not essential, that the assistant shall be able to move with freedom around the whole operating area. A

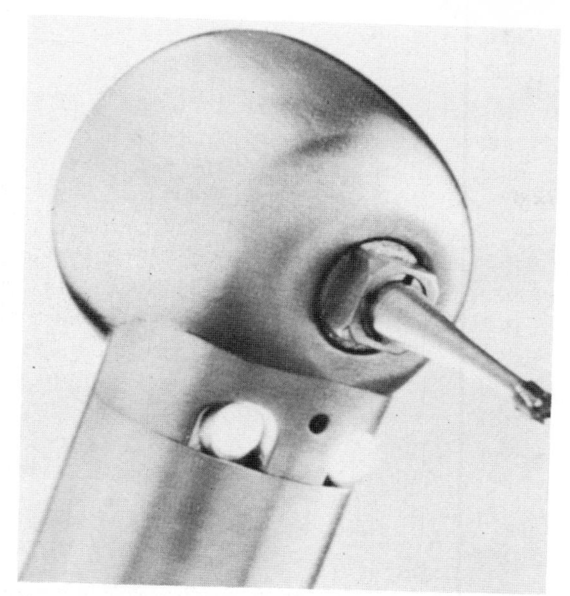

Fig. 1.5 Fibre optic system incorporated into the head of a turbine handpiece.

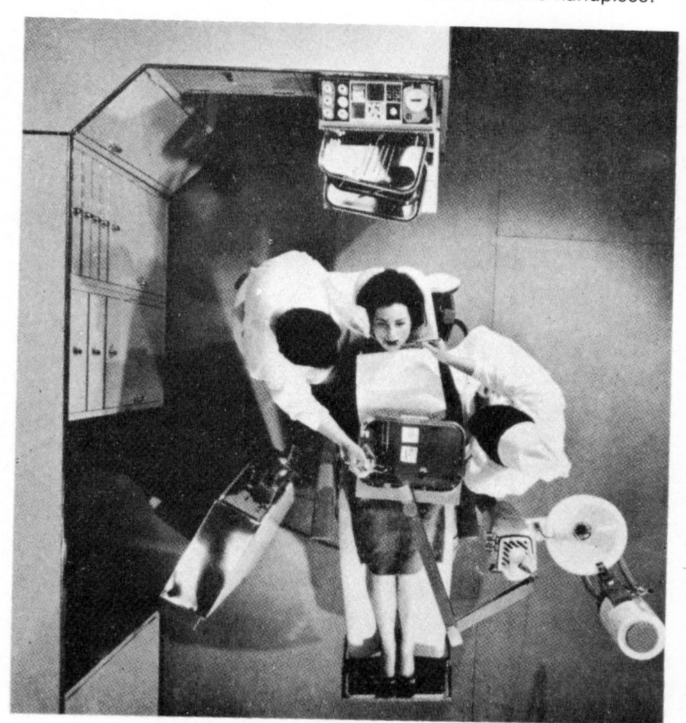

Fig. 1.6. Surgery lay-out showing patient reclining on chair. Operator and assistant are both seated on mobile stools. To the operator's right are air turbine, air motor, warm air, and spray; he has three working tops within easy reach. The assistant has access to the patient's mouth and controls the high-volume evacuator. For the purpose of the photograph, the operating light is omitted.

small trolley or mobile cabinet on the left side of, and half-facing, the patient (Fig. 1.6) is a very practical arrangement for her centre of chairside working. In this position she can see operator and patient, can assist with one or both hands, and is provided with all the commonly used materials and the instruments needed for their preparation and use.

There is a growing tendency to use the services of the dental surgery assistant much more intensively as part of an operating team. In her role as operating assistant located on the left of the patient, she can retract and control the tongue and cheeks, spray and dry the field, pass and receive instruments, prepare materials, and perform many minor but important tasks, closely co-ordinated with the operator. This integrated procedure between operator and assistant is often referred to as 'four-handed operating'. It calls, of course, for careful training and discipline not easily acquired, but leading to greatly increased efficiency.

Equipment

The disposition of equipment is also subject to variations of individual preference. Having regard to the factors of effective use, operator fatigue, and patient comfort, the following would seem reasonable.

1. Equipment used by the dentist should be mounted on the patient's right and should be easily mobile to allow use over a range of operating positions from 3 to 12 o'clock. This will include rotary equipment—air-rotor and air-motor—and a three-in-one syringe as a minimum.
2. Equipment used by the surgery assistant will be on the left of the patient, as detailed above.
3. The patient must of course be able to seat himself and get up again with ease, so his access to the chair must be unobstructed. Equipment should preferably *not* be mounted over the patient within his line of sight, with the exception perhaps of an easily mobile light-weight working tray which, when needed, can be drawn close to the field of operation.

The technique of operating on a supine patient, when the procedure in hand permits, is one which should be acquired (Fig. 1.3). This position is well tolerated by most patients, who rapidly find it restful. It is very important that protective spectacles be worn by the patient who, in this position, is seriously at risk of injury to his eyes from dropped or misdirected instruments and materials, and probably from infected droplets. The position shown allows the operator comfort, with good visibility and access for himself and his assistant. These methods have led to the ergonomic design of equipment and surgery layout. Instrument cabinets and washing facilities are often grouped behind and to the right of the patient so as to be immediately accessible to the dentist.

A solution acceptable to many operators is shown in Fig. 1.6. The operating lamp has been omitted; the air motor, air turbine, spray, and warm air syringe are at the operator's right hand and are on a mobile mounting. The spittoon and aspirator are also mobile and close to the assistant who sits on the left side. There are, however, many alternative arrangements made possible by the variety of equipment designs available. In the past decade considerable progress has been made in ergonomic design, that is to say, design consonant with the general principles stated above. Fig. 1.6 shows an example of their practical application to surgery layout.

Summary

Patient. Comfortably supported, legs, lumbar and dorsal spine. Head and neck slightly extended. Supine position suitable for many cases. Eye protection essential.

Operator. Seated whenever possible; thighs just above horizontal, weight evenly on both feet. Upper arms and spine vertical. Head slightly inclined. Free mobility between 9 and 12.30 positions.

Assistant. Similarly seated, with access to mouth, instruments, materials. Free mobility between 3 and 12 o'clock positions.

Lighting. Shadowless, quartz-halogen light commonest. Head-lamp and lighted mouth mirror very useful for special purposes.

Equipment. Used by operator, on right side. Used by assistant, on left side. Avoid equipment over patient when possible.

2

Hand instruments and rotary instruments

The design of hand instruments

In this context, a hand instrument—more correctly called a hand-held instrument—is the term used to indicate an instrument of light weight, usually constructed from a single piece of steel, used for operations on the teeth or on surrounding tissues, and for the manipulation of filling materials and other therapeutic substances.

Such instruments used in operative dentistry fall under the broad headings of probes, chisels, excavators, scalers, condensers, plastic instruments, and carvers. Their designs are numerous and their purposes multifarious. There are, however, certain general principles underlying their design and these should be understood by the operator.

The majority of hand instruments have three component parts (Fig. 2.1 (a)): the blade (which may be called the 'nib' in non-cutting instruments, or in the case of a probe, the 'tine'), the handle, and, connecting these two, the shank.

A straight chisel (Fig. 2.1 (a)) is a good example of a simple and effective instrument, which is used for cleaving and paring enamel and, to a lesser extent, dentine. It has a strong blade, connected by a rigid shank to a hexagonal handle. Such a straight instrument as this is the most efficient, and easiest in use, of the basic designs shown. Forces applied to the handle are transmitted, *unaltered in magnitude and direction*, to the working surface; for this reason the operator finds such an instrument effective and easy to control.

In order that an instrument shall act effectively, the blade must meet the surface to be worked at an optimum angle. A chisel paring enamel, a scaler removing calculus, a carving instrument contouring an amalgam restoration, or a scalpel cutting gingiva—all these instruments have an optimum angle of incidence of the blade upon the working surface. In some instances this angle is critical, in others it may vary over a wide range, but in each case the operator's *access* to the surface must be such that the instrument can be applied in the particular manner appropriate to the instrument and the operation.

The dental technician, many of whose procedures are mechanically comparable to those performed in the mouth, uses straight instruments and tools exclusively, for in this branch of dentistry access to the point of application is direct and comparatively easy.

There are many areas in the oral cavity where direct access through the aperture of the mouth is impossible with a straight instrument. For

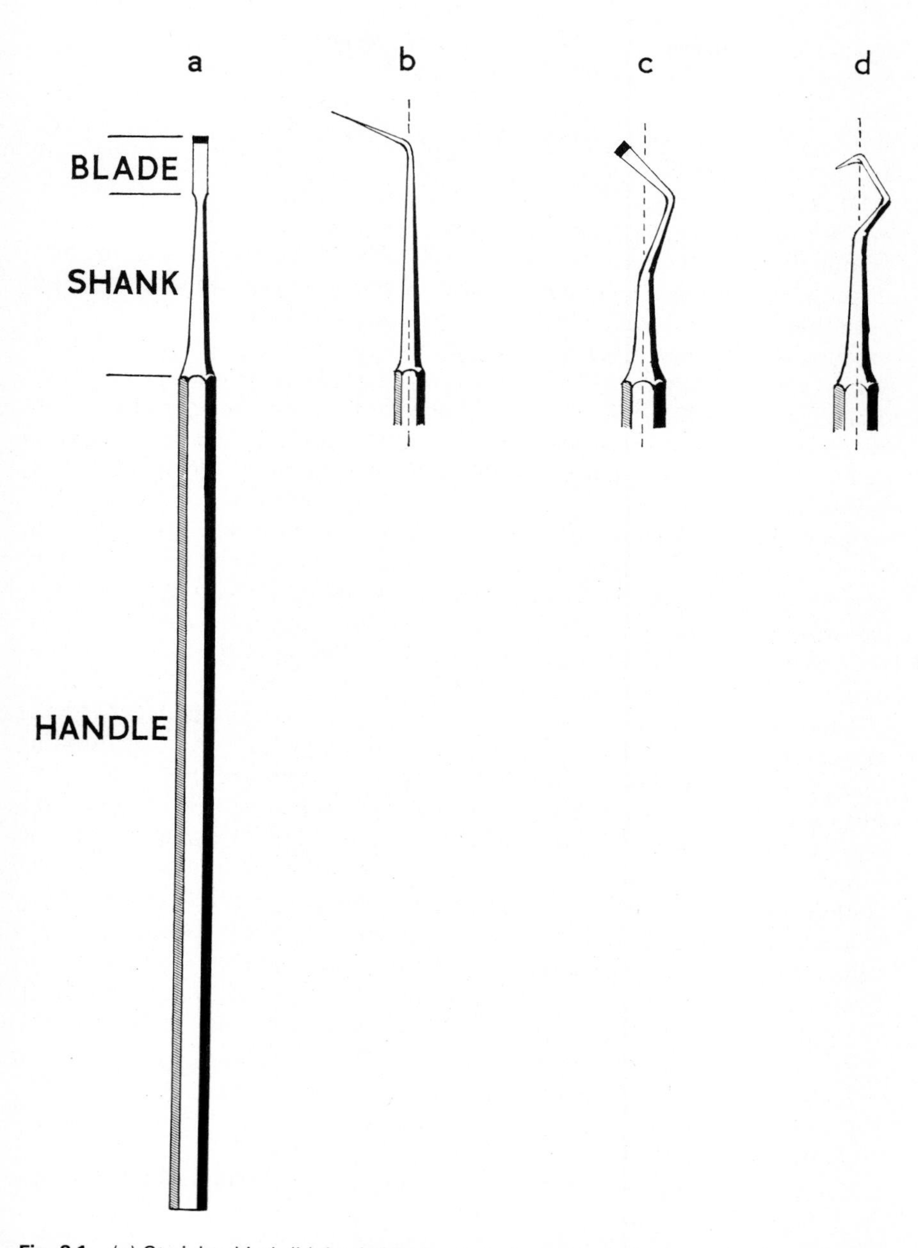

Fig. 2.1. (a) Straight chisel. (b) Angled probe: note the removal of the point from the axis of the handle. (c) Contra-angled or bin-angle instrument, with working point close to long axis. (d) Triple-angle modification of (c).

example, the distal aspects of posterior teeth. The field of usefulness of the straight instrument is therefore limited, and methods of allowing operative access to more difficult areas must be sought. To achieve this, the simple expedient of bending the shank of a straight instrument immediately allows the correct application of the blade to a number of surfaces which previously were inaccessible, and introduces the class of angled, or monangled, instruments (Fig. 2.1 (b)).

Of this class, the 'straight probe', that is one which has an angled shank but a straight tine, is a good example. This is commonly used for the examination of occlusal, buccal, and lingual surfaces of the teeth, and serves also to illustrate the shortcoming of the design. By the introduction of an angle, approaching 90 degrees, between tine and shaft, the point of the tine, which is also the point of application of the operative force, is now removed to a point *outside* the axis of the handle. This may be a serious disadvantage in the case of an instrument used with heavy pressure. In short, if heavy pressure is applied, the handle may tend to twist in the grasp and this tendency is greater the further the point is removed from the axis of the handle.

A second complication arises from the circumstance that forces applied to the handle are no longer transmitted unchanged in direction and magnitude to the worked surface; they have resultants which are more difficult to relate to the force applied by the operator. For these reasons it will be seen that angled instruments, though giving additional access, are not as easily used as those of straight design.

There remains a further important modification. The shank may be given another bend, which brings the cutting edge, point, or nib, *back into the axis* of the handle (Fig. 2.1 (c)), or so nearly as significantly to reduce the turning moment of the handle. This provides the contra-angle design of the bin-angle type, one which gives modified access and great stability which, in turn, means ease of control. Access can still further be improved and the stability retained by the introduction of a third angle (Fig. 2.1 (d)), with considerable increase of convenience for some purposes. This could logically be called the contra-angle design of the triple-angle type. It should be observed that the angles referred to in (b) and (c) (Fig. 2.1) need not be in the same plane, and in some types of instrument are commonly in two planes angulated one with another (see Fig. 7.31 p.125).

It is possible to generalize by stating:

1. **that access is mainly determined by angulation;**
2. **that stability is mainly determined by the close relation of the working edge to the axis of the handle.**

A method of specifying instruments by the width and length of the blade and its angulation to the axis of the handle enables a great variety of instruments to be described and identified. In Britain the design of hand instruments is standardized by BS2965:1970 (British Standards Institution 1970).

The design of the handle must be closely related to the purpose of the instrument, which in its turn determines its mass, the magnitude of forces to be applied, and the direction of their application. Instruments used primarily for tactile and exploratory purposes are usually light in weight and have handles of small diameter; such a design is, for physical and physiological reasons, well suited to purposes of fine discrimination. At the other end of the scale, an instrument designed to transmit heavy pressure, as for example, an amalgam condenser, must have a handle designed to give rigidity and adequate frictional grip.

The shank, connecting blade to handle, partakes of the functional characters of both. Its dimensions, rigidity, and angulation are governed by the same factors. Its incorrect design can mar a good instrument.

The purpose of hand instruments

It is appropriate here to consider the purposes of some of the commoner forms of hand instruments used in operative dentistry, to which these principles of design can be applied.

Excavators. These are used for the removal of softened dentine; they usually have a discoid or ovoid blade, the margin of which is bevelled to a sharp cutting edge.

Chisels. Straight and contra-angled are used for cleaving unsupported enamel, for paring sound enamel and dentine, and for the establishment of sharp-line angles between adjacent internal planes in cavity preparation.

Hatchets and hoes. These are similar to chisels in having a straight, bevelled cutting edge, and their purpose is similar. In design they are always angled or contra-angled and their blades are smaller than those of chisels. They differ from one another principally in that the cutting edge of the hatchet is in the plane of the shank and the handle, whereas the cutting edge of the hoe lies in an axis at right angles to this plane. Their particular function is the establishment of plane surfaces and sharp internal angles in cavity preparation.

Condensers, or pluggers. These are used for the compressing and forming of filling materials, particularly silver–tin amalgam, and cohesive gold. Because they are used with heavy pressure they are usually of contra-angle design. The design of the nibs varies with the nature of the material for which they are intended.

Plastic instruments. Some of these are like condensers and have rounded or flat ends used for pushing plastic material into a cavity. Others have flat blades and are used for conveying materials and for the conforming and

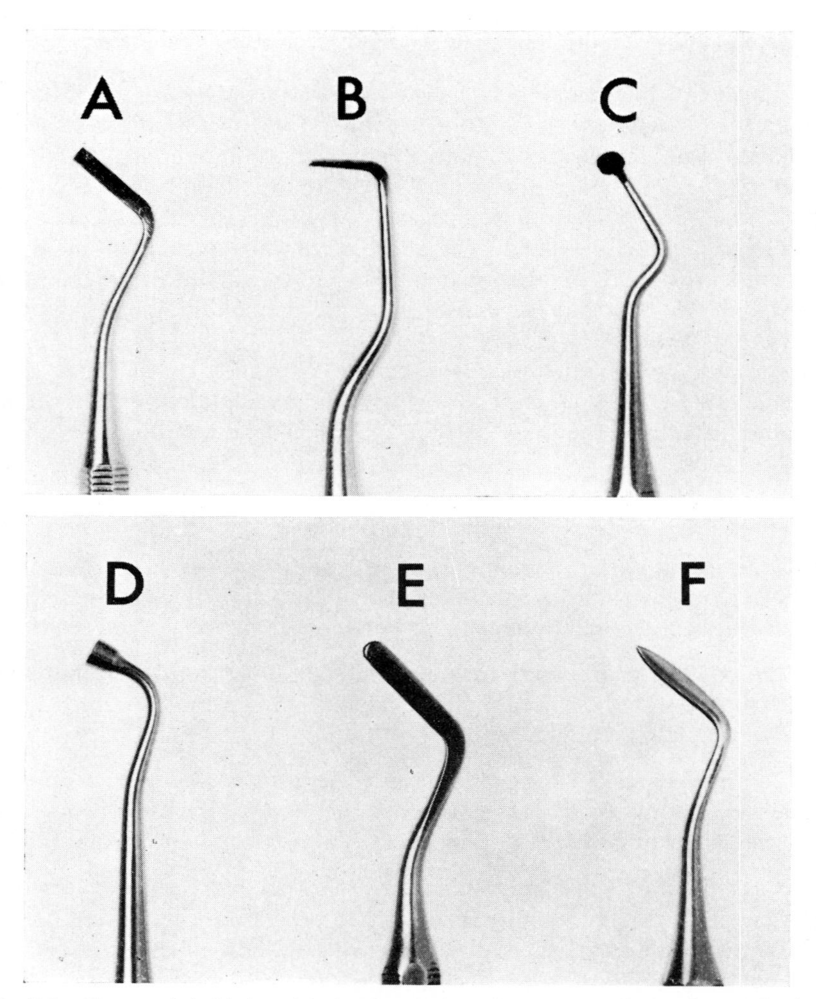

Fig. 2.2. Characteristic blades of A: hatchet; B: hoe; C: excavator; D: condenser; E: plastic; F: carver.

shaping of any plastic filling material, not involving the use of particularly heavy pressure. Rounded ends may be used for the burnishing of metallic restorations, in which case heavier pressure may be employed.

In using materials such as wax and gutta percha, and for other purposes too, it is necessary to heat the working ends of some instruments. *Heating damages both the surface and the temper of metal* and spoils the instruments for normal use. It is good practice to have a few instruments, say,

a carrying forceps, or tweezers;
a straight blunt probe;
a medium size flat plastic instrument, kept specially for heating purposes.

Carvers. These instruments have sharp, or semi-sharp, blades of various shapes, the purpose of which is the carving, by cutting, scraping, and paring, of plastic materials. The smoothness and sharpness of these instruments is of great importance and for this reason they should never be heated.

The maintenance of hand instruments

Instruments in the first three groups described above have one requirement common to all cutting instruments: for efficient use they must be kept sharp.

The acquisition of skill in the sharpening of instruments is essential to the good operator. A cutting instrument should be sharpened as and when required: the operator should be confident in his ability to achieve sharpness deftly and rapidly. This, together with good quality and well-tempered steel, is so important that the main methods of sharpening must be described.

Instruments having straight, bevelled edges such as chisels, hatchets, and hoes, may be sharpened upon a small flat Arkansas stone (Fig. 2.3 (a)), or upon a mounted Arkansas wheel in a handpiece (Fig. 2.3 (b)). Light machine oil is used as a lubricant and the instrument is so held that a bevel 30 to 45 degrees is produced at the cutting edge, according to the purpose of the instrument. A fine edge cuts more easily but is more rapidly lost; a less acute edge requires more pressure but is longer retained.

An alternative method (Fig. 2.3 (c)), also *suitable for excavators, scalers, and probes*, is the application of a fine abrasive disc, for example of sandpaper or vulcarbo. Light pressure is applied and if no lubricant is used, overheating of the metal and loss of temper must be carefully avoided; angulation of the blade must be suited to the edge to be covered. This method removes metal fairly rapidly and gives a coarser finish to the edge, but has the advantage of speed.

The introduction of tungsten carbide cutting edges for the more robust instruments such as chisels, cervical trimmers, periodontal scalers, and curettes has provided a much more durable sharp edge than has previously been attainable. With care these instruments remain usable for years, but they have to be returned to the makers for sharpening.

It is interesting to recall that until the early part of this century most dentists made many of their own hand instruments — from steel knitting needles, bicycle spokes, watch springs, and other sources of well tempered steel. Tools from other crafts like jewellery and watch-making were adapted for dental purposes. Some of our present instruments can be adapted, within limits, by small modifications in shape or angle of the blade or nib. For example, (Fig. 2.4). A flat plastic instrument with a square end is very effective for packing the gingival crevice; but when made slightly rounded and used at an angle of 70°, it can carve fissures in amalgam. A square-

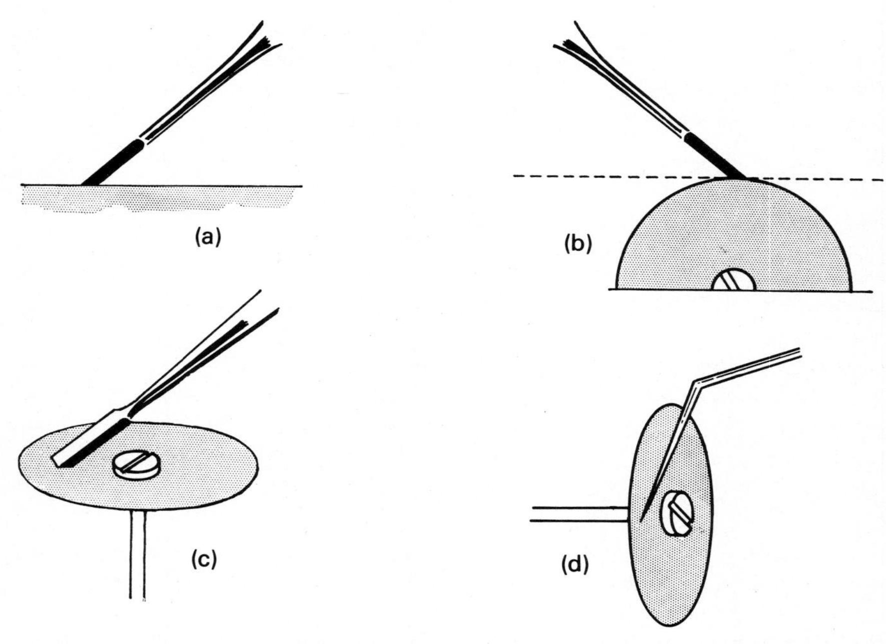

Fig. 2.3. Application of straight chisel to (a) flat sharpening stone; (b) mounted wheel; (c) mounted abrasive disc; (d) straight probe sharpened on disc.

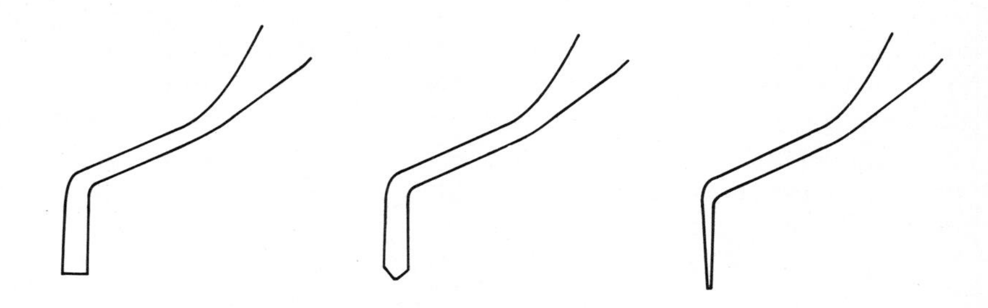

Fig. 2.4. A flat plastic instrument and a straight probe simply modified for special purposes.

ended probe is useful for teasing cements into small cavities; probes and trihedral scalers can be fined down to gain better access interproximally, and so on. There is scope here for ingenuity and inventiveness.

Stainless steel, an alloy generally unsuitable for cutting edges, has gained wide acceptability in recent years chiefly because it keeps a better appearance than carbon steel with repeated sterilization and also because cavity preparation is done more extensively with high-speed rotary instruments. The quality and temper of some hand instruments today leaves much to be desired and makes them inefficient in normal use.

The maintenance of non-sharp hand instruments is concerned with the preservation of their temper by the avoidance of over-heating in any form, and by the retention of a bright surface and smooth edges. A plastic material may be conformed with instruments of many shapes, but only those possessed of smooth and regular surfaces can effectively be used for finishing.

The handling of instruments

When using a hand instrument, it may be grasped in one of two ways:

1. The *pen grip* (Fig. 2.5 (a)) is self-descriptive and is most frequently used. It allows light and heavy touch and finely controlled movements over a fairly wide range. The middle and ring fingers are used as stabilizers.
2. In the *palm grip* (Fig. 2.5 (b)) the instrument is held between the thumb and forefinger and the handle lies diagonally across the palm, clasped by the remaining fingers. The thumb also acts as a stabilizer. This grip, used when operating upon maxillary teeth, provides heavy force in the axis of the handle over a limited range of movements. The *finger grip* (Fig. 2.5 (c)) is a modification of the palm grip in which the instrument lies over the proximal phalanges, grasped by the fingers. It is a grip of limited value, used when the palm grip fails to give the correct line of access.

Double-ended instruments with a working blade, tine, or nib at each end are very common. They save space in storage and reduce instrument exchanges in operating — a great advantage. Usually one handle will carry right and left, or mesial and distal patterns of the same blade, which makes for great convenience. The only disadvantage is that they must be handled with slightly greater care to reduce the risk of accidental injury.

It will be readily understood that the requirements of accuracy in fine movement, and those of safety in more forceful manipulation, demand that *all instrumentation be accompanied by finger support* upon adjacent firm structures, of which the firmest are the crowns of the adjacent healthy teeth. The thumb, and the third or fourth fingers of the hand holding the instrument are the finger rests most often used. This is a rule which must be closely observed; there are very few exceptions (Fig. 2.6).

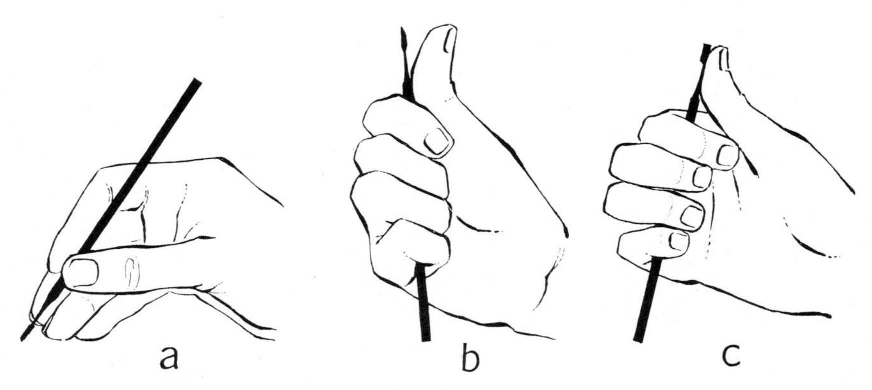

Fig. 2.5. (a) Pen grip; (b) palm grip; (c) finger grip.

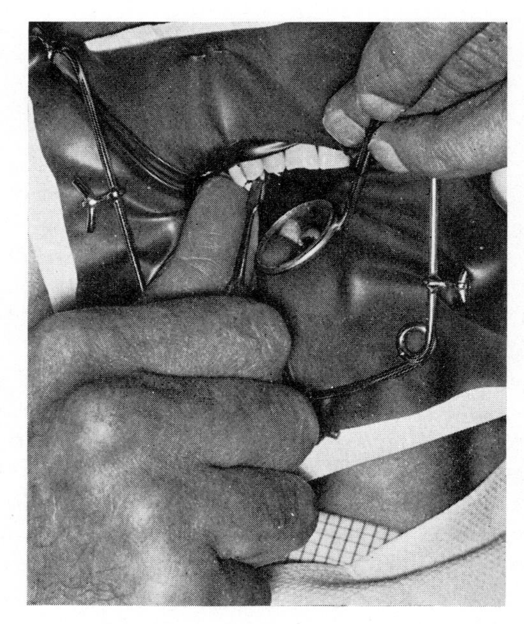

Fig. 2.6. Straight enamel chisel held by the palm grip, applied to a maxillary premolar. Note thumb rest.

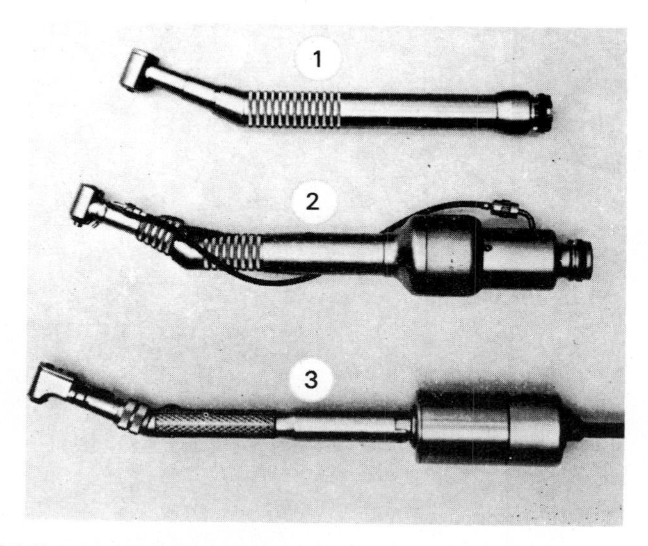

Fig. 2.7. (1) *Kavo* air-bearing turbine. The air turbine is in the head of the handpiece. Bur speeds up to 450 000 r.p.m. (2) *Kavo Dentatus* air motor. The air motor is contained in the base; bur speeds up to 100 000 r.p.m. (3) *Kerr Electro-Torque.* Low voltage DC motor; speeds up to 20 000 r.p.m.

Rotary instruments

Rotary instruments, consisting of burs, stones, and discs, are small instruments held in a chuck, called a handpiece. The instrument is rotated through the handpiece by power from an external source, either compressed air or more directly by an electric motor. There are three types of equipment used to provide the rotary power.

1. **The air turbine**, or air-rotor (Fig. 2.7 (1)), is the commonest and most widely used. It gives the highest speeds, in the range of 250 000 to 400 000 r.p.m. These very high speeds are achieved by a small air-driven turbine or rotor, mounted on air-bearings in the head of a contra-angle handpiece. The shank of the bur is inserted into the central axis of the handpiece and revolves with it. The handpiece also contains a system which directs finely atomized water at the cutting head of the instrument (Fig. 2.8).

2. **The air motor** (Fig. 2.7 (2)), is driven by compressed air by means of a three- or four-chambered motor placed at the base of a contra-angle handpiece and the rotation is transmitted by a shaft and gear, to the head of the handpiece. This apparatus covers the medium speed range, say, 4500 to 45 000 r.p.m. Another less common modification in the lower part of the medium speed range is a similar handpiece in which the air motor is replaced by a small DC electric motor (Fig. 2.7 (3)).

3. **Conventional 'dental engine'**. This equipment, introduced over 60 years ago was the chief source of rotary power in dentistry for over half a century and in a modern form, is still used in the dental laboratory. It con-

Fig. 2.8. Head of turbine handpiece showing spray outlets, and spray impinging on rotating bur.

sists of an electric motor from which the power is transmitted by a steel cable, or more often by a cord running over pulleys. In the surgery it normally covers a low range of instrument speeds, say, 1000 to 3000 r.p.m. This can be reduced or increased by geared handpieces. This type of 'dental engine' is obsolescent. Except for special purposes it is less and less used as the scope of the air motor is extended downwards to cover lower speed ranges, with high torque and ever improving control.

The speed ranges described above, namely,

High speed	45 000 to 400 000 r.p.m.
Medium speed	4500 to 45 000 r.p.m.
Low speed	1000 to 4500 r.p.m.

are entirely arbitrary. There are of course no sharp dividing lines but these speed ranges give a useful index of reference.

High-speed handpieces are invariably contra-angled and will therefore give access for use anywhere in the mouth provided that visibility and control are adequate. They require regular upkeep by way of cleaning, lubrication, and replacement of worn parts. Most turbines have oil droplets

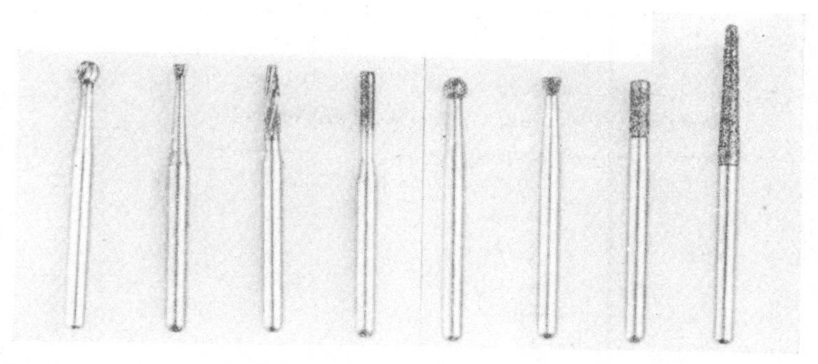

Fig. 2.9. Diamond and tungsten carbide burs of the friction-grip type, for use in high-speed handpiece.

suspended in the air supply. Not all of them will tolerate repeated sterilization because they include non-metallic parts in their construction, but all need regular cleaning and oiling. In these matters the manufacturers' advice should be closely followed.

Straight handpieces are more easily maintained because their construction is simpler, having no bevelled gear mechanism. They are used in the lower speed range, limited mainly to accessible places in the front of the mouth or to fixed and removeable prostheses where at this stage the surface to be worked upon can generally be removed from the mouth.

The small rotary instruments used in handpieces may be classified under the following headings.

Burs. These are small milling tools with blades of tungsten carbide, or cutting surfaces of fine diamond grit embedded in metal. The cutting head of the bur may be any one of the very wide variety of sizes and shapes of which the commonest are round, cylindrical, tapered, and inverted cone. The shanks of these burs, as shown in Fig. 2.9 are smooth and round-ended. This allows them to be forced by axial pressure into a plastic sleeve in the head of the handpiece. They are called 'friction grip' burs and are generally used in the high-speed range, but are also suitable for the medium range.

Stones. These are made of abrasives such as carborundum (green) or alundum (white or pink) which can be moulded into shapes similar to those described for burs. They are considerably larger than burs and are fixed directly to a shank or mandrel for insertion into a handpiece. These instruments are used over the low and lower medium speed range and the shank or mandrel is of the type used in the latch-type contra-angle handpiece, or longer for use in a straight handpiece.

Discs. These are used for surface abrasion or for edge-cutting. They are from 19 to 23 mm in diameter and may be made of steel, vulcanite, plastic, or paper. They carry abrasives that vary from diamond grit to finely divided cuttle-fish bone. Their capabilities, ranging from rapid removal of enamel and dentine to fine polishing of hard tissue, metal, and other materials, vary according to the coarseness of the abrasive and the speed. They are used mostly in the lower speed ranges and are mounted on various patterns of mandrel to fit both contra-angle and straight handpieces.

The term 'abrasive wheel' is also in common use and denotes a circular abrasive instrument, for example, of carborundum, carried on a fixed or removeable mandrel. It is usually thicker than a disc but smaller in diameter.

The commonest rotary instruments described here represent only a small portion of the very great variety of all shapes and sizes available, many for special and limited purposes, to be found in the catalogues of the major suppliers throughout the world. A study of these catalogues is both interesting and informative.

Characteristics of rotary instruments

In use rotary instruments have characteristics which vary greatly according to the speed of revolution and the torque of the cutting tool. They liberate heat proportionate to the work done. At low speeds this heat must be allowed to dissipate by intermittent cutting. At high speeds the heat must be dissipated by water or atomized water jets (Fig. 2.8). In the absence of an adequate means of dissipation of the heat evolved a harmful rise in the temperature of the tooth will rapidly occur, causing serious pulpal damage.

Noise and vibration are also produced and these are related to the nature of the cutting element, the concentricity of the tool, and its speed of rotation amongst other factors. In the lower speed range the finer the blades, or the grit, and the more concentric the tool, the less obvious is noise and vibration. Vibration is worst, that is to say least tolerable to the patient, in the range of 60 to 100 cycles per second, which are predominant over the range 3600 to 6000 revolutions per minute.

The concentricity of the bur head with its shank, and the axis of rotation of the rotor, are most important to accurate cutting. For example, if a diamond instrument has a tip diameter of 1.0 mm, and the neck or shank is bent so that the tip is eccentric by only 0.25 mm, the effective diameter of the tip then becomes 1.5 mm. This clearly means that the instrument cuts approximately half as wide again as the centred bur. Tungsten carbide instruments, being brittle, break rather than bend; diamond instruments are more likely to be bent, and if so should be discarded forthwith (see also p. 61).

At low speeds a bladed rotary instrument is liable to 'snatch' if it is allowed to rotate towards a sharp declivity on the worked surface, but will not do so if rotated away from such a feature. For this reason, instruments which cut equally well in both directions of rotation, such as diamond instruments, have a distinct advantage.

All rotary instruments produce detritus, in the form of dust, of the material being cut. This can reduce the cutting efficiency by clogging the

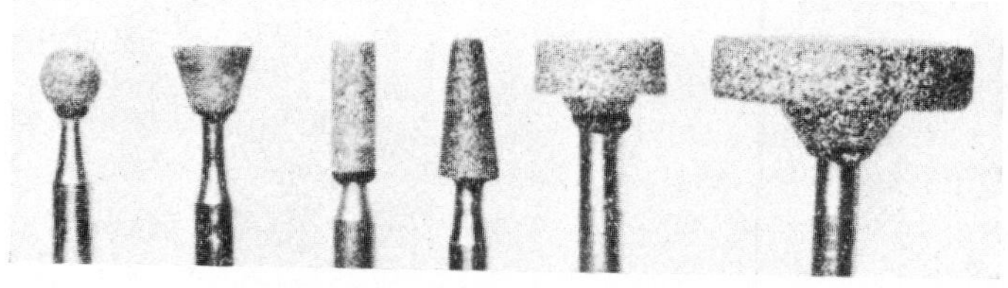

Fig. 2.10. Mounted carborundum 'stones' and wheels, used for removal of hard tissue and smoothing cut surfaces.

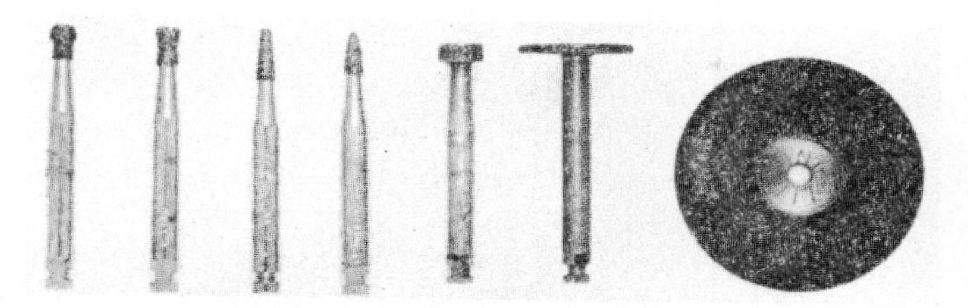

Fig. 2.11. Diamond instruments of various designs for use in a latch-type handpiece. Burs, wheels, and a side-cutting disc.

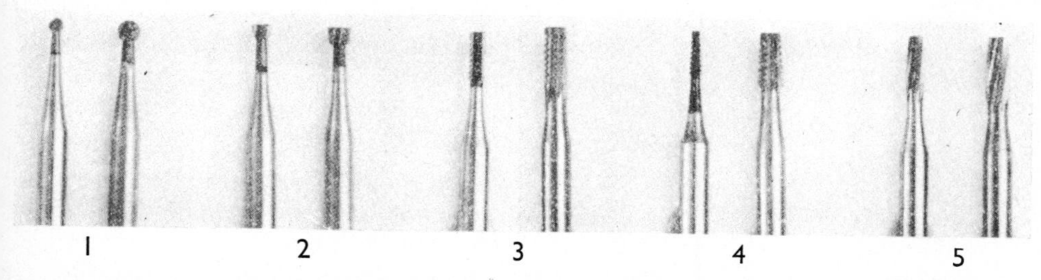

Fig. 2.12. Representative patterns of steel burs: (1) round; (2) inverted cone; (3) cylindrical cross-cut fissure; (4) tapered cross-cut fissure; (5) tapered plain fissure.

blades of the tool, and it obscures the operator's view of the cutting process. Both these disadvantages may largely be overcome by the use of suitable water sprays.

Such are the important characteristics of rotary instruments with speeds ranging from 1000 to, say, 15 000 r.p.m. Above this speed, the behaviour of these instruments undergoes progressive modifications which are very significant to the surgeon and to the patient.

High-speed instrumentation

For many years 'conventional' dental engines gave speeds of 1000 to 3000 r.p.m., and the majority of cutting was done nearer the lower limit of this range. Although the use of a water spray playing upon the cutting tool was borrowed from industry many decades ago, it is only in the last 20 years, as operating speeds have risen from 10 000 to 250 000 r.p.m. and higher, that water sprays have come into universal use.

At high speeds the tools must be of tungsten carbide or diamond; carbon steel loses its edge immediately. Completely effective cooling must be provided since the temperature at the cut surface may rise to the level of red heat in a fraction of a second. Atomized sprays, two or three in number, arranged to impinge as acutely as possible upon the centre of the cutting head, provide the most effective cooling systems for high-speed handpieces. Even so, care must be taken to see that sprays are used to full effect. For example, if the water content of the spray is insufficient, the tip of a long instrument may run dry though the remainder is well cooled. The head of the bur may be deeply placed in a cavity or in the interdental space where spray is unable to reach. Too heavy pressure on the bur may momentarily build up a wall of debris which will shield the point of cutting. All these may cause an unacceptable rise of temperature and damage to the tooth. The operator does well to remember a modification of an old saying—where there's a smell of burning there's fire!

In considering the properties of medium- and high-speed instruments, attention must be given to the torque and peripheral speed of the tools. By the *torque* is meant the *turning moment* of the tool. For instance, a cord-driven geared motor giving a bur speed of 40 000 r.p.m. has a high turning moment, and considerable pressure may be applied through the bur upon the cut surface without more than fractional slowing of the drive, though other factors such as cutting speed, heat, and vibration may rise conspicuously.

Peripheral speed means the linear velocity of the cutting blade and is derived by *multiplying the revolutions per minute by the circumference of the cutting head.* At a given pressure of the instrument upon the cut surface, the peripheral speed is a more reliable indication of cutting speed. A turbine at 200 000 r.p.m. has a very low torque and may be almost instantly slowed by comparatively light pressure. It uses a bur with a small cutting head of which the peripheral speed is of the same order as in the previous instance of the cord-driven handpiece.

Characteristics of the medium- and high-speed instruments

Cutting speeds. The speed at which hard tissues are cut is a function of peripheral speed and torque of the instrument. At speeds around 250 000 r.p.m. with moderate torque, sound enamel and dentine can be removed with ease and rapidity. Thus it is possible to make an extensive preparation quite quickly. So much so that the operator must be constantly on his guard to avoid over extension. *Excessive and unnecessary destruction of sound tissue is an irreparable loss and inevitably shortens the life of the tooth.* Detailed finishing of the cavity requires the use of lower speeds and slower cutting instruments.

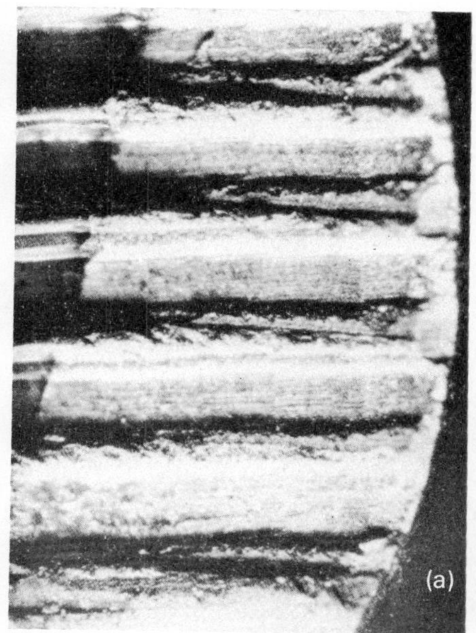

(a) The wall of an approximal cavity cut with a tungsten carbide cross-cut fissure bur at high speed (× 20).

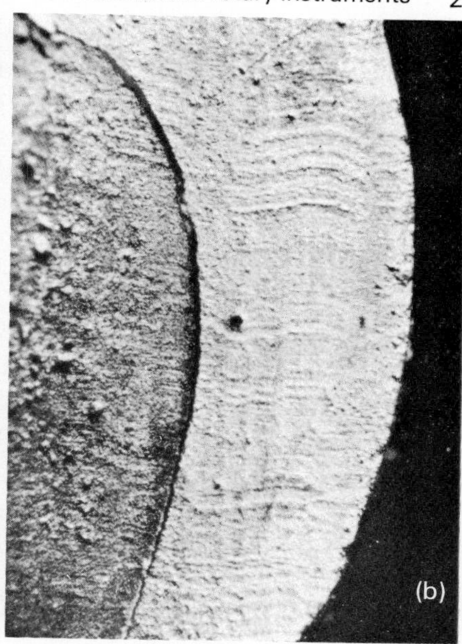

(b) A similar wall cut with a tungsten carbide plain-cut fissure bur at high speed (× 20).

(c) The same surface finished by planing with an enamel chisel (× 20).

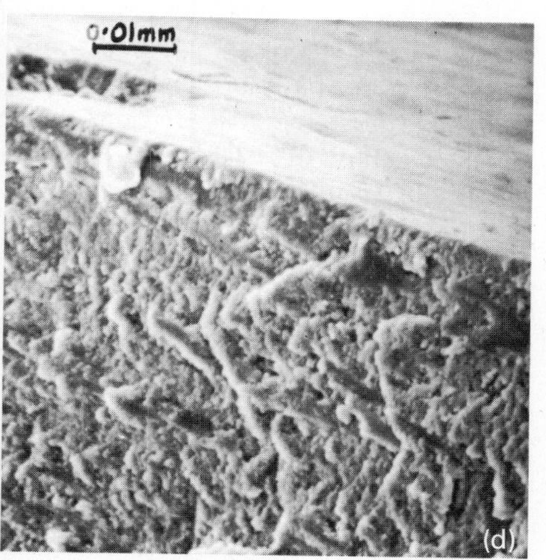

(d) A scanning electron micrograph at high magnification showing, *above*: the enamel surface, and *below*: the cavity surface cut by a plain bladed tungsten carbide bur at moderate speed.

Fig. 2.13. ((a), (b), and (c) by courtesy of Professor D. N. Allen; (d) by courtesy of Dr. I. E. Barnes.)

A surface cut at higher speed shows marked roughness and scoring particularly when diamond and cross-cut burs are used. To produce smooth surfaces and accuracy of detail it is necessary to resort to plain-cut tungsten carbide burs used at lower speeds.

Figure 2.13 shows very dramatically the roughness produced by a cross-cut fissure bur compared with a plain-cut fissure bur, both at high speed. This is compared again with the smoother finish produced by an enamel chisel. Figure 10.10 (p. 191) shows at higher magnification a scanning electron micrograph of a cavo-surface margin cut with a high-speed plain tungsten carbide fissure bur. This instrument gives the smoothest finish of any at present investigated.

Pain. Provided that cooling is adequate, pain experience is reduced as compared with low-speed methods without cooling. The reduction is, in most cases, insufficient to allow anaesthesia to be dispensed with, and it is therefore wise to use anaesthesia in all cases except for the briefest and most superficial operations.

Vibration. Lack of vibration at high speeds is one of the features most appreciated by the patient. Owing to the fact that the range of perceptible vibrations is limited, and the upper limit is in the region of 560 cycles per second, rotary instruments of speeds above 40 000 r.p.m. produce high-frequency vibrations above the range of perception. This applies particularly to turbines for, by the principles of their construction, they can be designed to contain only one moving part. The air motor and electric micromotor have more and heavier moving parts and bevel gears. These are susceptible to transmitted and harmonic vibration, and at the attainable speeds many of these, though well above the range of greatest discomfort, are not entirely outside the range of perception.

Pressure. At high speed the tendency of the tool to snatch, as at low speeds, virtually disappears and all tools work efficiently under minimal pressure. This allows delicacy of control by the operator and a marked reduction in the fatigue experienced by both operator and patient. On the other hand, discrimination between hard and soft tissues is much less than at low speed, and for this reason the removal of soft carious dentine with a hand instrument gives much greater safety. The high-speed instrument always seems to cut a little wider than is expected owing to its eccentricity, however slight and imperceptible, and due to the early stages of normal wear in the rotor bearings. This and the ease of cutting lead to over-extension in cavity preparation unless this is consciously counteracted.

Trauma. Histological studies indicate that high-speed cutting, even when effectively water-cooled, is not without a damaging effect upon the pulp. This appears to be comparable to that resulting from low-speed cutting, but it may be much more severe if cooling is ineffective. The tissue changes are characteristic of burning, with local destruction and disorganization of the cellular elements of the pulp, hyperaemia, oedema and capillary haemorrhage (Fig. 4.7; p. 56). The clinical manifestations are post-operative pulpal pain of varying degree and hypersensitivity to thermal changes.

There are many other aspects of high-speed technique which have to be taken into consideration. Sterilization, wear, lubrication, and general maintenance in good condition are of obvious practical importance. The method of cooling the cutting head which is sometimes deeply situated in a cavity or in the interdental space; the speed of the tool which throws the water centrifugally; the necessity for a mirror which keeps clear even when sprayed; the noise of the turbine — all these are features which modify technique and which to some extent detract from the efficacy of these instruments. These problems are, however, being solved. For example, the elimination of ball-bearing races at each end of the turbine rotor and the substitution of air bearings has resulted in a very marked reduction in noise and a simplification of the lubrication system.

The instruments and small equipment required for good operative dentistry in modern conditions will certainly exceed those briefly described in general terms in these chapters. For example, a reliable and effective saliva ejector; a source of atomized water spray in addition to that accompanying a rotary instrument and a low-pressure, high-volume aspirator; additional sources of light and heat for various purposes, and so on. Many of these are necessary to good practice, but the value which each operator attaches to individual facilities varies considerably. The student should learn to evaluate in a critical manner the value and efficiency of the various adjuncts to clinical practice, adopting only those which fit in with his methods of working to the greater advantage of his patient.

Summary

Hand instruments. Handle, shank, and blade (nib or tine). Single and double-ended. Straight, angled, and contra-angled. Angulation improve access but brings instability. Stability restored by contra-angle which brings working point closer to axis of handle. Holds, for general purposes, pen grip and palm grip. Finger grip occasionally. Thumb and finger rest on adjacent teeth essential. Maintenance. Cutting edges sharpened as soon as blunted. Non-cutting ends clean, polished, smooth edges. Avoid heating, except instruments kept for purpose.

Rotary instruments. Burs, stones, discs. Tungsten carbide and diamond cutting instruments needed for all medium- and high-speed range. Air turbine; air motor.

Effective water spray cooling essential above 4000 r.p.m. Cutting speed depends upon peripheral speed of tool and mechanical torque. Advantages of high speed; rapid cutting, lack of vibration, lightness of touch, good direct visibility. Disadvantages: risk of over extension, risk of thermal injury to dentine and pulp, poor mirror visibility. Precision cavity preparation and finishing need lower speeds and special instruments.

3

The examination of the patient

When a patient comes for advice or treatment, the dental surgeon undertakes to make a diagnosis, to prescribe treatment, and he may also form a prognosis of the case. The manner in which the diagnosis is achieved is of such importance that the principles of examination must be considered. The student may find it convenient to return to this chapter as a basis for further study when the approach to clinical work occupies his attention.

Diagnosis is the recognition of a disease. In order to come to a diagnosis the practitioner collects all available information, by questioning and by physical examination of the patient. The disease being recognized, the appropriate treatment is prescribed and carried out. At this stage a *prognosis* may be made. This is an informed forecast of the likely progress, duration, and termination of the disease. This prognosis may subsequently be modified as the disease and its treatment proceed.

The history

The object is to elicit from the patient every item of information which might have a bearing on the course of the disease up to date, and to avoid missing any factor which might have significance in the past development or the future progress of the condition.

The history, of which adequate written notes should be made, first establishes the patient's **immediate complaint**. This is a description, preferably in his own words, of the symptoms which have led him to seek advice. Such symptoms are frequently those of pain or soreness, of swelling, bleeding, stiffness of movements, and so on. An attempt should be made, without the use of leading questions, to encourage the patient to describe these as clearly and in as much detail as possible. For example, the question 'Does the pain in your jaw spread to your ear?' is one which could for a variety of reasons lead the patient to say 'yes', though the pain was not in fact felt in the ear. It is better to phrase the question, 'Do you feel the pain only in your lower jaw?' This is more likely to elicit an unbiased answer.

The next material point to be sought is the **duration and character** of the symptoms and care should be taken to establish, with accuracy, the occasion of the first onset of the particular condition. This frequently leads one further back in time than the patient at first appreciates. The progress of the disease may be traced by eliciting the character of the symptoms over this period; whether they have been constant or changing in character, continuous or intermittent, becoming more, or less severe, and the circumstances which have led to exacerbations or relief of symptoms. Thus, by careful questioning, an accurate mental picture of the previous progress of the condition is obtained by the examiner.

This by itself is insufficient, and attention should then turn to the **past dental history** which extends this picture further back and may reveal factors of general or immediate importance to the condition under investigation. Information on oral hygiene and dietary habits must be elicited at this stage.

The **past medical history** should also receive attention. Care is taken to seek information of past or intercurrent illness, of whatever nature, which may have a bearing upon the present complaint, or influence upon its treatment or prognosis.

Finally, in the investigation of the background, comes the **family history**, in which hereditary developmental factors, diatheses, and diseases having an important bearing on the condition under consideration may come to light.

The relative importance of these factors, which are summarized in Table 3.1, will vary with each type of case and with the age, intelligence, and cooperation of each patient. Assiduity and care applied to acquiring a detailed and complete history is well repaid, particularly in the more obscure forms of disease. A good history is fundamental to accurate diagnosis.

Table 3.1 Summary of history-taking

Name, age, sex, marital status.	
Address and telephone number, occupation.	
HISTORY	
Present complaint:	Patient's description.
	Onset and duration.
Details of symptoms:	Location, character, and intensity.
	Initiation.
	Time relationship: constant, intermittent, variable.
	Exacerbation, relief.
	Associated symptoms.
Dental history:	Regularity of treatment.
	Caries experience, periodontal disease.
	Extractions and other surgical intervention.
	Prostheses.
	Oral hygiene and diet.
Medical history:	General health, present and past.
	Throat, nose, ears, eyes, digestive, respiratory, and cardiovascular systems.
	Prolonged treatment of any form.
	Current drug treatment or allergies.
	Name and address of present doctor.
Family history:	Parents and sibs, alive or dead.
	Chronic or acute diseases; hospitalization.
	Dental experience of parents.
	Inherited and congenital conditions.

Having obtained, by this means, all possible information concerning the condition which has caused the patient to seek advice, and the general dental and medical background, the clinician frequently may find that he has already formed a tentative diagnosis of the condition. This is a natural sequence of events, but at this stage great care should be taken to keep an open mind, in order not to prejudge the condition. The physical examination should be approached with complete impartiality, lest in pursuing a line of thought which may lead to a premature and possibly erroneous conclusion, signs of diagnostic importance are overlooked.

During the course of taking the history the clinician will have had a good opportunity to observe the general appearance, attitude, behaviour of the patient, and these may be very informative. In the average patient, a *thorough general inspection* of the face, neck, and mouth, and of the affected area in detail, precedes the examination with instruments.

The physical examination

The basis of good clinical examination lies in the full and critical use by the clinician of all his senses and in the accurate interpretation of what he sees, feels, hears, and to a lesser extent, smells.

In the examination of dental and periodontal tissues the use of instruments such as a probe, a mirror, and cotton wool forceps is demanded by the special nature and position of the area under examination. The power to discriminate by sight and touch may become highly developed as the result of correct training and prolonged practice in the use of simple observation and of exploration with a probe. The sense of hearing contributes perhaps more than is generally recognized in the purely dental field, especially in the lower ranges where sonic and subsonic frequencies merge. The sense of smell only rarely contributes, but such a contribution can be significant.

The clinician will recognize that *no dental examination is complete which does not embrace the masticatory apparatus as a structural and functional whole and relate this to the patient's well-being*. The technique of complete examination must be understood and mastered.

So far as the **local examination of purely dental lesions** is concerned, the starting-point is a visual examination of the oral cavity, including the vestibule, the tongue and sublingual area, the hard and soft palates, and the faucial and pharyngeal areas. To this should be added an inspection of centric occlusion. These are essential and form a convenient base from which more detailed examination proceeds.

Examination of the clinical crowns of the teeth, the gingiva and periodontal attachments follow in detailed and regular manner. The occlusion and occlusal excursions are reviewed and the effect of any prostheses which may be present in the mouth noted.

Thus far the examination can be conducted with the simplest aids, a good light, a mirror, probes, and carrying forceps for cotton wool. In many conditions of dental origin such an examination may suffice to allow a tentative but correct diagnosis which further detail confirms. Occasionally this is not enough and the clinician must resort to **special methods of examination** of greater complexity and requiring much more extensive equipment and facilities. These procedures include the following:

Vitality tests of the teeth, by thermal or electrical stimulation.
Radiographic examination, bitewing or periapical techniques.
Transillumination, of teeth, alveoli, and nasal sinuses.
Occlusal registration and analysis.
Bacteriological examination, by culture and bacterial counts.
Haematological examination, cell counts, serology, and blood chemistry.
Biochemical examinations of saliva.
Biopsy and histopathological examination.

Of the above tests, radiography and vitality tests are commonly used, and occlusal recording and analysis are becoming more and more necessary to an understanding of the relationship of hard and soft tissues under functional conditions.

The last four categories named above require facilities found only in hospital. The possible need for them and the part they may play in the elucidation of disease must be borne constantly in mind.

The present nature of dental practice is such that we encourage all patients to present themselves for routine examination at regular intervals of about six months, and more frequently in the case of children. This being so, patients frequently present with no specific complaint and the clinician's object is to carry out preventive procedures and to eliminate the possibility of early or hidden disease so far unappreciated by the patient.

It must be emphasized that *most common conditions can and should be detected by the dental surgeon before the patient becomes aware of them.* When faced with difficulty in day-to-day diagnosis we should have in mind a useful aphorism, **the commonest diseases are the most common**. This is a reminder that only when we have eliminated the possibility of one of dentistry's two commonest diseases being the cause of the condition confronting us, should we *then* consider rarer and exotic causes. Because of the frequency with which he should see his patients, particular responsibility devolves upon the dentist with regard to the early recognition of premalignant conditions, as well as those of systemic diseases in which oral manifestations are of early occurrence.

The periodical routine examinations, even by an experienced clinician, should follow in all respects the general form of the most detailed examination, but it is obviously less time-consuming to confirm a state of normality than to detect and assess abnormality. Furthermore, experience teaches where and how examination may safely be abbreviated or omitted in detail, whilst the mind is attentive and on the alert for the possible occurrence of the abnormal.

There are also, in practice, cases of dental pain in which the diagnosis appears all too obvious and in which early relief of the symptoms is the immediate object of the patient's visit. Such cases can be a trap for the unwary. *A thorough comprehensive visual examination of the whole oral cavity is essential* and this may be followed by as much or as little of the routine examination as is required to establish a diagnosis beyond all reasonable doubt in the circumstances. There can be very, very few cases in which there is a valid excuse for not having had 'a good look', and unless the examiner is quite satisfied he should look again!

Table 3.2 Summary of routine examination

General visual inspection	Facial appearance, expression, skin colour and texture, sweating.	
	Bony contour, soft tissues, musculature.	
	Eye movement, sclerotics, pupils.	
	Breathing, cervical pulsation.	
	General bodily characteristics, stature, demeanour, gait.	
General examination of oral cavity	Mucosa of: vestibule, dorsum of tongue, sublingual area, hard and soft palate, fauces, oropharynx.	
	Teeth present and missing. Existing restorations and prostheses.	
	Periodontium — recession, mobility.	
	Centric occlusion — masticatory efficiency.	
Detailed examination of:	Mucosa.	
	Clinical crowns.	
	Periodontium.	
	Occlusion (prostheses).	
Accessory methods	Radiography — bitewing, apical intra-oral, extra-oral.	
	Pulpal response; thermal, electrical.	
	Transillumination.	
	Bacteriological.	Requiring special facilities
	Chemical.	Requiring special facilities
	Haematological.	Requiring special facilities
	Histological.	Requiring special facilities

All the findings in routine examination, and those disclosed by special methods, must be systematically recorded upon the patient's record card or notes. These will include a chart of diagrammatic or pictorial form on which much information may be graphically recorded. Further space is allowed for descriptive details which cannot be adequately portrayed.

To denote the permanent dentition the most commonly used scheme is the Szigmondy–Palmer notation in which teeth are numbered from the midline from 1 to 8 in each quadrant. The figures when written must carry vertical and horizontal lines to indicate the quadrant, thus:

6	
	4

There has recently been advocated a two-digit notation which has gained much international support. In this the quadrants are numbered 1 to 4, clockwise from the right upper, thus:

1	2
4	3

This figure is added to the serial number of the tooth, thus:

18 17 16 15 14 13 12 11	21 22 23 24 25 26 27 28
48 41	31 38

In this notation

6	
	4

becomes 16, 34 (one-six and three-four). It is claimed that this 'international notation' is simple, clear, and easily learned and, perhaps more important in an international context, it can be transcribed for computer programming, which is important in epidemiology.

It must be stressed that though the material recorded must be carefully selected and concisely expressed, these notes must be a full and informative account of the history, examination, treatment plan, and subsequent progress of the patient. In practice, administrative and financial details are generally kept with these records.

The detection of caries

Operative dentistry consists of more than the repair of the ravages of dental caries, but in the foreseeable future, prevention and early detection of this disease will continue to occupy an essential position in this aspect of dentistry.

The presenting characteristics of caries of hard dental tissues are discoloration, softening due to the progressive loss of mineral content, and the resulting loss of surface substance. Carious cavities over a certain size can be seen and felt with a probe more, or less rapidly, according to their position. The detection, however, of very early cavities and sometimes of recurrence of caries around existing restorations is another matter and may call for a high degree of discrimination.

In all cases of routine examination for caries the procedure should follow a regular pattern. For example, the operator might start his inspection at the left lower molars, following round the lower jaw to the right lower molars. Attention to the upper teeth would then follow in a like manner. The important point is that a constant routine should be adopted. It is also important that, for thorough examination, the teeth should be isolated with cotton-wool rolls and dried, segment by segment. *Refraction and reflection at a wet surface can easily disguise small differences of colour, form, and texture.* The area must, of course, be well lighted and the explorers used should be sharp, fine, well-tempered, relatively light in weight, and of such variety of pattern as will give adequate access to all aspects of the clinical crowns.

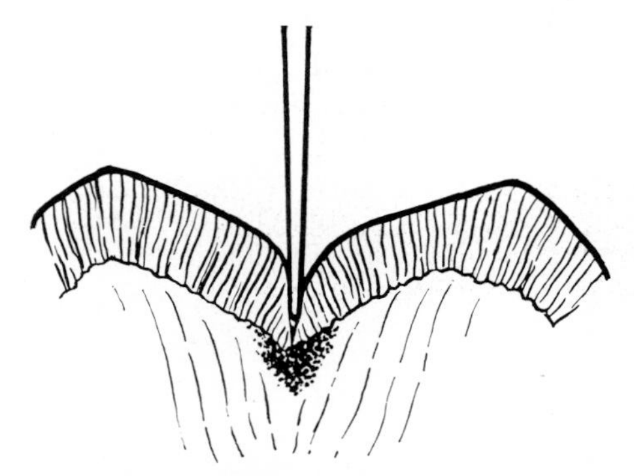

Fig. 3.1. Pressure of the probe against slightly decalcified walls of the early carious fissure results in the sensation of stickiness on withdrawal.

Caries starting in an anatomical pit or fissure, or on a plane surface, is best appreciated by a certain 'stickiness' to probing. In the case of a fissure this arises from a surface decalcification of the enamel walls by caries which, more often than not, has at this stage already involved the amelo-dentinal junction (Fig. 3.1). The compression, by the entering probe, of this softened region results in a slight frictional resistance to its withdrawal which is characteristic and fairly easily perceptible with experience.

The fact that interproximal caries starts near the contact area and in the area of close approach surrounding it obviously gives rise to difficulties in early detection. By the use of the fine curved or hooked probe it may be possible to insinuate a point into an incipient enamel defect, or at least to detect a breach of normal surface contour or texture. The ability of a tine to explore the contact area must depend upon the quality of the instrument, the fineness of its point, the size and shape of the contact area, and upon the amount of force which may be used without discomfort to the patient.

It is unusual for the interstitial lesion, of such size as to give rise to diffi-

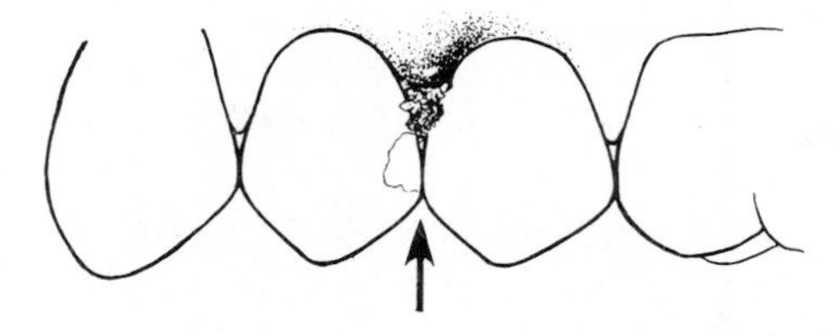

Fig. 3.2. Destruction of the contact point allows food-packing and injury to the interdental papilla.

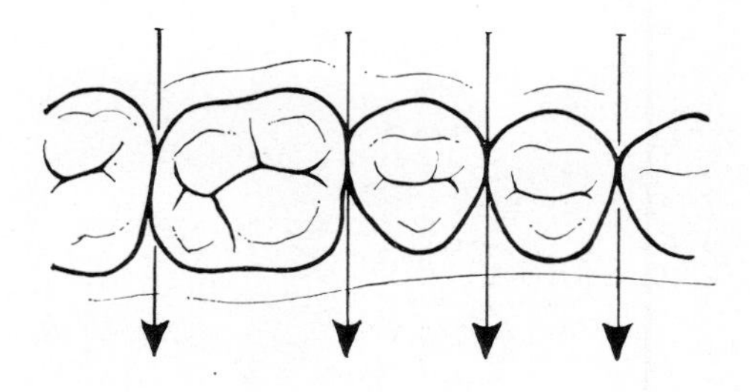

Fig. 3.3. For bitewing radiographs the rays are directed tangentially through the contact points.

culty in detection, to be the cause of symptoms. There are, nevertheless, a few patients who consistently complain of typical sensations of pain with sweet substances, heat, and cold, in the presence of very early lesions, but they are rare. Normally, the cavity which causes such symptoms is easily seen on clinical examination.

A history of food-packing between teeth, or evidence of damage to the interdental papilla as a result of this, is evidence of the destruction of the contact area and usually points to the presence of a well-established cavity (Fig. 3.2). The fraying of a piece of dental floss when it is passed through an apparently normal contact area is contributory evidence and may suggest that the area is the site of early carious destruction.

Bitewing radiography

In the detection of early interstitial caries the bitewing radiographic technique contributes powerful aid. In this technique the central beam of X-rays is positioned to pass at right angles to the long axis of the tooth, and tangentially through the area of contact (Fig. 3.3). A resulting radiograph

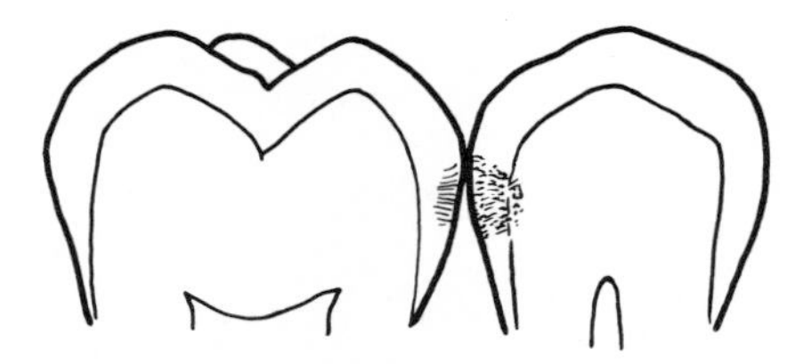

Fig. 3.4. Representation of a bitewing radiograph showing an early lesion confined to enamel and one just involving dentine (see also Fig. 3.5).

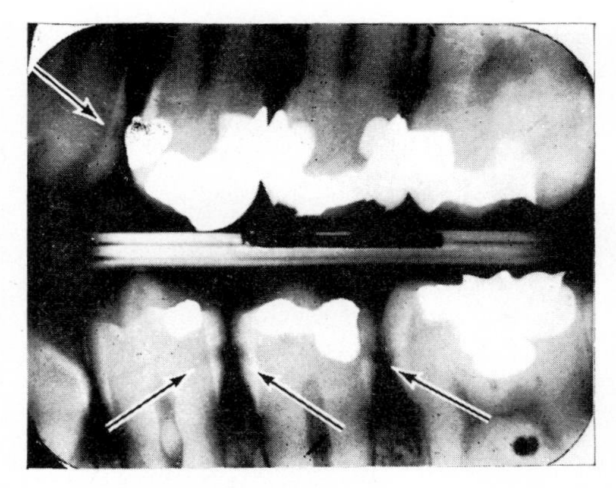

Fig. 3.5. Bitewing radiograph showing early enamel lesions of Class II and Class III cavities. Early involvement of dentine is just visible on the distal of the lower first premolar.

demonstrates a profile view of the contact areas (Fig. 3.4), in which interproximal enamel and underlying dentine may be distinguished in detail, because the outline interproximally is regular and the density minimal. By this means early lesions of enamel and dentine may be detected (Fig. 3.5). It must be emphasized that any overlapping outlines must be rejected. Only a clear tangential view is reliable for diagnosis.

It is generally conceded that, by use of the bitewing radiograph, early carious lesions can be consistently detected some time before they become detectable by clinical examination with the probe. Opinion may vary as to the stage at which some of these lesions should be treated by filling. The factors under consideration are the extent of the lesion radiographically demonstrated, the caries experience of the patient and the frequency of his visits, and the personal predilection of the clinician.

A sound general rule is that *the first certain involvement by caries of the underlying dentine*, shown in such a radiograph, is an intimation that treatment by restoration is desirable. It should be remembered when examining bitewings that most carious cavities when excavated to the boundary of sound dentine seem to be half as large again as they appear on the radiograph.

When, in a radiograph, an enamel lesion *alone* is shown, that is to say with no visible dentinal involvement, it may be desirable or judicious, except in patients with a high caries rate, to leave the matter for review at the next visit. In some patients, especially in those past the peak of caries incidence, *such enamel lesions may remain unchanged for years.*

The recurrence of caries at the cervical margin of a filling can be difficult of diagnosis with certainty. The additional space available in the cervical region of the interdental space helps, however, to simplify the problem of access with a probe in this position.

The evidence supplied by radiography of carious recurrence around the cervical margin of an existing restoration is frequently less decisive, due to a number of factors, the more important of which is the radio-density of a metallic restoration (Fig. 3.6). This may mask, in part or in whole, carious encroachment into adjacent dentine. Careful history, examination, and radiography, however, do not commonly leave the issue long in doubt.

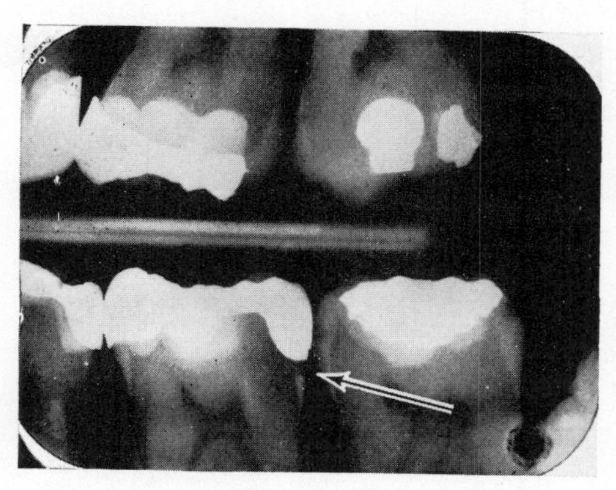

Fig. 3.6. An extensive carious recurrence at the cervical margin of a distal restoration in a mandibular first molar.

At another extreme is the grossly carious cavity, the result of delay and neglect. Such a cavity is immediately obvious on inspection and generally the important question to be answered in these cases is whether the pulp has been invaded by the carious process. **A simple procedure of considerable diagnostic value is performed by clearing the cavity of loose debris by a stream of warm atomized water, followed by drying the tooth and cavity with cotton wool. A small, fairly tightly rolled pledget of wool, held in forceps, is then pressed gently but firmly into the cavity in the direction of the pulp. If the pressure is definitely painful to the patient, a carious exposure and a vital pulp may safely be assumed in the great majority of cases.**

If no pain is experienced upon pressure, either the pulp is necrotic, in which case closer inspection will reveal the carious opening into the pulp chamber, or the pulp has not yet been involved. There remain, however, a small proportion of cases where, though pain cannot be elicited by direct pressure, caries has approached the pulp so closely that the underlying pulp is already lightly infected. For clinical purposes this is a carious exposure and is treated as such.

Although a small number of carious exposures occur without pain, *severe 'tooth-ache' is usually a result of extensive destruction of the crown and well-established pulpitis.*

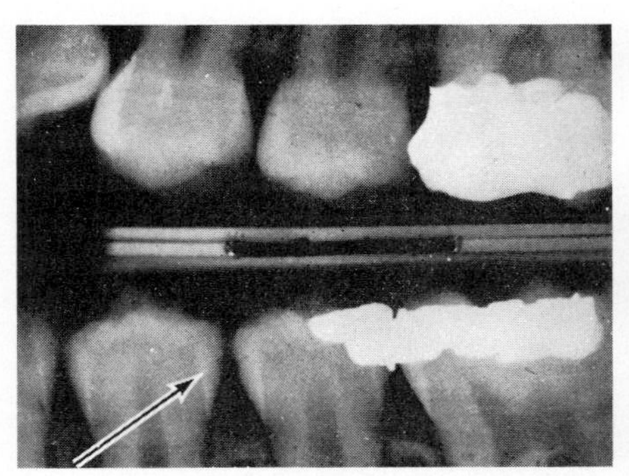

(a) Early enamel lesion.

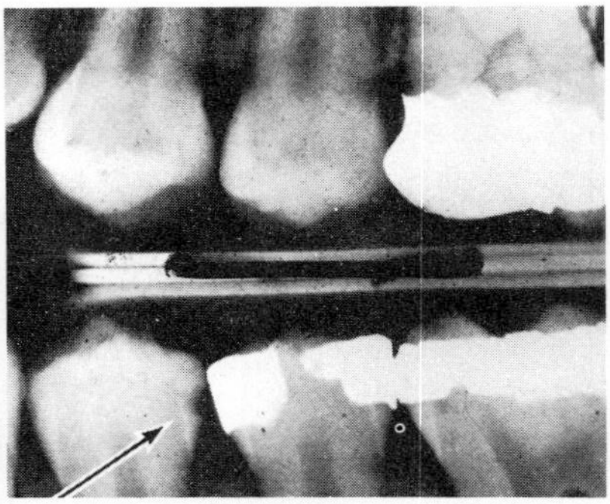

(b) Nine months later – late enamel lesion.

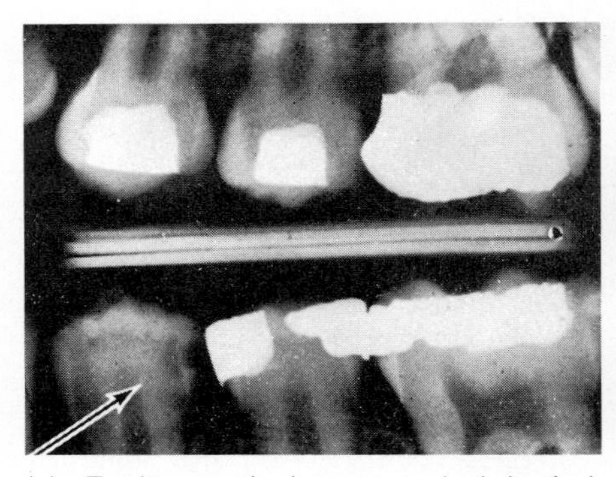

(c) Twelve months later – marked dentinal spread.

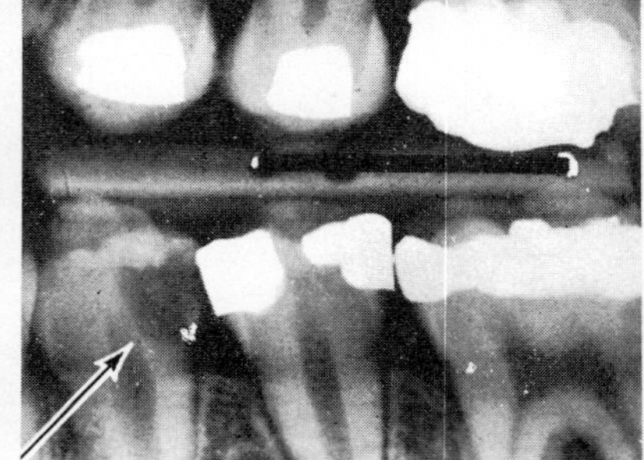

(d) Eighteen months later – approaching carious exposure.

Fig. 3.7. The radiographs record the progress of a Class II cavity in a mandibular first premolar, over a period of eighteen months in a patient aged 15–16 years.

The detection of caries on smooth and accessible surfaces, as for example in the buccal and lingual cervical regions, presents much less difficulty. Discoloration, softening, and loss of substance are all present in varying degrees. This, and the fact the surfaces are visible and accessible to instruments, means that it is only necessary to examine carefully with mirror and probe to find a cavity. In the diagnosis of this type of cavity, as in the occlusal type, radiography usually does not help. Nor is the bitewing radiograph helpful in the *early* detection of fissure caries. In this case the enamel and amelodentinal junction are irregular in form and the radiodensity of the crown is at its greatest. Fissure caries should be detectable to clinical examination long before it becomes visible with certainty on the radiograph.

Black's classification of cavities

For convenience of description we have adopted a simple form of classification devised by G. V. Black, based essentially upon the site of onset of the carious process, and on the relative frequency of the various sites involved.

CLASS I cavities are those originating in anatomical pits and fissures.

CLASS II cavities originate on the mesial and distal aspects of premolar and molar teeth.

CLASS III cavities originate on the miesial and distal aspects of incisor and canine teeth, but do not involve the incisal edge.

CLASS IV cavities originate as do Class III, but they are bigger and the incisal angle is involved in the cavity.

CLASS V cavities are those originating in the cervical third of the buccal and lingual aspects of all teeth, excluding cavities commencing in anatomical pits.

In practice it will be found that, if hypoplastic and traumatized teeth be excluded, there are very few cavities which do not fall into one or other of the above classes. The classification is an effective and convenient abbreviation and most uncertainties in use are resolved if it is remembered that it is based upon the *site of inception of caries*, and not upon the ultimate shape of the cavity prepared for restoration.

Exposed dentine

Pain may arise from areas of exposed dentine in the cervical regions of teeth subject to abrasion or operative trauma, from any area subject to erosion, and from the occlusal surfaces in cases of severe attrition. Pain which is of dentinal origin has a characteristic and penetrating quality, and is commonly provoked by cold, sweet substances, and by brushing, and to a lesser extent by acid fluids like fruit juice, and by heat.

The importance of diagnosis of this cause of pain lies in the fact that the pain is similar to that which may arise from carious involvement of dentine. Since the treatment of the two conditions is different, it is important to distinguish between them and to localize either as a preliminary to treatment (p. 101).

Exposed areas normally become insensitive with the passage of time as a result of the formation of dead tracts and areas of sclerosis. With the aid of some localization by the patient and detailed examination with a probe, sensitive areas may be discovered, although on occasions this may be very difficult to achieve with certainty. It is useful to remember that any exposed area is more likely to be sensitive around its periphery, because this is the site of the dentine most recently exposed, whilst the central portion is more probably isolated from the pulp by sclerotic dentine and dead tracts. These are the commonest causes of pain arising from dentine and the dental pulp. Together with chronic pulpitis they account for the majority of cases of this type.

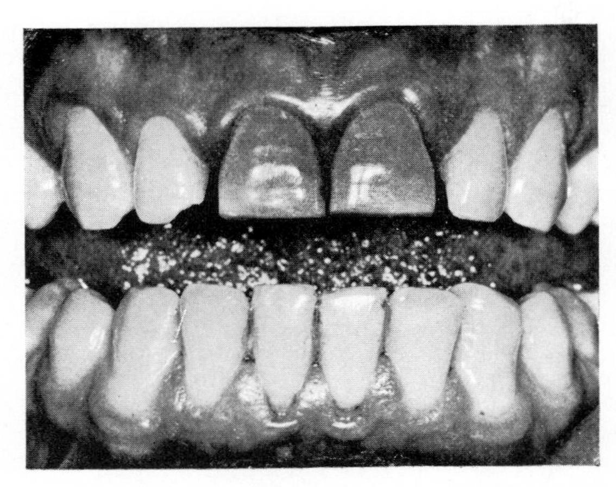

Fig. 3.8. Non-vital maxillary central incisors showing characteristic discoloration of marked degree.

Investigation of vitality of the teeth

Colour. A non-vital tooth is always, after a short period, darker in colour and less translucent than the corresponding vital tooth. The changes may be very slight or very marked (Fig. 3.8). If the change is slight, it is frequently difficult to detect small differences of colour, particularly when the tooth is a molar or premolar and when, as often happens, it is heavily filled. In the case of incisors the comparison is more easily made.

The first stage, therefore, in investigating the vitality of a tooth is to compare its colour and translucency with that of the similar tooth on the opposite side. If the latter is missing or is otherwise unsuitable, the comparison should be made with the nearest normal tooth of similar shape and size. Translucency can be appreciated by transillumination by a mouth lamp from the lingual aspect, in a darkened room. An alternative but less critical method is to allow the operating light to shine on the labial surfaces while the lingual aspect, in the shadow, is inspected with a mouth mirror.

Presence of a sinus. The existence of a sinus over the apical region is strong evidence in favour of a diagnosis of necrosis of the pulp with periapical abscess formation. The lip or cheek should always be retracted and the alveolar mucosa examined, remembering that the upper lateral incisor in particular, may point in the palate.

Thermal reaction. The normal vital tooth usually responds unmistakably to extremes of heat and cold; for practical purposes the non-vital tooth is unresponsive, though there are circumstances in which sensations may be evoked from adjacent tissues. These may mislead the patient, but should not deceive the operator.

The common method of testing thermal reaction involves the application, after isolation and drying of the tooth, of a pledget of cotton wool soaked in ethyl chloride and cooled by rapid evaporation. Heat is applied by warming a suitably sized portion of temporary gutta-percha to a temperature just short of the point at which visible disintegration occurs. This is applied to the tooth in similar manner, taking care in both cases to keep the application away from the gingivae in order to avoid the possibility of confusing sensations from that area.

A normal tooth is first tested as a basis for comparison, and with a perceptive and co-operative patient it is possible to gain a rough idea of the degree of reactivity of normal and abnormal teeth. The response to cold is more reliable than that to heat, but on occasions the response to both is quite indecisive and no conclusions can be drawn.

Electrical reaction. Dentine gives a characteristic lancinating pain when stimulated by the passage of an electric current of suitable intensity. There are several ways of achieving this as a means of assessing normal and abnormal pulp conditions.

One type of tester consists of an electrode which the patient holds in his hand, another which, moistened with some conducting paste, is placed upon the dry crown of the tooth. A direct current is passed and increased to a point on the scale at which the patient is just aware of pain. Thus one can obtain a quantitative estimate of the response which is reproducible and which, though not very precise, is more accurate than the results of thermal tests.

Neither thermal nor electrical tests are infallible. Not infrequently no reliable response can be obtained to either of these stimuli. Further, the more extensively filled the less reactive the teeth become, and it is these cases in which the pulpal condition is more often in doubt.

Percussion. This is performed by firmly tapping the crown of a tooth with a light instrument; the end of a mouth-mirror handle is commonly used. The crown should first be tapped in an axial direction and then obliquely, on buccal and lingual surfaces, and mesial and distal surfaces if these are accessible.

This yields *no direct evidence* of the pulpal condition. Though heavy percussion is unpleasant, it is painless in a normal mouth; if pain does result it is strong confirmatory evidence, in the absence of recent trauma and in the presence of other positive findings, of apical periodontitis arising from necrosis of the pulp.

Radiography. When apical radiography shows the presence of true periapical rarefaction of whatever degree, the death of the pulp must be assumed until it is disproved by positive clinical evidence. Inflammatory resorption of the bone is shown at its earliest stage by the disappearance of the *lamina dura* surrounding the apex.

Instrumentation. In the last resort, the operator may have to cut into an area of dentine which he knows from experience should be normally sensitive, usually in the cervical region of a Class II or Class V cavity. Such a step is rarely necessary, however, it being usually preferable to keep a suspected tooth under observation.

The periodontal tissues

Thorough investigation of the periodontal tissues is an essential part of the general examination of the mouth. It involves inspection of the investing mucosa and gingival margins and examination of gingival crevices, with an assessment of the amount of plaque and calculus deposited. In addition, the mobility of the teeth and the extent of pocketing must be explored. The bony changes can only be properly assessed with the help of intra-oral radiographs and this is an essential step in complete diagnosis. Similarly, the evaluation of plaque-control by use of disclosing agents and the performance of a detailed dietary analysis are indispensable, for a high caries rate and poor hygiene, if uncorrected, will lead to the failure of the best plan for restorative treatment.

Functional occlusion

The object of all treatment is to maintain the teeth and jaws in a condition in which they can function naturally and efficiently. It is therefore important to examine in some detail the relation of the dental arches in centric occlusion, in protrusive and in right and left lateral positions. Loss of teeth, irregularities due to drifting and tilting, over-eruption, premature contact, the efficacy of any dentures the patient may be wearing — all of these, together with the appearance of the anterior teeth, form a picture of the functional efficiency of the dentition. This is really the basis on which treatment is built.

It is mainly in cases of abnormal function of the temporo-mandibular joint, instances of severe malocclusion, and in patients needing extensive restorative treatment that models and occlusal registration will be needed. In these it is an essential adjunct to sound diagnosis and treatment planning.

Treatment planning

With the completion of all investigations and the establishment of a correct diagnosis it will be found helpful in all but the simplest cases to make a treatment plan. Many, perhaps even most, of the cases seen in practice involve a relatively short course of conservative treatment. Here it may be helpful to note the order in which the teeth should receive attention. In more complicated cases involving, for example, surgical, prosthetic, orthodontic, or other treatment, it is necessary for effectiveness and economy that the procedures shall follow a precise order. This sequence is arrived at by consideration of all available clinical evidence and all the therapeutic and technical details of the course of treatment prescribed.

A proportion of treatment plans will start with the relief of pain since this may be a matter of urgency. The completion of others may depend upon the outcome of investigations such as those required to decide upon the treatment suitable for a heavily filled and possibly non-vital tooth in a strategic position.

A typical treatment plan might run in this fashion:

1. Open and dress |6.
2. Scaling and cleaning.
3. Occlusal equilibration.
4. |47 gold inlays.
5. Partial lower denture.
6. Remaining restorations.

or

1. Scaling and cleaning.
2. |6 amalgam.? savable.
3. Extract |de remnants.
4. Upper fixed appliance.
5. 64|4 amalgams.

The true importance of making such a plan derives from these considerations:

1. **It ensures that the clinician reviews the treatment of the case in the light of all available evidence** and of each aspect of treatment in relation to the others.
2. **It is a valuable record for later reference**, particularly in complicated cases and after a lapse of time.
3. **It avoids the risk of disorderly and ill-advised management** of the case, which could lead to unnecessary discomfort for the patient, uneconomic use of the time of both parties, and even to ineffectual treatment.

The objective, when making the plan, is to draw up a scheme of procedure which takes full account of the patient's best interests and the accomplishment of the treatment with greatest efficiency. It is, moreover, generally wise, and greatly appreciated, if the matter is discussed with the patient and *explained in simple language*. It is important to remember that, however familiar the terms and procedures of dental surgery may be to the dental surgeon, most patients are ignorant of all but the simplest and commonest methods of treatment. Any course of treatment will be most successful if it evokes the patient's understanding and co-operation. To achieve this **it must be clear to the patient that in all the arrangements made for his treatment, the patient's best interest is the first consideration.** It must also be apparent that some appreciable progress is achieved at each visit, and the course of treatment as a whole must not be allowed to lose impetus, to drag out into a series of irregular, ineffectual visits and broken appointments, finally to founder in the frustration and loss of interest of all parties concerned.

Where a long and complicated course of treatment is required it may be better in some cases to choose a lesser objective and be sure of its achievement. In other cases it may be wise to divide the course into two or three shorter periods each with a well-defined aim, but separated from one another by periods when the patient can count himself free from the necessity of frequently recurring visits.

Reassurance and encouragement must also be given in a form suitable to the patient's age and understanding. However unemotional and placid he may appear, nearly all patients approach dental procedures with a degree of apprehension, most of which arises quite unnecessarily from ignorance of what they may be expected to endure. *It should never be necessary to inflict severe pain*; nevertheless most dental operations are unpleasant in one way or another, a fact which must always be borne in mind. A few quiet and sympathetic words at an early stage may well relieve these fears. At the end of the appointment a little appreciation of the patient's help and forbearance assists in building the very necessary trust and confidence which is the essence of the ethical relationship.

Summary

Diagnosis, recognition of disease. Prognosis, forecast of likely progress.

History. To elicit all relevant information from patient on previous course of disease. Immediate complaint; duration of symptoms; past dental history; past medical history and family history if relevant, see p. 32.

Physical examination. General inspection, face, neck, mouth, and affected area. Inspection of all mucosal surfaces. Methodical examination of clinical crowns and periodontium. Special tests, some by dentist, others require laboratory facilities and skills.

Routine and emergency examinations differences in procedure aimed at relief of emergency. Recording and charting, Szigmondy–Palmer and International notations.

Detection of caries. Use of mirror and probe. Importance of dry crowns. Early fissure caries and dentinal spread; early interproximal caries; frayed dental floss; food packing. Bitewing radiographs; X-rays tangential to contact areas; early enamel caries and earliest visible spread to dentine. Recurrent caries. Diagnosis of carious exposure by pressure. Radiographs no help in diagnosis cervical and fissure caries.

Black's classification of cavities. Classes I to V, based on site of inception and frequency of occurrence.

Exposed dentine. Important to distinguish from caries. Pain commonly to sweet and cold on freshly exposed area. Localized often at periphery of areas of abrasion and attrition.

Non-vitality of teeth. Loss of colour and translucency follows loss of vitality. Compare contra-lateral teeth. Sinus in apical region. Absence of thermal response suggests non-vital. Possible increased response in chronic inflammation. Electrical reaction; calibrated response. Percussive tenderness as evidence of apical periodontitis. Apical radiograph; irregularity of periodontal space and loss of lamina dura.

Periodontal examination, mucosa, crevices, plaque, calculus, mobility, pocketing, oral hygiene, plaque control; dietary analysis.

Occlusion. Centric and lateral. Drifting, tilting, over-eruption, premature contact. Dentures. Functional efficiency as a whole. Appearance of visible teeth.

Treatment planning. Plan ensures review of all treatment, provides record for reference, guides management of case. Importance of full explanation to patient, whose best interests are first importance. Reassurance and encouragement.

4

The control of moisture, pain, and trauma in operative procedures

The presence of moisture on or near the field of operation is a constant source of difficulty in restorative practice. Even when a moderate flow of saliva is temporarily diverted, the effect of capillary action along the gingival trough, and of seepage of blood and tissue fluid from adjacent gingivae have still to be guarded against. The need to control the access of moisture to the immediate field of operation arises for three main reasons:

Visibility

It has been mentioned that reflection and refraction at a film of moisture reduces the visibility of surface details. A more important factor in this context is that moisture retains in the immediate working area the detritus resulting from the use of rotary and other cutting instruments. This partly or completely obscures the precise point of working and necessitates frequent interruptions in order to clear the field of debris. This serious difficulty has been mitigated by the introduction of water sprays, and water and air jets in various forms. These were adopted primarily as a means of dissipation of heat resulting from the use of higher rotary speeds; the improved visibility, though secondary, has also proved important. Atomized water sprays directed at the cutting head of the rotary tool, are most effective in removing saliva and detritus. When volume and pressure are correctly adjusted the visibility of the working area is extremely good.

Contamination of materials

The inclusion of moisture has a detrimental effect upon nearly all filling materials, temporary and permanent. In some cases, such as amalgam composites and silicate cement, the properties are severely affected. It is essential that, for certain periods at least, cavities be kept dry if a good restoration is to be performed. The local application of drugs is less effective if the drugs are diluted and dissipated by saliva. This is another reason why moisture must be controlled, if not eliminated.

Contamination by infection

In certain techniques, especially those of endodontics, the preservation of a sterile field and the avoidance of extraneous infection is essential to a satisfactory result. This means that saliva and exudates from the mouth must be completely excluded, since these contaminants are invariably infected. Complicated techniques involving antiseptics and antibiotics are of no value whatever as long as there is a possibility of casual reinfection from the prolific organisms which abound in the oral cavity in health and in disease.

Methods of control

The rubber dam

There can be no doubt that the method which gives most complete control over moisture in the mouth is the use of the rubber dam, and its introduction over a century ago marked one of the significant advances in the history of restorative dentistry. It is the most certain of all available

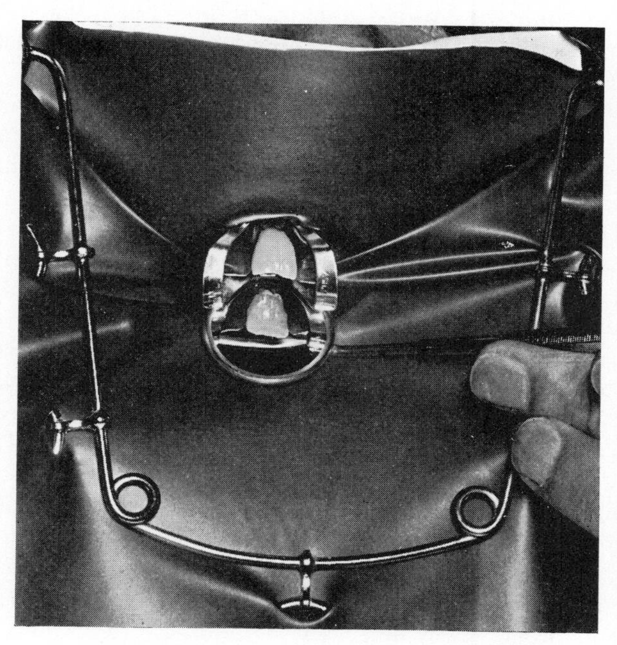

Fig. 4.1. Central incisor isolated by rubber dam, retained by cervical clamp.

methods and is the only acceptable method in procedures where absolute control is essential to the technique.

Reduced to its simplest form, it consists of making a hole of suitable size in a thin rubber sheet and slipping the hole over the crown which is to be isolated from moisture. This crown is thus situated on one side of the sheet, whilst the mouth with its moisture and infection remain on the other (Fig. 4.1). The line where the rubber encircles the cervix of the tooth can be made leak-proof, and the remainder of the sheet is supported and retracted. The isolated crown may be operated upon free from oral contamination and, equally important, the oropharynx is protected from the ingress of any small body which might otherwise fall to the back of the mouth.

Facility with rubber dam technique is not easily acquired and its expeditious and effective use is a good test of operative ability. No attempt will here be made to describe the diverse modifications suitable to the various conditions in which the technique is used. These have been well described and are only to be acquired by diligent practice. There are, nevertheless, some important points which can be easily missed.

Though it may be used to isolate many teeth at once, the dam should exclude from the mouth only as many crowns as are essential to the correct performance of the operation in question. Every additional hole made, every additional cervix included, introduces further possible sites of leakage; these should be avoided. For example, in endodontic treatment of a single tooth, only the crown of that tooth should be isolated. If an MOD restoration is to be done on ⌈6, then ⌈57 must certainly be included otherwise contact areas cannot be restored.

The contact points of all teeth to be included in the dam should be tested by the passage of dental floss before application of rubber is started. Tight contacts, and those obstructed by defective fillings, can frequently be eased with abrasive strips of metal and polished with fine linen strips. Others can be opened temporarily by the insertion of a small wooden wedge at the cervix. At all events, it is of greatest importance to know before application of the rubber is attempted whether the contacts are easily passable, difficult, or impassable.

There are teeth upon which it is virtually impossible to place or retain an effective dam. This may be by reason of the conical shape of the anatomical crown, the shortness of the clinical crown owing to incomplete eruption, or to the inaccessible situation of the crown in the dental arch. These limitations must be foreseen and in dubious cases a trial application of a suitable clamp or ligature, which would be used to retain the dam in position at the cervix, *before* the dam is prepared, will assist in its correct siting. It may often prove the feasibility of a procedure which looked impossible. With increasing experience the number of cases where the placing of a dam is impracticable progressively decreases.

It is clear that the subgingival cavity will present considerable difficulty and sometimes may, for practical purposes, prevent the successful use of the dam. Specially designed cervical clamps such as Ferrier's with which the rubber and the gingiva may be forcibly retracted in the bucco-cervical region may assist in some of these cases.

The exact conformation of the rubber margin at the cervix is a matter of primary importance if that margin is to remain leak-proof. First, the holes punched in the sheet rubber must be of the correct size to encircle the tooth cervix tightly, but not small enough to be torn during placing. Then the spaces between the holes must be big enough to allow the rubber to lie unstretched between the necks of the teeth, otherwise the holes are distorted and leakage of saliva readily occurs.

The distribution of the punched holes follows the outline of the upper and lower arches; it saves time and helps in placing if this pattern is stamped on the sheet in advance. The position of the holes can be varied between the larger and smaller arches and modified to adapt to irregularities of the teeth.

Contrary to what might at first sight be expected, the free edge of the circular hole, stretched to embrace the cervix, must be inverted into the gingival crevice (Fig. 4.2). The clamp or the ligature used to retain the rubber is placed in a position immediately coronal to the point of inversion. It does not include the rubber at any point but retains it simply by virtue of the fact that it prevents the rubber from leaving the cervix and the gingival crevice. Manipulation of this margin is greatly facilitated if the crown and the crevice are first carefully dried before placing the rubber.

The free edges of the rectangular rubber sheet must be supported and retracted to give clear operative access. This is done by the use of a wire frame (Fig. 4.1), or of an elastic headband; each device and each design has

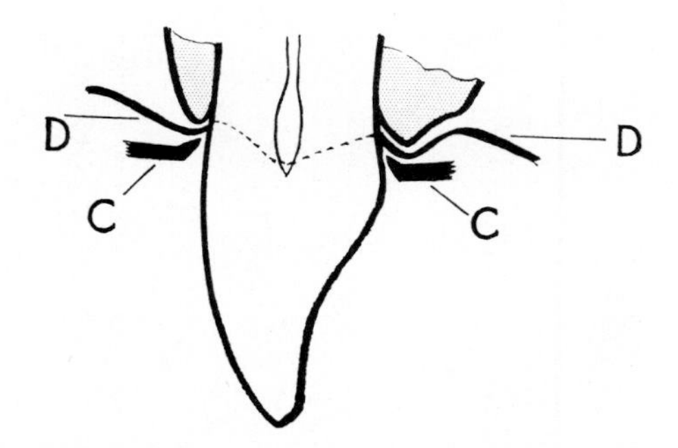

Fig. 4.2. Relationship of rubber dam (D) to cervix and clamp (C).

its advantages, but in all cases it is of importance that just sufficient tension shall be used, in the right direction, to give the best result in that particular case, having regard to the requirements of the operation and the patient's comfort.

The saliva ejector

This is an essential item of equipment. It is a water suction pump, of the venturi type common in laboratories, and is connected by a flexible tube to a curved endpiece (Fig. 4.3) the open end of which lies in the sublingual area. The suction required is of a low order, about 10 to 12 cm of mercury, and the design of the mouthpiece, of which there are many patterns, has been the subject of much ingenuity. There are several designs which merit attention. One type incorporates a metal flange to protect the tongue and a self-retaining device consisting of a spatulate metal member, adjustable to clamp lightly under the chin. Another type comprises a mouthpiece tube which is bifid, one branch designed to lie in the buccal sulcus and one in the

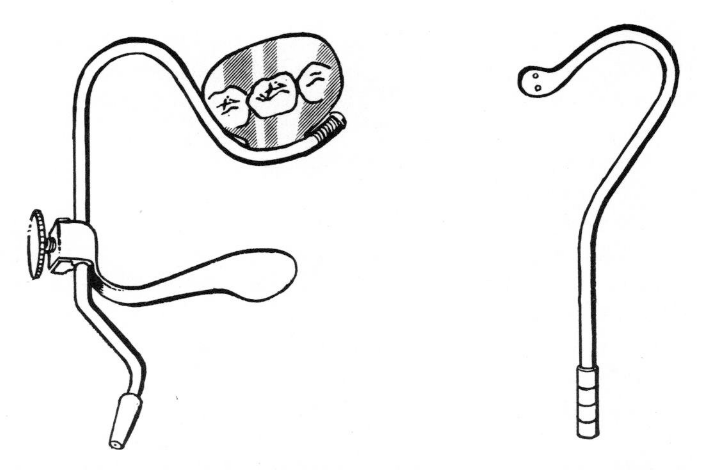

Fig. 4.3. Types of saliva ejector mouthpiece, one with tongue guard and retainer.

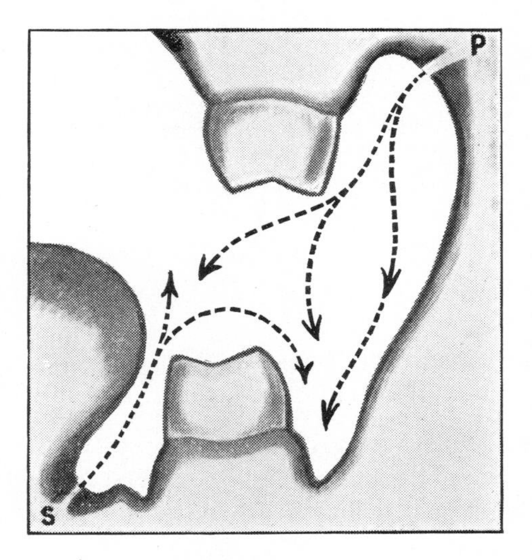

Fig. 4.4 Directions of flow of saliva, downwards from parotid duct (P) and upwards from sublingual and submandibular ducts (S).

lingual, and a number of patterns incorporate flexible parts in plastics. Disposable plastic mouthpieces which can be bent to suitable shape are also in common use.

The problem is to allow an adequate uptake of saliva, which may be very viscous, and of water, which may be copious with present cutting techniques, to avoid blockage by debris and, at the same time, to avoid aspiration of the fine mucosa of the floor of the mouth. One answer to this problem is the 'washed field technique', in which the large volume of water used in high-speed cutting is immediately removed by an aspirator which takes in a large volume of air at low negative pressure. The wide-bore nozzle of the aspirator is held close to the operation area by an assistant. It is very effective, but noisy in operation. Correct adjustment of water and aspiration results in a remarkably clean field with excellent visibility. This type of aspiration is commonly used when operating on a patient in the prone position.

Saliva enters the mouth (Fig. 4.4) from the parotid, sublingual, and submandibular ducts. From the orifice of the parotid duct it flows on to the maxillary teeth, down the buccal mucosa of the lower sulcus, and also across the retromolar region to both lingual and buccal aspects of the mandibular teeth. From the submandibular and sublingual ducts the sublingual area is gradually filled with saliva. By movement of the tongue and floor of the mouth, particularly in the act of swallowing, this saliva can well up to overflow into the buccal vestibule. In the prone position saliva from all ducts flows into the retromolar regions and thence across the angle formed by the dorsum of the tongue in contact with the palate.

The accumulation of saliva in the buccal and lingual sulci will obviously be greatly influenced by the position of the head. A moment's thought will confirm that a saliva ejector must be so placed that its opening is as near as possible to the lowest part of the area to be drained. Placed behind the lower incisors, an ejector will aspirate nothing but air if the patient's head be well tipped back. Placed in the left sublingual region, it will have little effect if the head is tilted to the right side. It is normally more effective in the sublingual sulcus than in the buccal sulcus.

It will therefore be apparent that head position should be considered not only for access to the area of operation, but also with regard to saliva control. With the rubber dam, the ejector is used for the patient's comfort, but many prefer to do without it. Most adults find it easy to swallow with mouth open when the dam is correctly adjusted. Generally speaking, the ejector is uncomfortable to the patient if used for long periods, and in any case, if it is ineffective, it should be discarded and another means of control adopted.

In four-handed operating with the patient supine, saliva control depends more on the constant use by the assistant of the triple syringe—one which delivers on demand a jet of water, air, or spray—and the evacuator. Her duty is to keep the prepared cavity clean and visible, to avoid the collection of water at the back of the mouth, to keep the mirror clear, and occasionally to retract lips or cheek. The commonest positions for the tip of the evacuator are opposite the first molar, either lingual or buccal, and the retromolar area. It is fairly easy for the supine patient to make an effective seal between dorsum of tongue and soft palate. Should any leakage occur, swallowing is not difficult.

The use of antisialogogues has never found general favour in conservative dentistry, though their efficacy is clearly shown when operating under general anaesthesia: atropine is almost routinely given as pre-operative preparation for the bed patient. The disadvantages are that it takes time to act, even when injected, and, more important, it is unselective in action. It paralyses not only the secretomotor nerves to salivary glands but all parasympathetic nerve endings. This can give rise to discomfort of some duration which, though not severe, is generally considered unacceptable when other methods of control are normally available.

Cotton-wool rolls

These are cylindrical in shape, 3 cm or 15 cm long, and available in three diameters. They are the commonest form of absorbent used in the mouth. Rectangular portions of compressed cellulose are also useful and some operators claim them to be more effective.

They are placed in the sulci; in the upper buccal sulcus close to the orifice of the parotid duct, in the lingual sulcus close to the orifices of the submandibular and lingual ducts, and in the lower buccal sulcus to absorb the flow which may escape either of the former. Their correct insertion is a matter of some importance.

When inserting a roll into either upper or lower buccal sulcus, the cheek should be retracted with a mirror, and the roll placed in the depth of the sulcus with a slight axial twist, outwards and downwards in the upper, outwards and upwards in the lower (Fig. 4.5). In the case of the upper, however, it is necessary to see that the roll is not *above* the duct opening, as can sometimes happen.

In the sublingual region, the roll should be held against the lingual cusps of the cheek teeth whilst the patient is asked first to protrude and then to relax his tongue. In relaxation the roll is made to follow the floor of the mouth back into normal position, where the lateral aspect of the tongue covers and retains the roll.

Several retainers for cotton-wool rolls are available. Of these probably the most effective is the winged clamp chiefly for use on lower teeth (Fig. 4.6), but any such device must allow easy exchange of the rolls which, even with the use of an ejector, may rapidly become saturated. A 15-cm roll threaded lengthwise on 13 cm of soft stainless steel wire is conformable yet semi-rigid and can be adapted to either upper or lower buccal sulci or to lingual and lower buccal sulci.

The cavity on which the operator is working may be dried with pledgets of cotton wool or other cellulose. Not only the cavity itself, but also the crown of the tooth and the gingival embrasures in the vicinity should receive this attention, and a short blast of warm air at low pressure completes the operation quickly and effectively. Drying, in this context, must be defined as the removal of free moisture from the surface in question, and it is to be sharply differentiated from dehydration of the tooth substance and, in particular, of dentine. Any attempt at dehydration of a cavity is noxious to the pulp and dentine, and the use upon vital dentine of dehydrating agents such as absolute alcohol has no place in operative dentistry. The surface

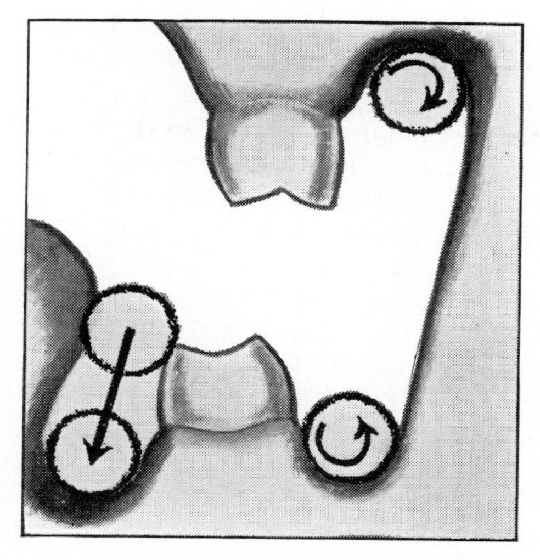

Fig. 4.5. Directions of axial twist used in placing cotton-wool rolls in the buccal sulci, and the path followed by the lingual roll into the sublingual position.

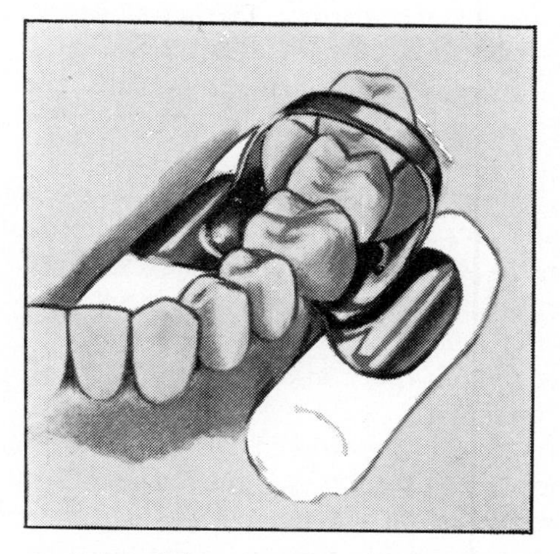

Fig. 4.6. The use of winged clamp to retain cotton-wool rolls in the lower sulci.

should be dried until it loses its reflective film of moisture and is just seen to blanch, and no more.

Haemorrhage and exudate

Both of these may occur from the gum and are frequently unavoidable when the cavity or area of preparation extends to the gingival margin or below. Haemorrhage usually ceases spontaneously after instrumentation ceases, but the damaged area continues to exude tissue fluid and this is of sufficient amount to be important, particularly as it occurs where visibility and access are difficult.

The primary rule for haemostasis anywhere—stop the trauma and apply pressure—applies here. Firm pressure with a pledget of cotton wool is occasionally effective in the mildest cases after a short period, but it is often unavoidably necessary to proceed with the trauma, for example in cavity preparation.

If, as is often the case, a local anaesthetic is already in use, the injection of a fraction of 1 ml close to the bleeding point controls it very rapidly by pressure and vasoconstriction. Isolation of the area with rolls and the topical application of 1:1000 solution of adrenaline hydrochloride on a cotton pledget packed on to the area is also very effective, and stronger solutions up to 1:200 may be used in this manner and, *for this purpose only*, without detriment.

Chemical styptics and the actual cautery are sometimes advocated for the control of bleeding. Some chemical styptics are unsatisfactory in that they leave behind considerable blackened eschar which, when disturbed, gives rise to recurrence of bleeding. Except in expert hands, the actual cautery carries some risk to adjacent tissues. Diathermy is very effective but requires the use of special apparatus not always possessed by the private practitioner. When, however, this apparatus is available the removal of small areas of gingiva, after the coagulation of bleeding points is usually rapid and very effective.

Seepage of exudate from a small area can be overcome in two ways. Where time permits and instrumentation is completed, the insertion of a temporary dressing to allow an interval for the area to heal is helpful and at the next visit the problem is solved by careful avoidance of gingival trauma. *It may be justified in some cases to perform, at the end of the initial visit, a very local gingivectomy in order to achieve this state of affairs at the next visit.*

Alternatively, the area being isolated with rolls and thoroughly dried, 20 per cent trichloracetic acid is applied to the raw area by the use of a pledget of cotton. In a matter of thirty seconds a white eschar forms, and this prevents exudate from escaping for some four to six minutes. The styptic may be reapplied if required. Zinc chloride solution of 10 per cent strength may also be used.

A general caution must be given concerning the use of escharotic drugs in the mouth. *They must always be used with the greatest care and circumspection. No more of the drug should ever be carried into the mouth than the minimum necessary to achieve the desired results.* The applicator, whether a pledget of cotton wool or the end of an instrument, must be watched all the way from the supply in the Dappen's glass to the point of application so that accidental contact with face, lips, cheek, and gum is avoided. Contamination of the instrument, thence the operator's fingers and the patient's person, must be carefully avoided. Clear labelling of containers and a routine check between assistant and operator that the drug asked for is in fact the one proffered — these must become a matter of routine strictly observed, in order to reduce to vanishing point a risk to the patient which is always present, always detrimental, and which may be disastrous.

The control of pain and trauma in operative dentistry

Some operative procedures are painless, some uncomfortable, but the majority are painful. Most of the more advanced techniques of operative dentistry and endodontics would be impossible without adequate means of pain alleviation.

The ministrations of the dental surgeon are still closely allied in the public mind with painful experience, in spite of the fact that effective anaesthetics and anodynes have been available for decades. The most serious implication of this is that fear of pain is still by far the most potent factor preventing the public from seeking adequate treatment at a time when treatment can economically be applied.

It is a primary responsibility of the profession at all times to use available methods for the avoidance and elimination of pain. It may be agreed that a patient should not be expected to submit to anything more than discomfort of moderate degree and short duration, and even this only when the measures required to avoid it would entail greater discomfort, or risk to his general well-being.

Avoidance of pain starts with careful handling of soft tissues. Excessive retraction, particularly of the labial and buccal frena, excessive drying of mucosal surfaces, accidental trauma by instruments to lips, cheek, tongue, and periodontium, are sources of pain which can be avoided by gentle and deft operating. That they are the criteria by which patients are apt to judge a dental surgeon's ability is an indication of the importance to the patient of this aspect of pain elimination.

Local anaesthesia is unquestionably the most important technique at the operator's command for elimination of operative pain. An anaesthetized tooth can, however, be seriously damaged. The commonest injury, because it is so easy to commit, is the incorrect use of rotary instruments, resulting in a sharp rise of temperature of the order of some hundreds of degrees in the immediate area of the cutting surfaces. This injures the odontoblasts, of which the fibrils are destroyed. In severe cases the damage to the parenchyma of the pulp may be sufficient to cause general hyperaemia, and local or generalized necrosis of the pulp.

Figure 4.7 (a) shows the pulp immediately below a very deep cavity cut at low speed under a continuous stream of water; the tooth was extracted shortly afterwards. Here the pulp shows no detectable signs of injury.

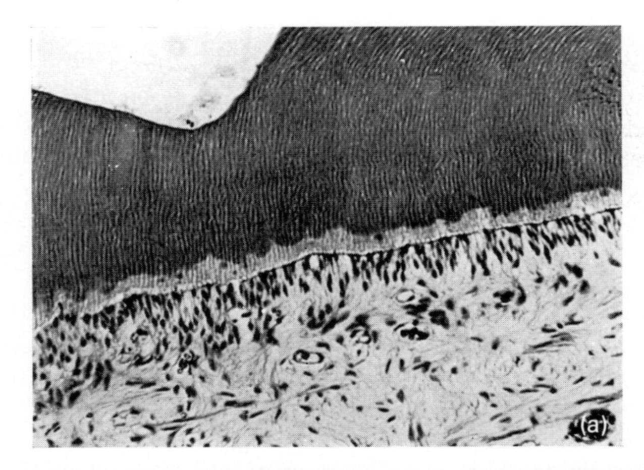

Fig. 4.7. (a) Pulp below the deepest part of a cavity prepared at 5000 r.p.m. under a continuous stream of water. (b) Localized destructive lesion of the pulp adjacent to a cavity prepared with an air-rotor with full spray. (By courtesy of Professor D. S. Shovelton.)

Figure 4.7 (b), by contrast, shows the pulp condition eleven days after the preparation of a cavity by air-rotor with normal spray. The destruction of the odontoblast layer is clearly seen, with all the cellular and vascular changes of acute inflammation.

The evidence suggests that with all mechanical methods of tooth cutting there is concomitant pulpal damage of some degree. Within usual limits this is capable of resolution. Discomfort of varying severity after filling is indicative of a reactionary hyperaemia.

When using all rotary instruments trauma can be reduced by having sharp instruments so that friction is reduced. By using low speeds of about 1000 to 1500 r.p.m., with moderate pressure of about one kilogramme and by cutting at intervals with short rests, heat is allowed to dissipate.

At rotary speeds above a level which may be arbitrarily put at 4500 r.p.m., it is no longer permissible to cut without some effective form of cooling device; this can be used with advantage at lower speeds. At high speed, as mentioned in Chapter 2, effective cooling presents several problems, the adequate solution of which is essential to the use of the method.

A further factor of importance is the extent of the operative injury. Under local anaesthesia much more extensive procedures are undertaken, in the process of which considerable areas of normal dentine may be invaded. This is undesirable and should be avoided for several reasons, of which one is the pulpal damage involved, with its associated pain, and the attendant threat of pulpal necrosis. When it is unavoidable it must be accepted as a calculated risk. Discomfort may be alleviated by giving the patient a simple anodyne after the operation.

The range of local anaesthetic substances and of 'non-pressor' vasoconstrictors, together with improved techniques made possible by disposable needles and syringes, places local anaesthesia in the forefront of analgesic procedures. It could be argued for example, that wherever the pain of an operation is likely to exceed the discomfort of injection and possible later soreness, in the absence of medical contraindications, local anaesthesia should never be withheld. It must be understood that such a broad statement presupposes knowledge of the drugs administered, sterile and well-maintained instruments, and close attention to technique. Only in these conditions can the inherent risks to the patient be reduced to acceptable proportions.

In these circumstances the nature of the proposed procedure is assessed and, if the case warrants it, as do a very high proportion of operative cases, the local anaesthesia is administered before the operation is started. **It is bad management to be forced, by the rising pain experience of the patient, to administer a local anaesthetic late in the course of an operation.**

General anaesthesia has a place in this field, particularly with the availability of intravenous drugs, rapid in action and elimination; methohexitone is the drug most commonly used. It is properly used by the expert only in those circumstances where full facilities are available for correct supervision of the unconscious patient and for meeting those anaesthetic emergencies which, though rare, can be fatal unless properly handled. Its particular indications are for those patients who are highly apprehensive and intolerant of operative procedures and who are yet in need of considerable treatment. It is often the method of choice for intractable children, for the physically handicapped, and for those of subnormal mentality.

The use of sedative drugs is important and could with advantage perhaps be more frequent than it is. The administration of barbiturates and compounds containing codeine assist in allaying anxiety and depress the pain response. Their judicious use can make all the difference between a refractory and intractable subject, and one who is quiet, submissive, and free from undue apprehension. A number of new techniques are now available for the use of heavy and light sedation, by the administration of drugs intravenously or by mouth. One of the most successful at present in use is diazepam, which allays anxiety and gives some relaxation and a degree of amnesia. This and other combinations of drugs are used in conjunction with local anaesthesia and have the advantage that, in contrast to full anaesthesia, the protective reflexes of coughing and swallowing are not abolished. The patient remains to some extent co-operative and on recovery remembers little or nothing of the operation. This is a development of significance for operative dentistry and one which offers promise for the future.

Relative analgesia by the use of inhalation anaesthetics, such as nitrous oxide-oxygen mixture and trichlorethylene vapour, to the point of abolition of pain perception, but short of loss of consciousness, has had its vogues, but is has various severe limitations. Suggestion under hypnosis is also effective in some subjects, but is again considerably limited in application. None of these methods can be compared in universality and simplicity of application, nor in freedom from undesirable side-effects, with local anaesthesia for the general run of operative work.

Control of minor emergencies

There are inevitable risks in all operations; dentistry is no exception. The dentist works in the limited space of the mouth, a cavity covered and surrounded by soft and mobile tissues, and in close proximity to the origins of the respiratory and alimentary tracts. Sharp hand instruments, high-speed rotary instruments, and the manipulation of small objects in awkward positions provide opportunities for possible injury and mishap. In this section minor emergencies which might arise simply from operative procedures will be considered. Some are almost trivial, some simply uncomfortable, but some hold serious implications for the welfare of the patient or the operator. Care and forethought in operating will greatly reduce inherent risks, but they cannot be entirely eliminated. In the event of any mishap it is very important that the operator should know what steps to take to minimize the injury.

Obviously, deft and skilful operating is of primary importance. The insertion of sharp or rapidly rotating instruments to the correct point of application, and their withdrawal without touching lips, cheek, or tongue is a basic requirement. The use of finger- or thumb-rest *at all times* and insistance upon sharp instruments is essential good practice. *Blunt instruments require more force and are likely to slip. All equipment, and particularly handpieces, should be regularly serviced.* Apparently trivial injury to soft tissue may upset a patient to an extent out of all proportion to its severity. Such incidents tend to undermine the confidence of the patient in the operator's manual dexterity.

A major consideration is the position of the patient. If he is almost upright a small loose object will fall fairly safely on to the anterior part of the tongue, floor of mouth, or the vestibule. On the other hand, if the patient is supine such an object will fall to the back of the mouth where the tongue and palate usually form an effective seal. If it is not retrieved the risks of inhaling the object are minimal. *In the reclined position, with the body at an angle of 45 degrees, the chances of a small object falling straight backwards into the oropharynx are certainly greatest.* Complete protection against this type of accident is provided by a properly applied rubber dam. This may be considered a primary reason for the use of the dam in endodontics, when a number of small sharp instruments are regularly in use.

When a small object is lost in this manner the first action is to examine quickly but thoroughly the faucial region, the sublingual and vestibular sulci, in that order. Do not disturb the tongue and palate too much, for the object may be retained in the oropharynx for a short time before ingestion. If a high volume aspirator is in use the possibility of the object's having been sucked up must be assessed.

Then get the patient out of the chair, as far as possible without his remaining in the upright position for too long, bend him sharply at the hips and give him one or two firm slaps on the back. This may result in the immediate delivery of the lost object. If not, and examination of the aspirator filter fails to disclose it, then the possibility of inhalation or ingestion must be considered. Sometimes the patient knows that he felt 'something going backwards' and may even know that he has swallowed it. Often the patient is unaware of anything untoward. Quite large objects such as a full molar crown can fall straight through the glottis into the bifurcation of the trachea without the patient's being aware of it.

The next step is a thoracic radiograph to localize the foreign body in the aesophagus or stomach or, more seriously, in a bronchus (Fig. 4.8). In the latter case bronchoscopy and early removal is called for to avoid the certainty of pulmonary collapse. If the object cannot be removed at bronchoscopy, then thoracotomy will be necessary.

The inhalation of any foreign body also carries the risk of laryngeal spasm requiring tracheostomy; every surgery should be equipped and every operator prepared for this rare but lifesaving intervention.

Most objects of dental origin inadvertently swallowed can be relied upon to pass through the alimentary tract without incident. Progress should be monitored by abdominal radiographs and if necessary by faecal examination. Surgical intervention is fortunately only very rarely necessary, but of course careful supervision is necessary till the risk is past. Objects of low radiopacity carry an additional risk due to difficulty in their localization by radiography and this is more serious when they are inhaled. Radiopaque dental plastics are available and should perhaps be used more often than they are.

In all dental operating injury to the patient's face is a risk which must be

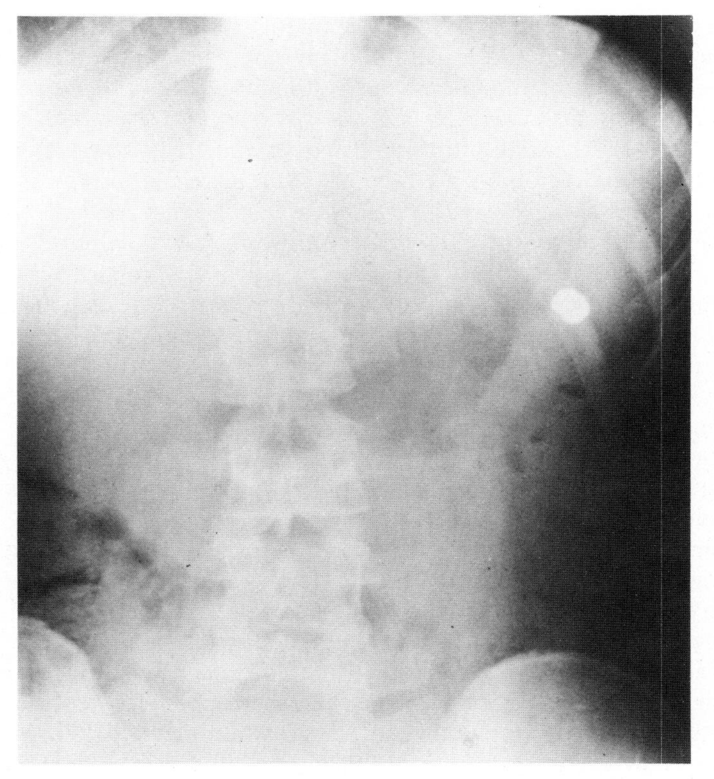

Fig. 4.8. Radiograph showing a gold crown in the stomach. (By courtesy of Professor A. H. R. Rowe.)

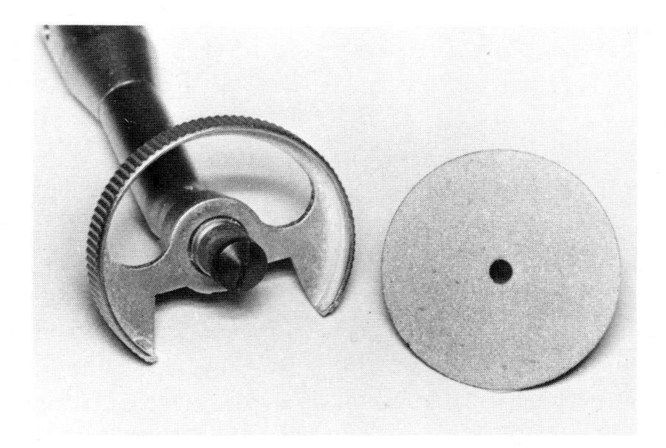

Figs. 4.9. Disc guard and disc.

totally avoided. Here the supine patient is particularly at risk. *No instruments or materials should be passed, manipulated, or adjusted, over, across, or near the patient's face.* If the patient is not already wearing his own spectacles, large enough to afford protection, *he should be provided with plastic protective spectacles*, lightly tinted to reduce glare but not dark enough to obscure his eyes from the operator. Ophthalmic injuries, fortunately rare, can be very serious but they are for all practical purposes entirely avoidable.

Concentrated and escharotic drugs are not commonly used these days. However, the warning given in this connection on p. 54 must be borne well in mind. It applies also to hot instruments which are occasionally used and which can cause a burn on the mucosa unnoticed until a local anaesthetic has worn off.

When operating on the lower teeth inferior dental or mental anaesthesia may leave the lower lip anaesthetic for some time after the patient has left the surgery. *He should be warned against eating food before sensation recovers*, for quite severe laceration of the lower lip can occur should he indulge in vigorous chewing whilst the lip is still anaesthetized.

Rotary instruments are another source of trauma to soft tissues. Fast cutting and abrasive discs can cause severe laceration and should *never be used in the mouth without a disc-guard*, of which there are several patterns (Fig. 4.9). Some coarse sand-paper discs can be damaging. In the use of all these, careful retraction and protection by operator and assistant of the tongue, lips, and cheeks is required. The tissues should be gently but firmly retracted with spatulae, retractors, mirrors or the flattened end of the aspirator tube, with care taken to avoid pressure on the alveolar mucosa or excessive retraction of the labial and buccal frena. Cuts of the mucosa, should they occur, may need one or several fine thread sutures to bring their margins together and speed healing.

High-speed instruments are also a source of risk because of their high kinetic energy, especially if they are bent and rotating eccentrically. **The rules are**, that the shank must be firmly held in the chuck, incapable of working loose; that bent instruments must be discarded immediately and, particularly, *that no attempt be made to re-straighten them*, for this leads to sudden fracture and a high-speed projectile; that long tapered diamond instruments are particularly prone to bending and should therefore be examined before use. This is best done by holding the handpiece away from the patient and giving a short tap on the foot control so that the concentricity of the tip of the instrument can be checked. Surgery assistants should be trained to avoid damage to these instruments and to recognize it when it does occur. Dropping a handpiece or even putting it down heavily is the commonest way of causing such damage.

If the head of a tungsten carbide bur breaks at the neck it usually does so when cutting, and therefore within the confines of the cavity from which the head can be recovered. The fragment is usually small and so of low kinetic energy, particularly if the bur has been slowed by pressure at the time of fracture. If it cannot be found it may be wise to radiograph locally to exclude the possibility of its being embedded in adjacent soft tissue. When using the air rotor in the vicinity of a breach in the mucosa, for example a gingival flap, or close to the orifice of an empty root canal, it is well to remember that surgical emphysema can be caused locally by the stream of compressed air forming the spray. This is important because of the possible spread of infection into deeper tissue planes. The risk can be minimized by avoiding directing the jets into these areas.

There are small but constant risks affecting the operator and dental surgery assistant. Broken burs, fragments of detritus flying off the bur, infected droplets of spray, all these make some form of eye protection highly desirable, if not essential for safety. In the event of injury to the eye copious irrigation with warm saline is the only immediate treatment which can be carried out in the dental surgery. The patient should in most cases be referred to an ophthalmologist to assess the extent of possible damage.

The aphorism that prevention is better than cure is nowhere more true than in the field of accident and misadventure. In the event, the dentist's first concern must be for the patient's welfare. He should explain what has happened but not admit liability or offer to compensate the patient. He should make a careful record of the incident at the time, and he may consider it prudent to inform his defence society.

Risks may be reduced to an acceptable level only by strict habits, by careful forethought and skilful operating, and by proper training of all assistants. The patient has the right to expect the dentist to follow closely those precepts.

Summary

Control of moisture. To improve visibility, avoid contamination of materials. Control infection.

Rubber dam: pre-check on contacts and application of clamp or ligature.
Include as few teeth as necessary to proper performance.
Inversion of rubber into cervical crevice essential.
Saliva ejector: venturi type water pump, low pressure, low volume. Disposable plastic mouth pieces adapted to lingual or buccal sulci. Position of head important.
Cotton wool rolls: used in sublingual and vestibular sulci; method of placement.
High volume aspirator: used by assistant.
Haemorrhage and exudate: control by local anaesthetic infiltration, vasoconstrictors and styptics. Diathermy. Care in use of escharotic drugs.

Control of pain. Avoidance of trauma to soft tissue, especially periodontium, gentle retraction.
Local anaesthesia: most effective. Sterility essential, disposables desirable. All forms of anaesthesia carry risk of thermal injury to pulp and dentine; high speeds, proximity to pulp, extent of cavity.

General anaesthesia: only with full anaesthetic facilities; chiefly for highly apprehensive, mentally and physically subnormal.
Sedation: pre-operative, alone or with above techniques. Post-operative anodynes when indicated.
Relative analgesia and hypnosis, effective but of limited value.

Minor emergencies

Inevitable risks: demand correct treatment when accidents occur. Importance well maintained instruments, equipment, especially handpieces.
Patient position: 45 degree reclined most hazardous. Inhalation and ingestion of foreign bodies. Function of rubber dam. Immediate treatment on loss of small object.
Radiographic localization.
Possible surgical intervention, cases of inhalation.
Radiographic monitoring of ingested object.
Protection of face and eyes: supine patient most at risk; no operative procedures over or near face.
Protective spectacles essential.
Avoidance of burns: lacerations; risks of high speed instruments.
Surgical emphysema.
Risks to operator and assistant.
Patient welfare: first consideration.
Records. Defence Society. Reduction of risks to acceptable level.

5

The principles of cavity preparation and lining

In the first decade of this century G. V. Black formulated stages in the preparation of a cavity for the reception of a filling and he also established the main principles governing the design of such cavities.

So sagacious was his understanding of the problem that his formula is still, more than half a century later, the basis for teaching on the subject. The principles (though some have been modified by more recent concepts in histopathology and by the emergence of new techniques), are fundamentally unchanged and have yet to be seriously challenged.

Carious cavities may occur in numerous sites and are presented to the dentist in all stages of development. The objectives of the restorative procedure are to eradicate diseased tissue, to prevent recurrence of the lesion, and to restore the appearance and function of that portion which has been lost. In order to prepare a cavity for the insertion of a filling, carious tissue must be removed and the cavity so shaped that a filling of the intended material, amalgam, gold, cement, may be built and retained in the tooth.

That it is possible to formulate principles of wide applicability in a subject apparently so diverse, derives from a knowledge of **the structure of teeth, the natural history of the disease process, and the properties of the material to be used for the restoration**. It is worth while recognizing the relevance of these factors before proceeding:

1. Much depends upon the fact that we are dealing with predominantly calcified tissues, the hardest in the body. Enamel, considerably the harder, relatively inelastic, and liable to fracture in the direction of its prisms; dentine, hard, tough, and elastic. The efficacy of the bonding at the amelodentinal junction of these physically dissimilar tissues is often overlooked. So, too, is the physiological unity of the dentine–pulp complex, with its susceptibility to injury and its limited power of response. *The fact that enamel and dentine cannot be regenerated should remind us never to destroy them unnecessarily.* These characteristics influence cavity design.

2. The fact that caries is a degenerative disease commencing in areas of stagnation and liable to recur in those areas; that its penetration of enamel is slower than its progress in dentine, and that enamel therefore becomes undermined; that it presents a threat to the integrity of the pulp and evokes characteristic responses in dentine and pulp. All these factors must be taken into account.

3. All properties of the restorative material are of direct consequence. The appearance — that is to say, its colour, refractive index, and

translucence — its physical properties such as hardness, plasticity, edge strength, conductivity, as well as many of its chemical properties—all these may have a bearing upon details of design.

4. There are mechanical problems which arise from the fact that most restorations are subject to considerable strain by reason of the masticatory forces which they bear. These strains are repeated at frequent intervals over long periods of time. The mechanical problems involved are extremely complicated. Fortunately the most important ones can be simplified to an extent that they are comprehensible to the dentist, and are still generally applicable.

Cavity preparation

The steps in cavity preparation are as follows:

1. The establishment of outline form.
2. The establishment of resistance form.
3. The establishment of retention form.
4. The treatment of residual caries.
5. The correction of enamel margins.
6. The toilet of the cavity.

No stage can be considered except in relation to the others; there is a certain degree of overlapping, and principles affecting one frequently have a bearing on another. It is necessary to study these steps in detail and to follow the application of some general principles involved.

For the purpose of preliminary study, let us assume that the carious lesion under consideration is of an optimal size for treatment. *This implies an early lesion which is just detectable by mirror and probe or with the help of a bitewing radiograph.* In the case of the more advanced lesions the principles still apply, but their application is complicated, to a varying degree, by the actual extent of the lesion.

Outline form

This term refers to the outline of the cavity, and its meaning is most easily visualized as the line of junction between the surface of the completed filling and the adjacent surface of the tooth. It has certain important general characteristics.

Clearly, the outline must enclose the extent of the carious lesion, and must include areas of enamel which are unsupported, due to destruction of the underlying dentine. It should follow a sinuous course, avoiding sharp angles which would be points of mechanical weakness, either of the filling margin or of the enamel margin (Fig. 5.1).

Not only must the lesion be included in the outline, but any adjacent areas not at present carious, but liable to become so in the foreseeable future, should be included. By so doing, the risk of subsequent carious

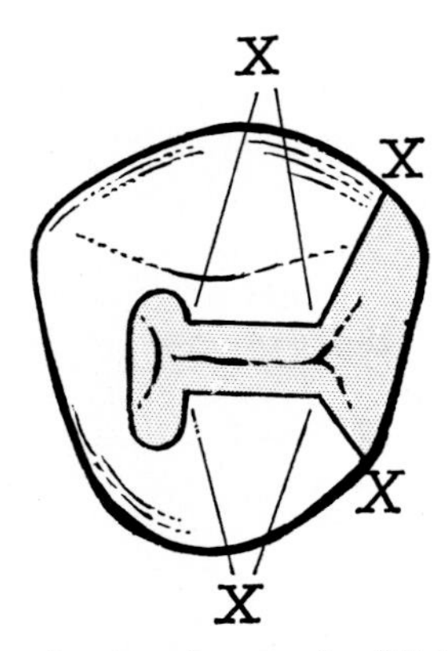

Fig. 5.1. A faulty outline form showing sharp angles (X) the points where filling or cavity margins are liable to fail.

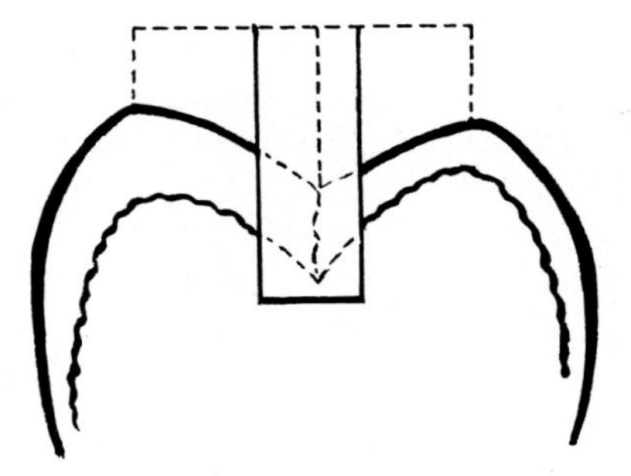

Fig. 5.2. When a fissure is cut out, the cavity margin should be about one-third of the distance up the slope of the cusp.

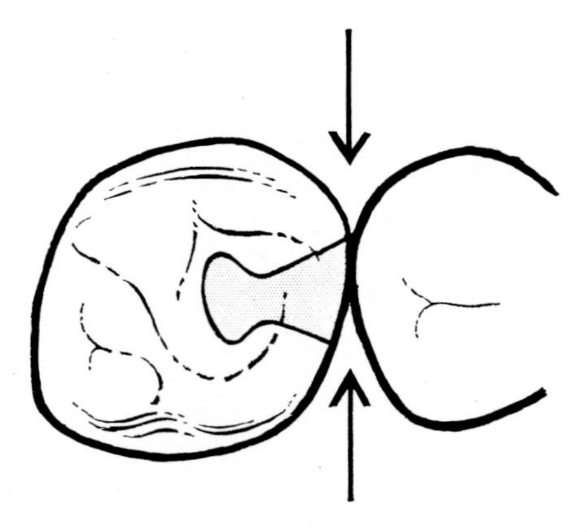

Fig. 5.3. The axial margins lie in self-cleansing areas when *just* visible laterally.

recurrence at or near the cavity margin is reduced. This is the principle of *extension for prevention*, and it is one which should be applied judiciously, in the light of the operator's experience, and with due regard to the caries liability of the patient at the time and in the future. The principle can only be justified on the assumption that a fully effective restoration can be inserted; otherwise the caries liability of the area may well be increased by the existence of a defective filling, rather than reduced.

The outline must lie in areas which are self-cleansing and not excessively exposed to occlusal trauma. For example, where occlusal fissures are included, the outline should lie approximately one-third of the distance up the slope of the cusp (Fig. 5.2); in this case the bottom of a natural fissure is a potential stagnation area, and the peak of the cusp an area of excessive occlusal strain.

In interproximal areas the axial margins, that is to say, the margins parallel to the long axis of the tooth and situated on the buccal and lingual aspects of the interproximal surface, must be placed far enough laterally to lie in an area subjected to natural or artificial cleansing. In effect, this is achieved if the margin of the filling is *just* visible when viewed normally from the buccal or lingual aspect (Fig. 5.3).

The cervical margin, which connects the axial margins of a Class II cavity below the areas of contact and of close approach, cannot be designed in a way which promotes natural cleansing or any significant degree of artificial cleansing. this margin is therefore placed below the stagnation and plaque formation occurring in the areas of contact and close approach. This, in an early cavity, may establish the margin just above the gingival crest where less irritation of the gingiva can result.

Until recently it was thought that recurrence of caries at the cervical margin was less likely if the margin were placed within the gingival crevice. The belief, based upon little more than clinical impressions, has been called in question. Experimental evidence and controlled investigation favours the view that the risk of carious recurrence depends more directly upon other factors such as the complete removal of carious dentine, an effective marginal seal and the smoothness of the surface of the restoration. The irritation of the crevicular epithelium by the margin of any restoration may be mild. It is acceptable only because unavoidable, and at worst, is very destructive (Fig. 5.4). *Unfortunately the size of most cavities when first treated is such that there is no choice but to place the cervical margin within the crevice but when conditions allow it should be above the free gingival margin.* Elsewhere on the crown the position of the outline is determined by balancing the opposing factors of stagnation and undue functional strain.

It will be clear from these considerations and from others to follow that a certain sacrifice of sound tissue is unavoidable. Over-extension of the cavity can, however, readily occur because it is easier to work in a large cavity and the inexperienced operator is likely to want to err on the safe side. *Over-extension is irretrievable*: it inflicts needless damage and shortens the life of

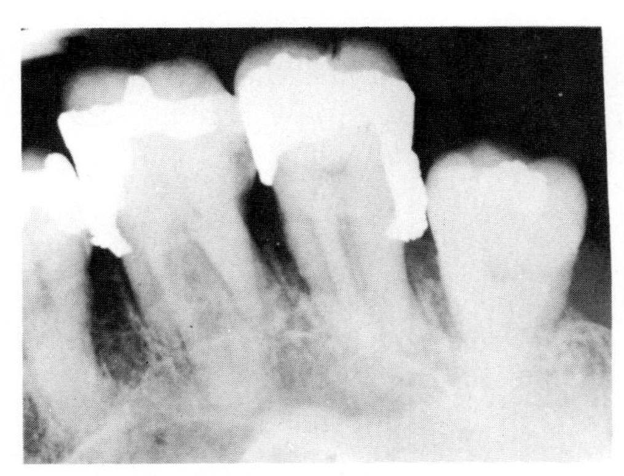

Fig. 5.4. Radiograph showing alveolar destruction caused by gross excess of amalgam at cervical margins.

the tooth. On the other hand an inadequately extended cavity also invites early failure of the restoration. The correct choice of design comes from an appreciation of the factors involved and experience in their application.

Resistance and retention

These stages of cavity design are related to one another and are usually achieved concurrently; they have a direct bearing upon the outline form. The resistance form is represented by those features of the internal design of the cavity which are intended to meet the forces of occlusion to best advantage. The resultants of occlusal forces experienced by fillings vary in magnitude and direction according to the portion of the crown which they restore, and the direction of functional movement of the mandible.

A simple example is an occlusal restoration of the Class I type (Fig. 5.5). Here the occlusal force is predominantly at a right angle to the occlusal plane, and the flat floor of the cavity opposing this force constitutes, in this case, the resistance form. The application to more complicated cavities is necessarily more complex.

Cavity surfaces which impart resistance form are not in themselves sufficient to ensure that a restoration is not dislodged. Other features of design must be included which will resist dislodgement under the various strains of occlusion. These features constitute the retention form and they are achieved by three means:

1. **The cavity may be undercut**. Generally, any cavity whose greatest internal diameter is greater than the diameter of its aperture in the same plane is self-retentive though there are some special geometrical exceptions to this definition. This provision can be met by undercutting opposing walls of the cavity (Fig. 5.6). The possible disadvantages of this method of retention are

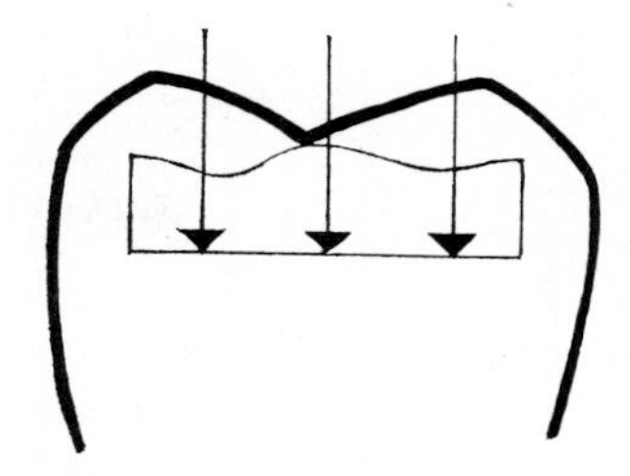

Fig. 5.5. Sectional view of a Class I restoration in which the floor of the cavity represents the resistance form of the cavity.

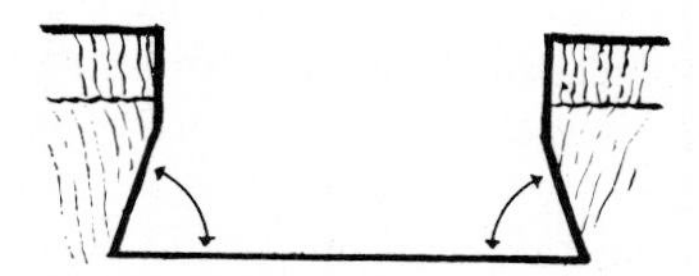

Fig. 5.6. Diagrammatic representation of undercut dentinal walls used as retention for a simple restoration.

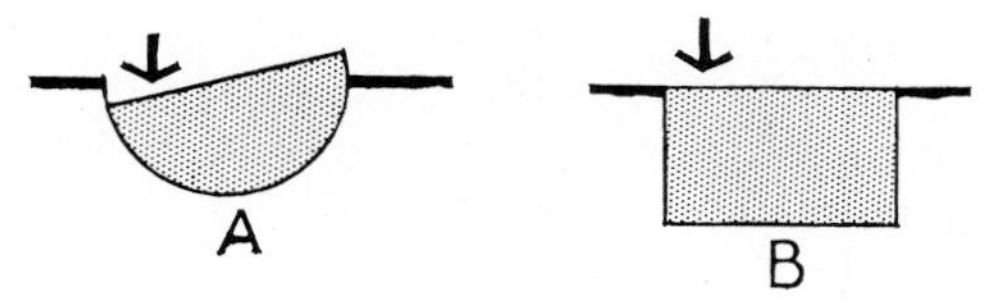

Fig. 5.7. An amalgam restoration placed in a hemispherical cavity (A) could easily be displaced. A restoration placed in a parallel-sided cavity (B) could only be removed by destruction of the filling or the tooth.

that, if excessive, the undercuts open an inexpedient amount of normal dentine, and if they are placed too close to the aperture of the cavity, the margins are weakened by removal of the dentinal support of the enamel.

2. **Parallel walls**, when on opposite sides of the cavity, provide excellent retention if the material of the restoration can be relied upon for very close contact with those walls, as is the case with silver–tin amalgam (Fig. 5.7). In these conditions it is only necessary to prepare walls which are exactly parallel or, since this requires great precison, to allow only the slightest internal divergence. The principle of *near-parallelism* has applications to retention in many aspects of restorative procedure. For example, straight walls, flat floors, and sharp line angles in the internal design of a cavity, are derived directly from this principle.

3. **The dovetail lock**, which is a common method of achieving retention, should be recognized as a form of undercut cavity, usually in an occlusal plane (Fig. 5.8). Its position is such that it opposes forces which tend to dislodge the filling in a mesial or distal direction. *There are two important things about a dovetail.* First, it must be sited in sound tissue without

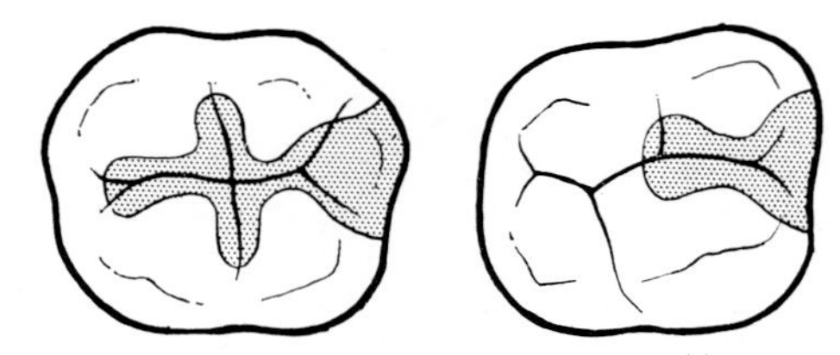

Fig. 5.8. Examples of occlusal locks related to the morphology of the respective teeth.

weakening the general tooth structure. Second, the dimension at the 'neck', the narrowest point, must be sufficient to allow the material of the filling adequate strength to withstand the forces to which it is likely to be submitted. It should be remembered that the strength of the neck is more effectively increased by deepening the cavity than by increasing its width, but in cutting a deep lock, particularly in premolars, the cusps are deeply separated and the remaining crown is thereby weakened. It will be apparent that in the process of establishing retention by cutting a lock, modification of the outline form is unavoidable.

The term *convenience form* is sometimes used in connection with cavity preparation. It is referable chiefly, though not exclusively, to cohesive gold technique and is used to describe a modification of cavity form to allow either easier preparation or easier insertion of the restoration. Reference will later be made to it.

Residual carious dentine

In the case of an early lesion, such as has been presupposed, it sometimes happens that, when the first three steps have been completed, there remains little or no carious dentine still to be dealt with. This is a very significant fact from which several important deductions may be made.

It implies that *with lesions of this size the extent of the lesion has little bearing upon the design of the cavity.* A corollary to this is that the cavity at this stage has been designed upon the theoretical consideration of the properties of tooth tissues, the aetiology of caries, and the properties of the filling material. It follows that for each type of restoration there is a minimum size below which the restoration will be unsatisfactory in some respects.

Usually at this stage there is a small residue of stained and softened dentine in what was the deepest part of the cavity. Since this is an early lesion, there can be no risk of involvement of the pulp, and softened dentine is removed until the whole cavity surface is composed of dentine, normal in colour and consistency. In large cavities with extensive carious involvement there are other factors to be considered. These will be discussed in Chapter 6.

It is of particular importance — and it is almost impossible to overemphasize this — that the periphery of the cavity be scrutinized with the greatest care. Carious enamel is easily visible, white and opaque, but caries

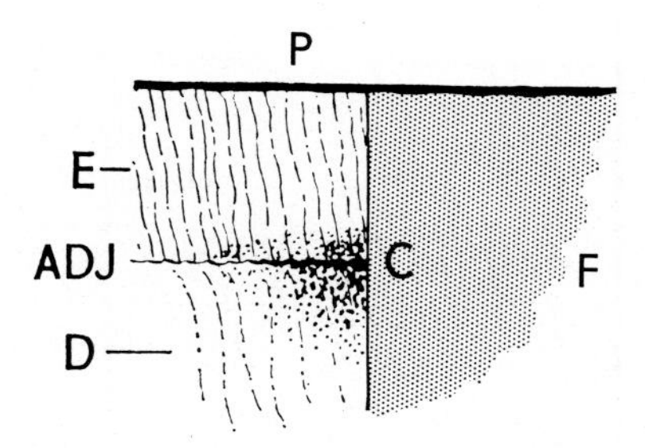

Fig. 5.9. E: enamel; D: dentine; F: filling; C: caries at ADJ; P: marginal enamel prisms deprived of dentinal support.

at the amelodentinal junction, particularly in the region of the cervical margin and under the peaks of cusps, may escape detection. **The periphery of the cavity must be indubitably caries free, this is one of the few rules in operative dentistry to which there can be no exceptions.**

The reasons, almost self-evident, are these. If an enamel margin were disintegrating it would obviously be impossible to create an hermetic seal between the filling and the enamel. If the enamel margin were intact and a small area of caries remained at the periphery of the dentine (Fig. 5.9), it is certain that the enamel prisms overlying that area, deprived of support and involved in disintegration at their proximal ends, would soon collapse and so destroy any marginal seal which had previously existed. *The integrity of the marginal seal is crucial to the durability of the restoration.* It can be stated without fear of contradiction that, in the long run, the great majority of all restorations fail on account of the failure of the marginal seal and the recurrence of caries which ensues.

There are many factors which influence the peripheral seal other than the presence of marginal caries at preparation. These will emerge later, but one can be discussed now.

Enamel margins

Following the removal of residual caries, the margins of the cavity should again be reviewed to ensure that there is *no overhanging and unsupported enamel*, and that there are no unsupported enamel prisms. Since the prisms are invisible to the naked eye, the latter is done by inference . The course of an enamel prism is not a straight line; it is nearest to being straight at the cervical region, and is at its most devious in the cuspal areas. It is possible to approximate by saying that, in general, the direction of the prisms is normal to the external surface of the enamel, but it must be remembered that this is only an approximation.

If the cavo-surface angle, that is to say, the solid line angle between the surface of the tooth and the adjacent wall of the cavity, were less than 90 degrees (Fig. 5.10 (a)), there would certainly be a considerable number of incomplete and unsupported prisms at the margin. The assumption is that

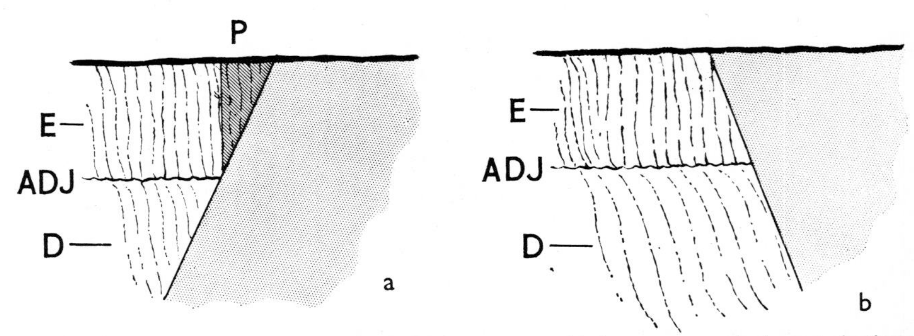

Fig. 5.10. (a) When the cavo-surface angle is less than 90 degrees, marginal enamel prisms (P) are unsupported. They will disintegrate under stress. (b) When the cavo-surface angle exceeds 90 degrees, incomplete enamel prisms are supported by dentine and not exposed to direct pressure.

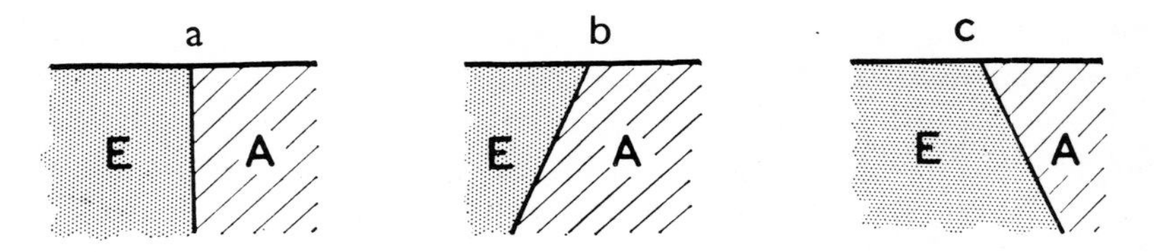

Fig. 5.11. (a) If enamel (E) and the restoration (A) are equally strong a cavo-surface angle of 90 degrees is desirable. (b) If enamel is *stronger*, the cavo-surface angle may be less than 90 degrees. (c) If enamel is *weaker* than restoration the angle should exceed 90 degrees.

these would be prone to failure. When this angle exceeds 90 degrees (Fig. 5.10 (b)), the probability is that any incomplete enamel prisms would be supported by dentine and would not receive direct occlusal stress. The liability to failure in these conditions may be presumed to be markedly less.

Another factor determining the cavo-surface angle is *the marginal strength of the filling material*. On theoretical consideration, if the edge strength of enamel and of the restoration were equal and subject to the same load, the optimum cavo-surface angle should be 90 degrees (Fig. 5.11 (a)). If the edge strength of enamel were greater than that of the restoration, the angle should be less than 90 degrees (Fig. 5.11 (b)); if the converse were true, then the angle should be greater than 90 degrees (Fig. 5.11 (c)), in order that the edges be of equal strength. It will later be shown that in most types of restoration we try to achieve a cavo-surface angle of approximately 90 degrees, but in the case of gold restorations the margins are bevelled at 45 degrees.

Cavity toilet

In this, the final step in preparation for the reception of a filling, the object is to clean and dry the cavity. In the course of mechanical preparation a certain amount of detritus collects and is liable to remain in the cavity. This is best dispersed by a warm atomized water spray: any debris which may resist this should be loosened with a sharp probe before the spray is repeated. In high-speed cutting, the water spray removes detritus very effectively in most circumstances. Even so, debris displaced from the cavity can fall into it again, and hand instruments also leave debris. After cleansing, the cavity should be isolated before it is recontaminated with saliva, and dried.

It has been noted that no dehydration of the cavity walls is permissible. From evidence now available and from clinical experience it appears that the use of antiseptics to 'sterilize' the cavity surface confers little or no advantage and may act as additional irritation. If the cavity has previously been otherwise contaminated, by wax or oil for example, a solvent such as chloroform may be used to clean the surface, but a 2 per cent solution of detergent is more suitable for most purposes.

Before proceeding to the detailed examination of techniques of the commoner restorations, it will assist to consider the types and purposes of cavity linings and the temporary fillings which are in frequent use.

Cavity linings

In a deep or extensive cavity it is usual first of all to place a lining of a different material than that of the eventual restoration. The lining may serve one or more of the following purposes: protective, therapeutic, or structural.

Protective linings

These are more frequently used than are linings of other types. Their purpose is to protect dentine and pulp against the irritant effects of thermal changes in the case of metallic restorations, and against toxic substances in non-metallic restorative materials.

Metallic restorations transmit thermal changes much more readily than do normal tissues, and the temperature range of modern diet may exceed 60 °C. Protection against such changes requires a relatively thick layer of a good insulator (Fig. 5.2 (a)) and it appears that dentine is a slightly better insulator than zinc phosphate, which in turn is considerably better than zinc oxide and eugenol. *It is therefore not justifiable to deepen a cavity by removal of dentine in order to line it against thermal shock.*

Zinc phosphate cement is strong enough to take transmitted compression, though it has the disadvantage of being acid in reaction and is therefore irritant. From what is known of the chemistry of this cement it appears that there is less free acid — and for a shorter period — than is the case with silicate cement. Clinically, zinc phosphate cement seems to exert very little of this irritant effect if it is mixed fairly thickly and placed in the centre of an old cavity, even though the cavity be deep. The reason lies in the fact that the centre of the cavity is separated from the pulp by areas of sclerotic and secondary dentine, the permeability of which is much reduced.

The periphery of a recently prepared cavity is the region where freshly cut dentine is chiefly to be found, and this is the part of the cavity most susceptible to chemical irritation. This irritation may arise not only from the effects of free phosphoric acid in zinc phosphate and silicate cements, but also from the free monomer of autopolymerizing acrylic resin. What methods are available for protection against these irritants?

Zinc oxide and eugenol. This is a simple mixture of zinc oxide powder with eugenol—in earlier days clove oil was used—to form a slow-setting mass of stiff, putty-like consistency. In former times it was universally used for temporary restorations and linings as it is virtually non-irritant to freshly cut dentine and of low thermal conductivity. It makes a good marginal seal and is fairly hard when mixed with greatest powder content. Its use as a cement has been largely supplanted by modifications which have retained its good properties but have shortened the setting time to a few minutes and increased the early compressive strength.

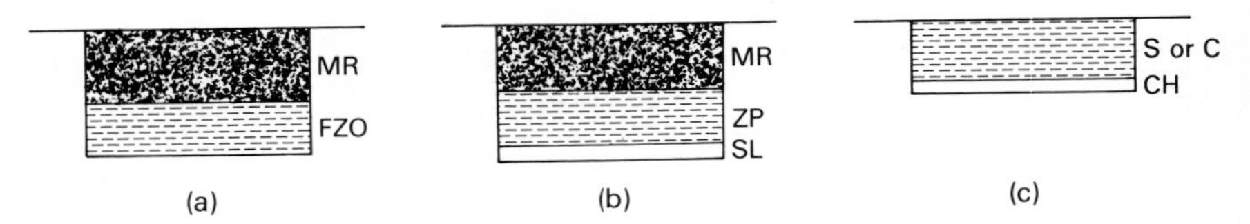

Fig. 5.12. Use of common lining cements: (a) Fortified zinc oxide lining, FZO, used as thermal insulation below a metallic restoration, MR. (b) A sublining, SL, beneath a lining of zinc phosphate, ZP, beneath a metallic restoration, MR. (c) A thin lining of calcium hydroxide cement, CH, as a protective lining beneath a silicate or composite restoration, S or C.

The eugenol in these cements retains some of its antiseptic properties when mixed with zinc oxide, but it is a plasticizer of acrylic resins and is said to discolour silicate cements. So it is unsuitable as a lining material for those restorations.

Fortified zinc oxide cements. A number of proprietary zinc oxide cements. These are generally of two types: one with the addition of polystyrene or a similar polymer as a binder, the other using *o*-ethoxybenzoic acid to replace two-thirds of the eugenol. These are often referred to as the EBA cements.

The powders may also contain finely divided silica and powdered natural resin, which greatly improves the physical properties of the set cement. Their particular qualities are a shorter setting time, about three to four minutes, non-irritancy and greater strength. They can be used for most lining purposes and will withstand the condensation pressures of amalgam within a short period after the initial set. Because of the eugenol content they are unsuitable under acrylic and composite restorations.

Calcium hydroxide cements. These are cements in which the powder is mainly calcium hydroxide, mixed with zinc oxide and bound with a salicylate chelating agent. They are used primarily for protection; their mechanical strength is poor. They are either neutral or fairly strongly alkaline in reaction, but are apparently of low irritancy and may be used for direct pulp capping (p. 239) and also in the deep carious lesion (p. 87). The cements containing phosphoric acid, namely silicate and zinc phosphate, when they come into contact with calcium hydroxide cement, form insoluble calcium and zinc phosphates which reduce the irritancy of the available free acid. Because they contain no eugenol they can be used under acrylic or other plastics restorations. To sum up, the calcium hydroxide cements can be widely used where a non-irritant lining is required.

Zinc polycarboxylate cement. This cement is essentially a combination of zinc oxide with polyacrylic acid. It was formulated to be adhesive to enamel. With a completely clean surface considerable adhesion can be demonstrated, but in practice this property is insufficient to give it a pronounced advantage in this respect. It is non-irritant and has a crushing strength comparable to or slightly better than the fortified zinc oxides and eugenol, so it has been widely used as a protective lining, and for the cementation of crowns when extensive sound dentine has been involved.

Zinc phosphate cement. This is the hardest cement used for lining but with the increasing variety of cements available, phosphate cement tends to be confined to cases where its irritancy can do little harm and its strength is of particular advantage. It is therefore often used with a sublining as mentioned above. It is also quite safe in large, well-established cavities such as have previously contained a restoration, where the pulp is well protected by dead tracts and areas of sclerotic dentine.

The various methods of protection by lining against chemical and thermal irritation can be summarized thus:

1. In anterior teeth, cavities to be restored with silicate or resin are best lined with calcium hydroxide or polycarboxylate cement. This need not exceed 0.5 mm in thickness but should cover as much as possible of the dentine directly overlying the pulp.
2. In the majority of cavities in posterior teeth, fortified zinc oxide in one of its several forms or polycarboxylate cement may be used for all purposes.
3. When zinc phosphate cement is used, freshly cut dentine should be protected by one of the sublinings mentioned.

For further consideration of the factors in cavity preparation which control the liability to chemical irritation see p. 148.

Structural linings

The structural function of a lining is usually combined with its protective function. For example, zinc phosphate cement, used primarily to insulate, is shaped to serve an aspect of the internal design of the cavity as in the Class II amalgam restoration (Fig. 7.21, p. 118). This cement may also, in correct conditions, properly be used to block out undercuts in some inlay preparations (Fig. 11.5, p. 211). It must not, however, be used to support otherwise unsupported enamel which could become stress-bearing; for this purpose its physical properties are inadequate.

Therapeutic linings

These are used for the purpose of applying medicaments such as chlorobutanol, carbolized resin, silver nitrate, etc., to underlying dentine; it is fair to say that, as a result of histopathological evidence, the use of concentrated drugs is less in favour than formerly. The medicament may be applied directly to the dentine and covered with zinc oxide and eugenol mixed with strands of cotton wool or, if compatible, may be mixed with these cements and applied as a sublining to a temporary or a permanent filling.

Temporary restorations

These are used either as a means of closing and protecting for a short interval between visits, or as a means of sealing in drugs as referred to above.

Fortified and accelerated zinc oxide and eugenol cements are in general the best for temporary fillings. They are easy to prepare and do not set so hard that they cannot be excavated from the cavity when required. The strength and durability can be increased by mixing to a firmer consistency and by the incorporation of cotton-wool fibres. Lack of irritancy and a water-tight seal at the tooth surface are the characteristics which make these cements the first choice for this purpose. No other cements equal them in this respect except EBA and polycarboxylate and these form too hard a mass to be easily excavated. It is an obvious and inconvenient fact that the more durable a temporary restoration is the more difficult is its removal.

There are times when a temporary restoration must serve for a long period, as when a patient is suddenly called away. A zinc oxide cement mixed firmly and with cotton wool added may last for two or three months but will be difficult to remove. Before inserting the cement the cavity may be lightly dressed with petroleum jelly (Vaseline), leaving a small excess in the corners and line angles, deep in the cavity. This will assist the eventual removal of the filling. With large Class II cavities temporarily restored for a long time, a copper band contoured to cervical and occlusal should be placed around the clinical crown to support a temporary restoration safely and in comfort (Fig. 5.13).

Simple zinc oxide with eugenol—or, of course, oil of cloves may be used—can be mixed, without particular care as to the manner of mixing, to an extremely stiff consistency. Such a mix as this has good lasting qualities.

Zinc oxide and eugenol is frequently said to be obtundent, and the cement mixed with clove oil even more so. Certainly it is non-irritant to freshly cut dentine, but it frequently leaves a cavity highly sensitive except when it has remained for some months. There are occasions when the deepest layer of cement overlying the pulp can be left in a position when the remainder is

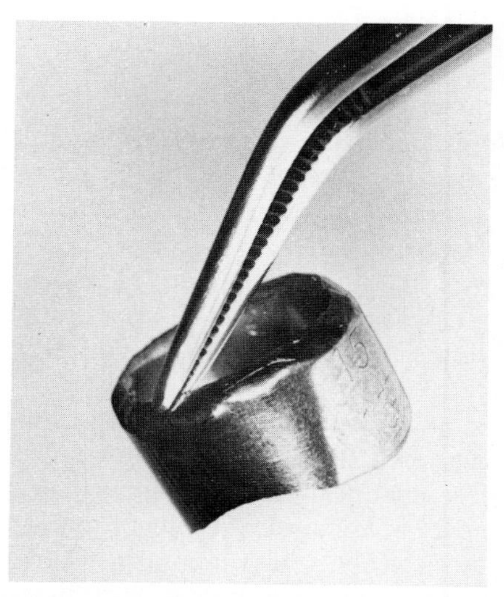

Fig. 5.13. Copper band contoured to gingival margin and in-turned occlusally, for retention of a large temporary restoration.

removed. This deep layer may then be used as a lining or sublining to the permanent restoration which follows.

The method of mixing zinc phosphate cement is of considerable importance to its properties and the technique of its various uses. Its setting time can be prolonged by 'slaking' the fluid, cooling the slab, and slow mixing. To slake the fluid a very small quantity of powder, about 3 or 4 pins' heads, are mixed with the fluid about a minute before the main mix starts. The setting time may be shortened by rapid mixing.

Preparation of zinc phosphate cement. A 0.6 cm (¼in) thick glass slab is adequate; it may be cooled. A good quality stainless steel spatula may be used. *Small quantities of powder are incorporated first, followed by larger quantities later. Each portion of powder is thoroughly incorporated before the next is added, spatulating with a circular motion over a wide area.* The setting time may be, to some extent, controlled by slow or rapid mixing.

This cement may be used as a thick creamy mix which will *just* not drip from the spatula, or as a thick mix of putty consistency. The technique of insertion in the cavity will vary accordingly. If a thick creamy mix is used, a blunt-ended probe is the most suitable instrument for teasing the cement, a small portion at a time, into the appropriate part of the cavity. With cement of putty consistency, a discrete portion of cement is carried to position on a small round-ended plastic instrument. When in position, the cement should be tamped firmly against the cavity surface and conformed to the correct shape with plastic instruments moistened with alcohol.

The preparation and use of this cement should be carried out with precision and care, and it is well worth the operator's while to acquire a standardized procedure for its use in various circumstances.

Gutta-percha. Temporary gutta-percha can be used for short periods in simple cavities, affecting one surface, or in compound cavities well enclosed and not exposed to excessive bite. It is *unreliable as a cavity seal*, but if the surface of the mass is made tacky by immersion in chloroform before insertion, it is probable that closer adaptation to cavity walls can be achieved. Its particular advantage is that it is easily removed but it often leaves the dentine of the cavity hypersensitive to touch. It appears that this drawback can be reduced by applying carbolized resin to the cavity before insertion of the gutta-percha.

These are the substances most commonly used as linings and temporary fillings. Care should be exercised in choosing the most suitable material and in using it in the correct manner. The results may make a great difference in terms of comfort to the patient and convenience to the operator. The reader will probably have noticed the wide range of usefulness of zinc oxide and eugenol in its various preparations. There is scarcely any aspect of dentistry in which this ubiquitous substance has no contribution to make.

Summary

Cavity preparation. Design depends on structure of enamel and dentine, pathology of caries, and properties of restorative material. Conservation important, for hard tissues never regenerate. Five steps in cavity preparation: Outline; junction of restoration and coronal surfaces; encloses lesion and areas at risk; sinuous; correct siting of occlusal, aproximal, and cervical margins.

Resistance and retention. Surfaces opposing occlusal forces constitute resistance form; retention by undercut, occlusal dove-tail. Parallel walls or near-parallelism.

Residual caries; in optimal cavity remove all caries; particular attention amelodentinal junction, cusps, and periphery generally. Marginal failure commonest cause of carious recurrence.

Enamel margins; no unsupported enamel; cavosurface margin depends upon restorative material. Cavosurface angle commonly 90 to 110 degrees.

Cavity toilet; warm spray, removal of all debris; light drying, no dehydration.

Cavity linings. Protective, structural, therapeutic.

Protective against heat and chemical irritation, free acid or monomer.

Zinc oxide and eugenol, non-irritant, good seal, non-conductor, now replaced by fortified cements, shorter setting time, early compressive strengthen; polystyrene and EBA modifications.

Zinc polycarboxylate similar in most properties, adhesive to enamel. Zinc phosphate hard and adherent but irritant to vital dentine.

Structural linings, usually zinc phosphate, serve an aspect of cavity design, e.g. blocking out undercuts. Mixing this cement important.

Therapeutic linings, to apply drugs to dentine; not commonly used.

Durable temporary restorations; stiffer mix with cellulose fibre. Larger temporary fillings, support and easy removal.

Gutta percha; quick and easy insertion and removal, no marginal seal, leaves cavity sensitive.

6

The technique of amalgam Class I and Class V restorations

It will be noted that the term *permanent* as applied to fillings is used in a comparative sense. The distinction between temporary and permanent fillings is arbitary, yet in practice it is clear enough. A temporary restoration may fulfil its purpose for days, weeks, or months, whereas a restoration whose life could not be measured in years could not pretend to permanency in the accepted sense. The temporary filling sometimes proves surprisingly durable and instances of permanent restorations which have remained functionally intact for several decades are not unusual.

Of the permanent restorations now in use, those of silver–tin amalgam account for a very considerable majority. This is due to research which has improved the properties of the material and the study which has been applied to the technique by generations of conscientious practitioners.

Amalgam is a material which can be severely mishandled and yet produce, on completion, a superficially passable result. This is perhaps its greatest shortcoming. Its failure, partial or complete, after a shorter or longer time is usually attributed to the inherent defects in the material and not to failures in technique, which is where the fault more frequently lies. The correct performance of the technique demands an understanding of the properties of amalgam and strict attention to detail, as well as an awareness of its limitations.

Amalgam is used to best advantage in the restoration of Class I, Class II, and Class V cavities, and this chapter will describe the preparation and filling of typical Class I and Class V cavities; Class II cavities are dealt with in the next chapter.

In the description of cavity preparation, the use of high-speed instruments will be described and where appropriate the use of lower speeds will be either assumed or described. High-speed rotary instruments are very versatile and easy to use; *they are accurate only when concentric with the rotor.* An eccentric tool can lead to rapid and destructive over-extension of a cavity. The operator must be constantly on his guard against this.

For purposes of simplification in the study of cavity preparation, it will be assumed in the first instance that the carious cavity concerned is of optimal size as previously defined. Modifications in the technique as applied to larger cavities will then be indicated. The use of local anaesthesia or other medication, according to the clinical indications, will be assumed.

THE CLASS I AMALGAM RESTORATION

Cavity preparation

The procedure in normal practice is to define the outline of the cavity with a high-speed instrument; this can be done rapidly and it is at this stage that over-extension can easily occur. The cut surfaces are then smoothed, line angles more clearly defined, the position of margins and bevels determined. Lower-speed instruments are more suitable for this because they cut more slowly and, when carefully used, need not be water-cooled, so that visibility of fine detail is improved. Finally, hand-cutting instruments may be used to sharpen line angles when necessary.

Rubber dam is not normally used for high-speed cavity preparation, but in the completion of final detail and the insertion of the filling material it is a great help to the operator without chairside assistance in the control of moisture and the provision of good visibility. In four-handed operating, however, these factors are capable of such control that the comparatively cumbersome and occasionally time-consuming use of rubber dam can no longer be upheld on grounds of efficiency.

Initial access to the cavity is obtained by the use of a No. 2 or 3 round bur applied at the point which appears to be most affected by caries, commonly the central fossa, or the marginal pit, on the occlusal surface (Fig. 6.1). The bur is allowed to penetrate to a depth of approximately 2.5 mm.

With a No. 3 cylindrical fissure bur in the case of molars, No. 2 in the case of premolars, at the average depth of 2.5 mm carious and stained fissures are now cut out from the point of penetration towards the periphery (Fig. 6.2). The bur should be used with the lightest touch, keeping its speed at the maximum, as the margin is cut back to an area of sound enamel and dentine. To conserve sound tissue it is better to choose small burs rather than larger. Bladed tools must, of course, be tungsten carbide and care should be taken to avoid chipping the blades. Fine-grain diamond instruments are also very effective, but because they can be slightly bent by a fairly light blow, they often become eccentric in revolution and therefore very destructive. They should be immediately discarded.

In establishing the outline form of the cavity the question of extension for prevention must be considered. *In order to include in the restoration adjacent areas which are likely to become carious and to leave the outline so placed that nowhere does it cross a stagnation area, it becomes necessary to extend the cavity to include the greater part of the adjacent fissures.*

This must be done with judgement and care. The steepness of the cuspal slope and the depth of the fissure affect the width and extent of the cavity. A shouldered bur may be used to limit the depth of penetration into a non-carious fissure (Fig. 6.3). Proceeding along the fissure, the extension should stop at the point where its margin enters a self-cleansing area. **The principle must be thoroughly applied to be effective, but excessive destruction of sound tissue is unwarrantable.**

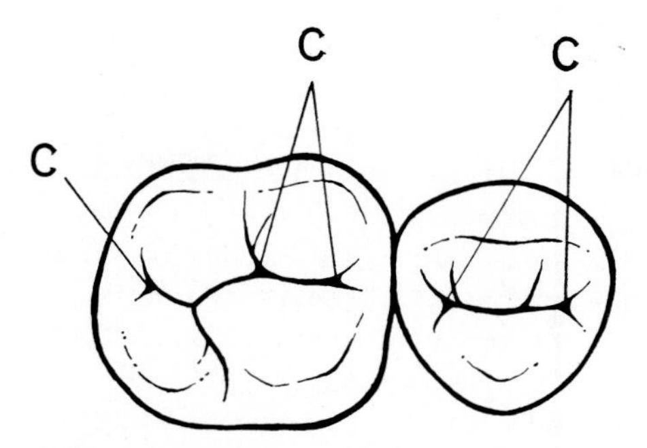

Fig. 6.1. Sites of initial caries, C, in upper molar and premolar teeth. These are the points of access in cavity preparation.

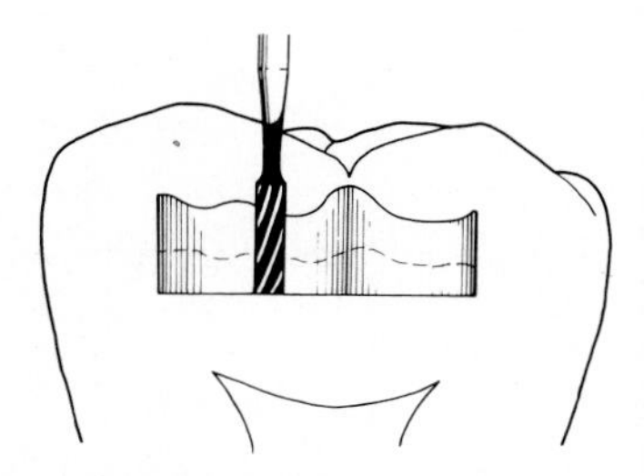

Fig. 6.2. Use of cylindrical fissure bur in Class I cavity to establish flat floor and vertical walls.

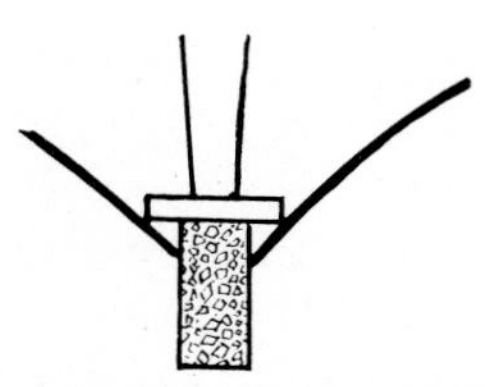

Fig. 6.3. A shouldered diamond instrument used for cutting out a fissure. The shoulder limits the depth of penetration.

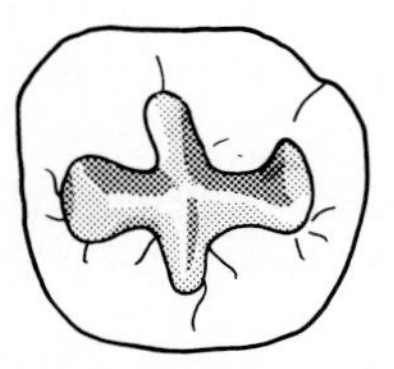

Fig. 6.4. Final outline of Class I restoration in lower molar tooth. This is extended to include all fissures likely to become carious.

When this stage is complete the cavity will possess an outline such as is shown in Fig. 6.4. The width of the limbs of the cavity is for the most part that of the burs used, but it is slightly larger at the points of confluence of fissures. At its mesial and distal extremities it approaches the marginal ridge and its underlying dentine. At this point it is necessary to see that the marginal ridge is not weakened by excessive removal of its substance.

The cavity should be of an average depth of about 2.5 mm, having a flat floor which in this case represents the resistance form. The walls of the cavity may be parallel to the long axis of the tooth. At most, they should have no more than a slight internal divergence, that is to say, showing slight divergence as they pass from the occlusal surface to the depth of the cavity. In these conditions no other form of retentive undercut is required.

There is frequently a small area below the point of initial access where the floor consists of carious dentine. If this is so, the softened dentine is excavated with a sharp hand excavator, or with a No. 4 round bur at low speed, with a light touch, until normal dentine is reached. Thus a depression, of size corresponding to the extent of the caries, then exists in the floor of the cavity and, though there may yet be a small area of lightly stained dentine in the centre of this, the floor and walls of the cavity should present a uniformly hard and resistant sensation to investigation with a sharp probe.

Using a sharp, well-tempered probe and heavy hand pressure of the order of 4.5 kg, sound dentine allows only barely perceptible surface penetration and, this is the crucial point, *there is no sensation of stickiness, or resistance to withdrawal of the probe.* The existence of even the slightest sticky sensation implies that it is softer than normal; it is therefore decalcified and, it must be assumed, it is infected. Such dentine should, in the case under consideration be removed since its removal implies no risk of exposure of the underlying pulp.

Following the routine steps in cavity preparation stated in the previous chapter, the enamel margins of the cavity are now reviewed, first to confirm that they are everywhere correctly sited, and second, to shape them in such a manner that they have the correct cavo-surface form. The general principles described on p. 71 are those which govern the best marginal enamel form and the correct cavo-surface angle. In the practical application of these principles it will now be apparent that certain difficulties arise and these will be discussed.

In certain limited regions of the outline of the Class I cavity described, the adjacent enamel surface of the tooth is for practical purposes flat and the wall of the cavity, extending upwards from the floor–wall angle, meets the surface at an angle of 90 degrees (Fig. 5.11 (a)). Here there is no difficulty in the application of the principle, for the enamel and amalgam margins are, for practical purposes, at maximal strength. The difficulty becomes apparent, however, when the outline form passes across a steeply sloping cusp shown somewhat exaggerated in Fig. 6.5 (a), where the axial wall of the cavity, as cut by a fissure bur, produces a cavo-surface angle A of 135

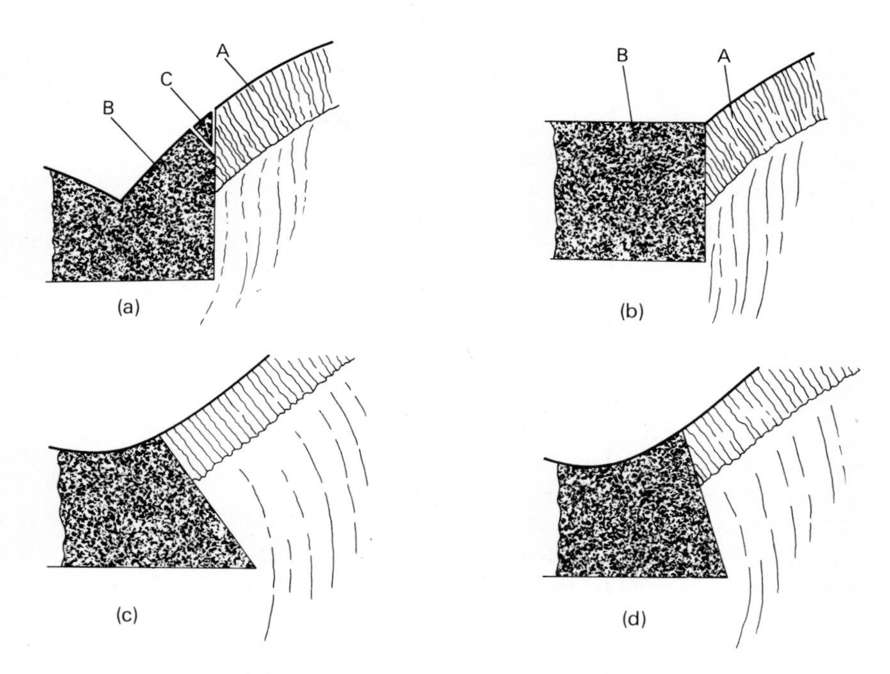

Fig. 6.5. Some factors affecting the finishing of enamel margins (see text).

degrees. This is a very robust cavity margin, but it carries with it the necessity for an amalgam margin B of 45 degrees, if the normal form of the fissure is to be restored. Such an amalgam margin would be fragile and susceptible to fracture (point C in Fig. 6.5); the region of the fracture would form a minute stagnation area, a potential nidus for caries, and the marginal breakdown of the restoration. This would be unacceptable.

Theoretically the problem may be met in either of two ways. The cavo-surface angle of 135 degrees may be retained and the surface of the amalgam made flat, with no attempt at fissure reproduction (Fig. 6.5b); in this case a relatively strong amalgam margin is obtained but the anatomical form of the occlusal fissure is lost or is modified, in a way which is not generally permissible, chiefly because it would form premature contact with an occluding cusp. Alternatively, the whole wall may be sloped away from the centre of the cavity, parallel with the enamel prisms and so produce a 90 degree cavo-surface angle (Fig. 6.5 (c)). Usually this would entail excessive destruction of dentine and for this reason a compromise must be found between (b) and (c).

The slope of the cuspal surface is usually about 30 degrees, sometimes less. The wall of the cavity is adjusted to give a cavo-surface angle of 100 to 110 degrees, involving only a moderate undercut into dentine (Fig. 6.5 (d)). The amalgam margin is now 70 to 80 degrees and this may be considered an adequate compromise solution, provided that, in the technique of insertion, every measure is taken, particularly in condensation, to produce amalgam of maximal strength. The operator must at all times, however, be aware of the factors which determine the detailed design of the margins and be prepared to modify the design as circumstances may dictate.

To achieve a satisfactory enamel margin a fine straight-cut fissure bur, either cylindrical or tapering (for example Meisinger 73, 73a, or 161), may be used. The bur should be run fast with water-cooling and used with a light touch to remove irregularities and smooth the outline into sinuous curves, and at the same time to establish, by the angle of its incidence, the correct cavo-surface angle. Recent studies of the results of various methods of smoothing enamel margins, using the detail shown by the scanning electron microscope, have indicated quite conclusively that a multibaded, or blank tungsten carbide bur (the Baker–Curson bur, see p. 192) used at high speed, water-cooled, gives the smoothest enamel surface of any (Fig. 10.10, p. 191). This is probably the method of choice.

There is another consideration in design which should be mentioned here. Provided that the cavity walls are generally undercut, that is to say inward sloping, nothing is gained by an acute floor–wall angle. On the contrary, a sharp line angle causes concentration on internal stresses in that area where cusp and restoration are loaded occlusally. In some positions this could initiate fracture of the cusp after years of heavy loading.

With completion of the review of the enamel margins the preparation of the minimal cavity is finished. Before proceeding to the technique of filling, the modifications required in the preparation of a larger Class I cavity will be considered.

The larger cavity

In this case there is an easily visible and palpable carious defect which, having started at the intersection of fissures, has involved the surrounding enamel to the extent of producing a perceptible hole some 2 to 3 mm across. The outline of the cavity is established with the high-speed bur or diamond, extending the margins back into sound tissue, thus eliminating unsupported enamel and carious dentine from the periphery of the cavity, but the floor of the cavity overlying the pulp chamber still consists of soft and discoloured dentine.

The factors governing outline form and enamel margins are those already discussed and have the same implications; the cavity has a greater area and involves more of the cuspal slopes (Fig. 6.6).

With the completion of the outline form carious dentine will still, with certainty in these cases, be found on the pulpal floor of the cavity, and this is removed with firm, deft strokes of a sharp spoon excavator, working from the centre of the cavity outwards. A hand-cutting instrument, such as an excavator, is the best instrument for this purpose because it allows the operator good discrimination between softened and normal dentine. The

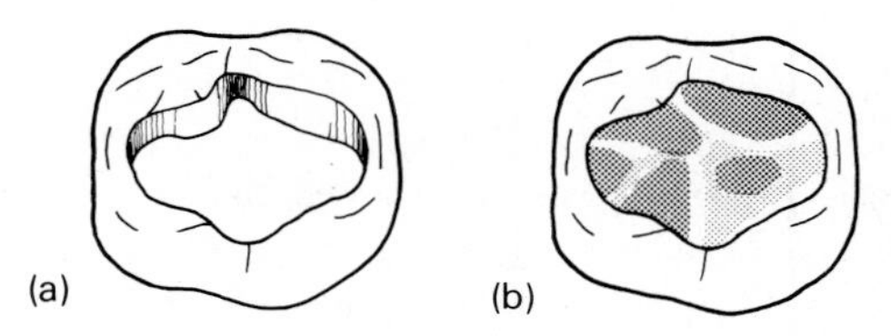

Fig. 6.6. Outline form (a) of extensive Class I amalgam cavity with lined base and a completed restoration (b).

cavity should have been laid open to a considerable extent before the removal of caries is undertaken, in order that the field may be clearly seen and easily reached. *Much time and effort are lost by trying to excavate a cavity through an aperture which gives inadequate access. When difficulty is unexpectedly experienced at this stage the operator should ask himself two questions: 'Can I see into every part of this cavity?' and 'Can I reach every part with this instrument?'* If the answer to either question is no, the need for further extension of the outline to include a carious area and to allow better access or a different choice of instrument is usually apparent.

Whilst the depth of the cavity is increased by excavation, the lateral extent is continually under review. In a large cavity, unexpected extension along the amelodentinal junction may be met; the undermining of a cusp may be considerable and may lead to necessity for further extension of the outline form by removal of unsupported enamel with chisel or bur.

In this way the stage is reached when the cavity margins are finally established in self-cleansing areas, the outline smooth and sinuous, and comprising sound enamel supported by sound dentine. Unsupported enamel is nowhere permissible and marginal ridges must remain well supported. In the depth of this cavity the dentine may well be found to be both stained and softened; in either case it must be presumed to be infected. The questions which now arise are 'How much dentine should be excavated, and with what risk of exposure to the pulp?' and 'Is it ever permissible to leave known infected dentine and, if so, how should it be treated and what is its ultimate fate?'

Treatment of carious dentine

The histopathology of dentine, the results of research, and the results of practical experience all confirm that in certain clearly defined conditions some carious dentine can and should be left at the *base* of a cavity.

In some cases of moderately deep cavities the clinically carious dentine, that is to say, dentine which is *softened* as well as being discoloured, may be excavated until the underlying dentine is apparently normal. More commonly, the dentine is as hard as normal but retains some degree of grey–brown staining, however slight. Histological evidence suggests that this dentine is not heavily infected and in fact it may be difficult to demonstrate bacteria in this tissue. It must, however, be accepted as infected since it cannot be proved otherwise. When excavation to this extent can be carried out without the risk of exposing the pulp, leaving what is frequently referred to as 'sound' dentine in all parts, it is both wise and expedient to do so. So far, few clinicians would disagree.

When, however, the carious process has proceeded beyond this, to the point where the pulp, though still uninfected, is closely approached, other factors must be considered. It is impossible to tell with certainty how closely the pulp has been approached and to exactly what extent a protective barrier of secondary dentine has been laid down. The ability of the pulp to respond by laying down secondary dentine and the rate of its deposition, normally a slow process, seems to vary within wide limits. It is, moreover, impossible to tell with certainty, in the absence of a macroscopic carious exposure, whether the pulp is, or is not, yet invaded by micro-organisms. The clinical history may give a lead here, but again, symptoms are not entirely reliable.

In these circumstances the operator relies upon his clinical judgement, that is to say, upon his assessment of the case based upon the experience of many other cases, the clinical details, and his knowledge of the pathology of the condition. A decision is made as to how much carious dentine may be removed without the risk of overrunning and destroying such natural barriers to invasion as may have been laid down. *In effect this usually means that when there is no clinical evidence of pulpal infection but where, nevertheless, the approach of caries threatens the pulp, it is permissible to leave a thin layer of carious dentine, of about 0.5 to 1 mm, over the pulpal area of the cavity.* **This does not in any way affect the statement that the periphery of the cavity must be unquestionably caries-free.**

It is inevitable that in occasional cases bacterial invasion of the pulp may have occurred without clinical evidence that this is so, and that this will proceed to pulpal necrosis and its sequelae. Such experimental evidence as there is suggests, and clinical practice certainly confirms, that these cases are few and far between. It is also apparent that many teeth can be conserved with vital pulps, that might otherwise be lost.

When a small area of carious dentine is left, it is reasonable to postulate that this area is, or will become, separated from the pulp by secondary dentine, and it will be segregated from adjacent normal dentine by the development of zones of sclerosis. If the orifice of the cavity is closed by a watertight restoration, any bacteria within the cavity are deprived of their metabolic requirements and die, or at least are incapable of reproduction and dissemination. The evidence is that many cavities become sterile within a short period and a few, though still capable of harbouring viable bacteria, do not extend in size over long periods of observation.

If this is true, it may be asked whether larger quantities of carious dentine should not be left, thus reducing the need for careful judgement and avoiding all risk of possible exposure. This extension of the principle is unacceptable on four counts, two of which are practical and two hypothetical:

1. Unless excavation is pursued thus far, it is not possible to determine with reasonable certainty the absence of a carious exposure.
2. The area of even a large cavity is relatively small and nothing may be allowed to increase the risk of leaving undetected caries at the periphery of the cavity.
3. Hypothetically, an extensive pabulum of carious dentine might provide conditions for sustained viability which could persist long enough for pulpal invasion to occur.
4. A soft basis such as would be formed by a considerable area of carious dentine would provide an unstable foundation for a stress-bearing filling which, if the slightest movement were to take place, would be subject to marginal failure.

Regardless of any specific treatment, the minimal carious tissue left behind in the conditions referred to undergoes characteristic physical change which can be seen when a filling with sound margins is removed. It darkens in colour, becomes very hard and eburnated, but shows no sign of carious extension. Many drugs have been recommended for the treatment of carious dentine, most of them with a view to sterilization. The application of strong drugs may do further injury to the dentine and pulp and should therefore be avoided. It seems that the choice of a non-irritant, oily, and mildly antiseptic dressing such as zinc oxide and eugenol is effective. The application of calcium hydroxide in a simple form would seem logical, since it would tip the reaction of the area towards the alkaline side, and produce conditions inimical to caries without causing tissue irritation. Many proprietary preparations based on these and other substances are available for this purpose. The procedure is sometimes called 'indirect pulp capping'.

In practice it seems to matter little what form of non-irritant dressing is chosen *provided that the restoration ultimately inserted is hermetically sealed and can remain so.* It is common practice to apply a lining of fortified zinc oxide and eugenol cement, which meets most requirements, or a sublining of calcium hydroxide followed by phosphate cement. The insertion of a permanent filling may proceed forthwith if the cement base is hard enough and the condition of the pulp not seriously in question. If either of these is open to doubt, it is preferable to insert a temporary filling and place the tooth on probation for a period of one to twelve weeks.

In the consideration of the large Class I cavity, it is clear that more and more of the occlusal surface may be involved, with the destruction of cusps and the weakening of the lateral walls. Thus the point is reached where one or more of these features must be removed and restored in amalgam, or alternatively, a more suitable form of restoration, for example, a gold inlay, should be chosen. The principles of design of the compound amalgam restoration are discussed in the next chapter.

Cavity toilet

If the cavity has been cut by any apparatus embodying a water spray or jet, there is little, if any, detritus in the cavity at completion of preparation. To make sure, however, the cavity should be washed with warm water or the atomizer spray before proceeding to isolation of the tooth. The necessary steps to isolation are then taken; this may be the application of rubber dam or the use of cotton-wool rolls. There can be no doubt that the use of rubber dam is an advantage for lining and filling procedures, unless four-handed operating can be relied upon to ensure complete moisture control.

When the area is isolated the cavity must be inspected to see that it is both dry and clean before proceeding. If a temporary filling has been previously inserted, the cavity surface should be cleaned with a pledget of cotton wool moistened with chloroform or with a solution of detergent, and re-dried.

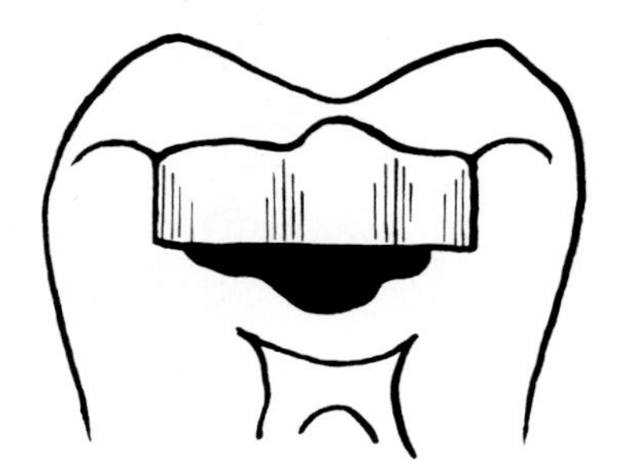

Fig. 6.7. Class I cavity prepared and lined.

Lining the cavity

In the case of a minimal occlusal cavity, fully extended and cut to the optimal depth of about 2.5 mm, it is unnecessary to insert a protective lining before the insertion of amalgam. Indeed, at this depth it is undesirable to place a lining, for this would reduce the depth of amalgam to a point where its strength might be insufficient to withstand the pressure to which it is liable to be subjected.

Where the further extension of caries has necessitated the deepening of the floor beyond the optimal depth, a lining should be placed to insulate the pulp and to restore the floor of the cavity to the optimal depth. The deeper the cavity, the more necessary is the lining for purposes of insulation. Polycarboxylate or EBA cement provides a non-irritant lining and a base which is strong enough to support condensation of amalgam (Fig. 6.7). A sublining of zinc oxide may be required or dispensed with for any of the reasons previously discussed, according to the indications. If used this is overlaid with zinc phosphate which restores the floor to its normal flat contour providing a firm and resistant base for the condensation of amalgam.

Insertion of amalgam

For many years alloys have conformed to the formulations of ADA specification No. 1 or the closely similar BSI Specification BS 2938/61. Recently, however, two important modifications have come into clinical prominence. One is the alloy powder of spherical particles, the so-called 'spherical alloy'. The other is the use of alloys having a high copper content, that is to say above the 6 per cent maximum previously thought desirable. These new alloys may contain from 12 to 28 per cent copper. It is claimed that the physical properties of the set amalgam are significantly improved in ways that are clinically important, for example, in decreased dimensional change and corrosion, an increased compressive and tensile strength.

These innovations have not required much change in usage. The spherical alloys (see p. 122) are considered to be less sensitive to the clinical variations common to most operators. The instructions supplied by the manufacturers

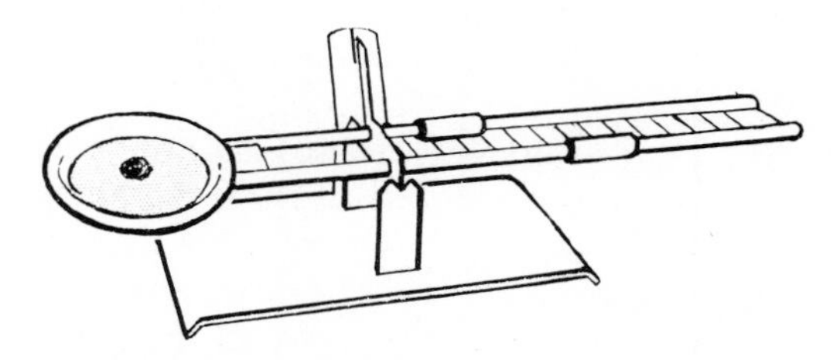

Fig. 6.8. A simple balance for the weighing of mercury and alloy powder.

of a reputable alloy should be closely followed according to the particular method of mixing recommended.

In most cases the powdered alloy and the mercury should be apportioned in the ratio of 5 parts of alloy to 7 to 8 parts of mercury, by weight. In selecting the quantity of amalgam for any given restoration, it should be remembered that enough is required to fill the cavity, to overpack the surface which is subsequently trimmed back, and to allow for an excess which is discarded with excess mercury during the process of packing. It is preferable to have a slight excess of prepared amalgam, rather than risk a shortage in the terminal stages of condensation. As an indication, a full extended Class I cavity in a lower molar will require approximately 0.8 g alloy and 1.2 g mercury.

Amalgam alloy powder can be presented in several forms; as loose powder; compressed into pellets; weighed and sealed in plastic sachets or contained with mercury in a plastic capsule. In each case except the last the appropriate amount of mercury must be added. The proportioning of alloy and mercury is most accurately achieved by using an amalgam balance (Fig. 6.8), a cheap and simple apparatus which is accurate enough within the required limits. For reasons of speed and facility many proportioners are available which measure the constituents by volume. Though convenient in use, these are subject to error unless used with great care. They must be clean, maintained with their reservoirs more than one-third full with alloy and mercury respectively, and they should be used in such a manner that their measuring chambers, usually embodied in a spring-loaded plunger, are completely filled before being released. If inaccuracy in proportioning is in any way to be condoned, it is preferable to have a slight excess of mercury, but this must be expressed from the plastic mass.

A mix of amalgam of too low mercury content is sandy in texture and difficult to work. It is true, however, that a correctly condensed amalgam when set contains slightly less than 50 per cent by weight of mercury. For this reason Eames and other workers recommended that the proportions of alloy and mercury in the mix should be equal. This avoids the possibility of a harmful excess of mercury and produces a set amalgam of good physical properties. The resulting mix is 'drier' than normal and calls for more than usual thoroughness in condensation. It is presumably for this reason that the method has not been widely or permanently adopted.

Amalgamation. The correct amount of alloy and mercury, correctly proportioned, may be amalgamated by one of three methods:

1. The pre-filled capsule may be chosen, or the alloy and amalgam measured and transferred to a capsule provided for use with the machine vibrator (Fig. 6.9). This provides rapid agitation for about 15 to 25 seconds measured on the time-controlled switch on the machine. The correct time for complete amalgamation will vary with the alloy and the type of machine. By running a series of trials of varying mixing times, a standard time is arrived at and in these conditions, it is claimed, a mix of constant character is obtainable.

2. The use of a pestle and mortar. The powdered alloy and mercury are triturated in the mortar under a light pressure, about 500 g weight, for a period of about 90 seconds, varying according to the type of alloy. The mix is complete when the mass assumes a smooth silvery consistency and shows a tendency to adhere to the sides of the mortar. This mass is then transferred to a 7.5 cm square of rubber dam, and kneaded between finger and thumb for a further 30 seconds, when a smooth, soft, and homogeneous texture is obtained.

3. The constituents may be placed in a thick rubber finger-stall and amalgamated by kneading between finger and thumb for a period of half to one minute, according to the type of alloy, and until the consistency described above is attained. This method is simpler since it is a single operation, and it also avoids the risk of over-trituration due to excessive pressure of pestle and mortar. If this procedure is to be delegated to an assistant, it is probably safer and more consistent in practice than the former (Fig. 6.10).

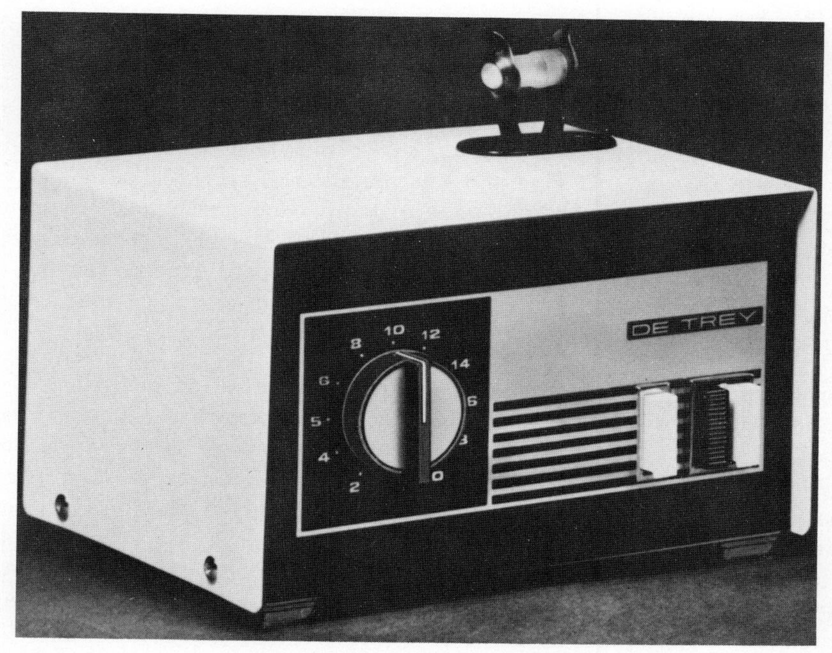

Fig. 6.9. A mechanical vibrator used for amalgamation of alloy or mixing cements contained in capsules.

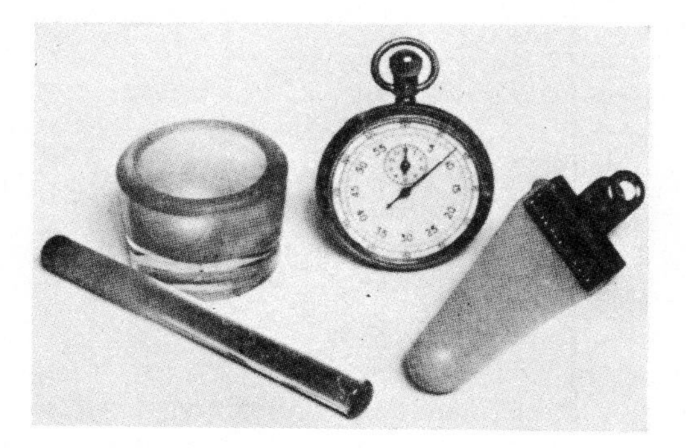

Fig. 6.10. Two methods of hand amalgamation. *Left* pestle and mortar; *right*, rubber finger-cot. A stop-watch or timer should be used.

The advantages of the simple methods shown above are that they are independent of any electrical supply and the equipment is cheap and easily replaced. In all three methods timing is important if over-amalgamation is to be avoided. If an automatic timer is not part of the equipment, and particularly if the operation is delegated, it is advisable to emphasize the time factor by using a stop-watch or other suitable timer.

Whatever method is used to obtain a satisfactory mix, the mixed mass is next transferred, avoiding contact with fingers, to a small square either of gauze of several thicknesses or of one thickness of chamois leather, and wrung between finger and thumb, or between the jaws of flat pliers, in such a way that excess mercury is expelled into a small receptacle. A little water in this will prevent splashing and dissemination of the expelled mercury. The mass is then ready for loading into an amalgam carrier for transfer to the cavity. The casual spreading of mercury about the working area must be avoided, for in some circumstances and over a long period, harmful concentrations of mercury vapour can build up in the atmosphere. Ventilation should be good; air extraction systems can be used.

Condensation. It is the habit of some operators to fill the carrier, for the first time only, with amalgam taken from the mass prior to expressing the excess mercury. In this way the first load transferred to the cavity is richer in mercury and, in the initial stages of condensation, the amalgam is more plastic. It is therefore thought to be more easily comformable to the deeper parts of the cavity. There would seem to be some advantage in this, but it must be emphasized that this excess must be eliminated early in the packing process, otherwise there is a risk that the deeper parts of the restoration may retain an excess of mercury and be defective in strength. For this reason the method is sometimes condemned.

The condensing instrument of choice for the Class I cavity under consideration is of the type represented by the double-ended, contra-angled plastic instrument No. 153. This has a flat, smooth condensing surface, and the contra-angled shank allows heavy pressure with stability (Fig. 2.2, p. 14). The diameter of the nib must be such that it can be accommodated on the floor of the cavity *at its narrowest point*, otherwise condensation of the deepest layers of amalgam is impossible in those areas.

Condensation proceeds by heavy pressure upon the amalgam mass in the centre of the cavity and the progression, by 'stepping', toward the walls of the cavity and the ends of the fissures. Heavy pressure is used throughout and as the instrument reaches the cavity wall, that portion of the surface amalgam which, being rich in further expelled mercury, has worked to the top and now lies around the nib, is removed from the cavity by an upward lift of the instrument against the wall (Fig. 6.11 (A)). This part of the amalgam is to be discarded and is allowed to fall away from the cavity. If the nib is shaped as shown in Fig. 6.11 (B), with a well-marked neck to the shank, removal of excess from the cavity is made easier.

When the first load is heavily condensed over the cavity floor and into the line angles and the excess removed, the second load follows and is similarly condensed, its mercury-rich excess being removed in similar manner. The operation proceeds using an outward stepping motion and discarding the topmost layers until the level of the cavity margins is reached.

Two aphorisms concerning the operation of condensation may be men-

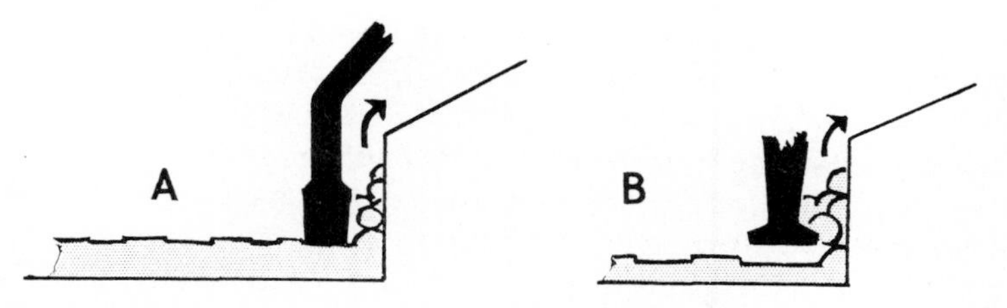

Fig. 6.11. The use of a flat-ended amalgam instrument with a sideways stepping motion and the elimination of mercury-rich amalgam against the wall of the cavity. For the purpose of lifting this excess from the cavity the nib shape shown in (B) is more effective.

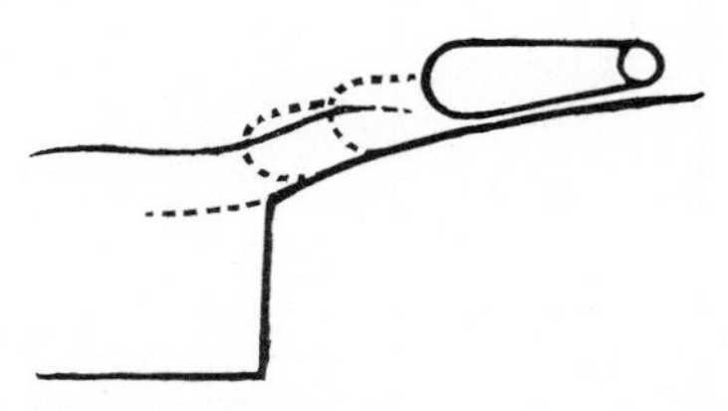

Fig. 6.12. The use of a flat plastic instrument, lying across the margin and moving along it, to define the outline form.

tioned. With a correctly prepared mix it is impossible to express too much mercury as the result of packing. If, at the conclusion of packing an amalgam restoration of any considerable size, the operator's fingers and wrist do not ache with fatigue, the restoration must be inadequately condensed. The point to be made is that the heaviest packing pressure which can safely be applied should be used throughout the operation. It is fallacious to suppose that the earlier layers inserted can later be condensed by heavy pressure in the terminal stages of packing. **Inadequate condensation is probably the commonest fault in amalgam technique**; it is one which cannot be remedied, but it can easily be avoided.

Carving. As the amalgam level reaches the cavity margins packing continues in the same manner to allow an excess to build up over the ultimate level of the finished restoration, overlapping the cavo-surface angle, and covering the contiguous occlusal surface. Condensation here must be particularly thorough, *for the strength of the amalgam margin and the integrity of the marginal seal* depend upon this stage of the operation.

When the margins are everywhere covered and fully condensed, carving the occlusal surface may be started. This is conveniently done by using a flat-ended plastic instrument, Ash No. 179. The amalgam surface is first attacked by defining the *cavity outline*, using the sharp blade of the instrument, lying at right angles across the direction of the outline. In this manner the configuration of the amalgam is trimmed to approximate to that of the adjacent enamel surface (Fig. 6.12). The blade, lying *across* the margin, moves *parallel* to it and so, by proceeding round the outline, the complete margin is defined. In a large restoration this leaves the centre incompletely carved and the delineation of the fissure is produced by the rounded end of the blade of the same instrument or of a smaller.

The reproduction of deep fissures such as may have naturally been present should not be attempted. The bottoms of the fissures should be rounded and capable of being burnished with a small round-ended instrument. It will be recalled that the steepness of the amalgam surfaces is a determining factor in their marginal strength (see p. 83), and an attempt to reproduce anatomical carving too closely may well have detrimental results in this respect. The main fissures should, however, be clearly represented and the marginal pits adjacent to the mesial and distal marginal ridges are of functional importance.

It is of great importance at this stage to realize that the attainment of a smooth and uniform surface, whilst the mass is still capable of being carved, has the greatest bearing upon the ultimate finish of the fully set and polished surface. Irregularities can only be removed later at the expense of considerable loss of surface contour. Sharp, smooth carvers must be used. Light burnishing of the surface with a small, smooth, rounded instrument is permissible to reduce fine surface irregularities. It should, however, be borne in mind that excessive burnishing may have the effect of transferring an excess of mercury toward the margins of the cavity. It is said that this excess can cause a loss of edge strength.

The occlusal surface should, before finishing, be checked by removing the rubber dam or cotton-wool rolls and asking the patient to place his teeth *lightly* together. If no abnormal contact is felt he may then be asked to grind his teeth together in centric occlusion. If the occlusal carving has been accurate, no abnormality of the bite should be perceived by the patient. If the restoration feels slightly high to him, a brightly burnished mark will be seen on inspection of the filling surface. This should be carved down until no abnormal sensation is appreciated by the patient. If necessary, articulating paper may be used to localize the high spot, but this is normally unnecessary since the burnished area is usually apparent on a surface which is otherwise of a matt texture.

Finally, a small, tightly rolled pledget of cotton wool may be used to remove the smallest irregularities and any small fragments which may lie upon the surface. The patient is instructed to rinse his mouth vigorously, holding the head well forward. In this way fragments of amalgam which may have found their way into the buccal and lingual sulci can be removed. The patient should also be told to avoid biting on this type of restoration during the next hour.

Polishing. No amalgam restoration is complete without final polishing which may be carried out at the next visit, after an interval of at least twenty-four hours. The surface should first be inspected to detect any burnished spots which may, even now, show areas where an excessive load is being exerted. Any such area should be reduced with a steel finishing bur which may also be used to run lightly over the whole area. Only the lightest pressure should be necessary and this is followed by a rubber-pumice wheel, of which the smaller size, the so-called sulcus disc, is particularly suited to polishing the fissure surfaces. As in all polishing procedures, success depends upon the use of a light uniform touch and the constant movement of the instrument so that the working area of the wheel impinges upon the surface from a variety of directions. If the wheel is kept aligned in one direction, grooves tend to develop and the character of the carved surface is lost. Care is exercised to avoid any risk of overheating the amalgam in the use of these abrasives.

When the surface presents a uniform 'satin' finish the final polish may be given with a soft cup-shaped brush and a fine abrasive. Some of the proprietary dentifrices are convenient, pleasantly flavoured, and capable of imparting a very fine polish to the restoration. Zinc oxide or stannous oxide moistened with equal quantities of water and spirit may also be used.

Modification of the Class I cavity preparation

The Class I cavity may, by definition, also occur in sites other than occlusal surfaces. The commonest of these are the buccal pit of the mandibular molar, the cingular pit of the maxillary lateral incisor, and the lingual fissure of the maxillary first molar.

The pit cavities, when small, may be treated as single surface restorations, roughly circular in outline, about 2 mm in depth, with flat floor and straight walls. Technically these are similar to, but smaller than, the occlusal cavity described. If caries has been allowed to progress, it may well be necessary to excavate these cavities to a much deeper level. They should then be lined, following the principles referred to on p. 88.

It sometimes happens with a molar occlusal cavity that the lingual fissure in the upper, and the buccal fissure in the lower, are also involved by caries. In these cases the affected fissure is included in the outline form (Fig. 6.13) and thereby becomes continuous with the occlusal restoration. The slopes of these fissures are not steep and, if the caries is minimal, they should not be widely extended. Their treatment follows in all respects the principles affecting fissures on occlusal surfaces.

The Class V amalgam restoration

Class V carious cavities occur as a result of stagnation and accumulation of plaque on the surface close to the gingival margin, more commonly on the buccal than on the lingual aspect. The cavity may frequently extend below the free gingival margin and if recession has occurred the carious area may extend into the cementum and dentine of the root. In the elderly the cavity

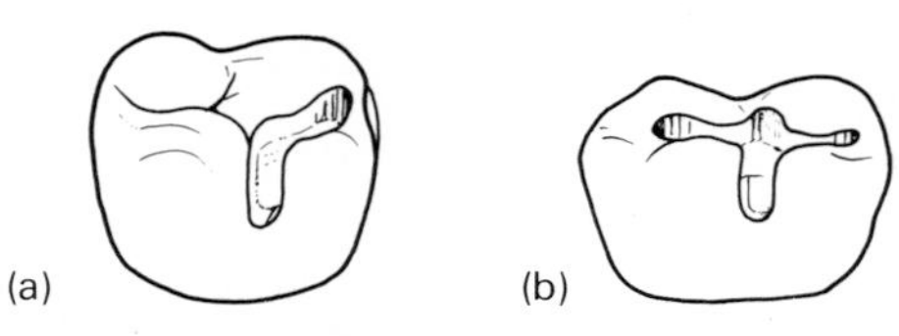

Fig. 6.13. (a) A distal occlusal cavity extended to include the lingual fissure of an upper first molar; (b) an occlusal cavity extended to include the buccal fissure of a lower first molar.

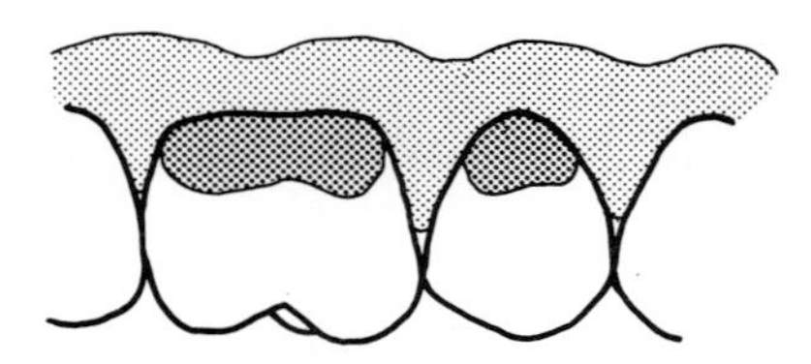

Fig. 6.14. The characteristic outline form and position of the prepared Class V cavity.

may lie very largely in the root dentine. As this is a lesion which starts on a 'plane' surface, in contradistinction to a Class I cavity, the extent of the lesion is more easily appreciated before preparation starts because its widest spread is mainly on the surface and the carious area tends to be saucer-shaped. The restoration of Class V cavities resembles in many ways that of Class I: there are, however, some significant and interesting differences.

It is worth noting that with cavities in premolar and molar teeth, the use of a reversed mirror head is a considerable help. This simple variation of the normal mouth mirror often gives better retraction of the cheek, reflection of light on to the cavity, and a reflected view of the working field, with considerably greater ease of manipulation.

Cavity preparation

The outline of this type of cavity usually conforms to the general shape shown in Fig. 6.14. The margin nearest the occlusal surface must be placed outside the normal stagnation area; this brings it to a position near the junction of the middle and gingival thirds of the buccal or lingual surface. The location of the gingival margin is usually determined by the extent of the caries and in practice this commonly lies below the free gingival margin, within the gingival crevice. There is, however, no doubt that anything less than a perfect restoration in this position will cause gingival irritation and a poor margin will cause severe chronic gingivitis. Theoretically the best site for the margin is at or slightly above the free gingival margin, but as already noted the extent of the lesion usually decides otherwise.

The rounded mesial and distal margins of the cavity present an insoluble problem from the point of view of stagnation. They must, wherever they lie, be within the gingival third of the coronal surface. To extend them mesially

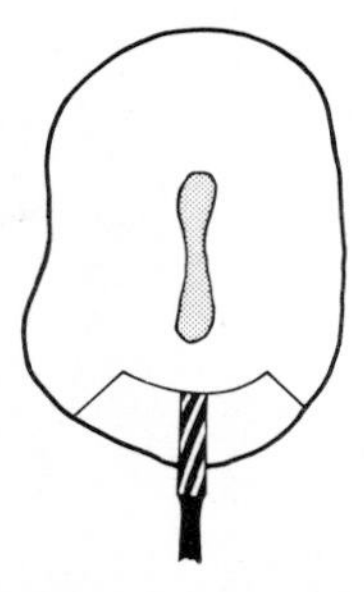

Fig. 6.15. Diagram of transverse section showing the use of a flat fissure bur in a Class V cavity. Note the curved floor and the radially disposed walls.

and distally towards the interproximal areas does not reduce the stagnation factor and introduces difficulty in cavity preparation and condensation of amalgam. The only positive requirement which must be met is that the outline must enclose the whole extent of the carious area, and, therefore, the margins must lie in sound tissue.

The cavity can be defined with high-speed burs but particular care must be taken, for the results of over-extension can be particularly difficult to deal with and rapid excoriation of the gingiva can occur with bleeding, which makes it even more difficult to see. Direct vision must be used whenever possible and access to cavities further back in the mouth usually demands the use of medium- and low-speed instruments. The completed cavity should be of uniform depth between 1.5 and 2 mm; this implies that the cavity floor is convex. In any case, completion of the cavity outline and finishing the walls should be done with lower-speed instruments, or with the special high-speed tungsten carbide instruments previously referred to. These burs are cylindrical in form, and should be kept in radial alignment when cutting, roughly at right angles to the tooth surface (Fig. 6.15). The cervical margin of the cavity, being close of the gingiva, is better prepared with a round bur at low speed and completed with a hand instrument such as a hoe, Ash 28, or hatchet, Ash 22. In this way injury to the gingival margin and the haemorrhage (see next page) are minimized. If the cross-section is roughly circular or elliptical, the floor of the cavity should lie on the circumference of a smaller concentric figure. Thus the floor-wall line angle is cut at 90 degrees and the cavo-surface angle also approximates to a right angle.

The mesial and distal walls of a cavity such as this are often widely divergent and it is difficult, sometimes impossible, to make them retentive in form. This is particularly so when the margin is placed at the mesio-buccal or disto-buccal angle of the crown. Retention must therefore be derived from the occlusal and gingival walls, of which the dentine may be given a 5-degree inward divergence along their entire length. Excessive undercutting with an inverted cone bur should be avoided as unnecessary. This restoration is not subject to direct occlusal stress and the forces tending

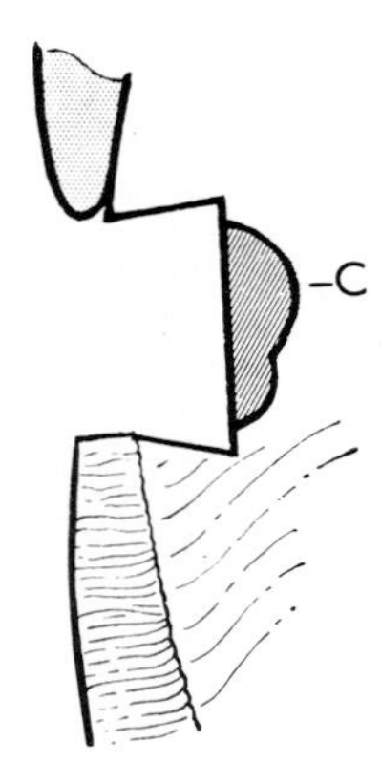

Fig. 6.16. Diagram of vertical section through Class V cavity showing undercuts on gingival and coronal walls and restoration of floor by cement C.

to dislodgement are minimal. The cavity must, of course, be self-retentive and it should always be cut at sufficient depth to allow this (Fig. 6.16).

In all but the shallowest cavities, the centre of the cavity floor may be deeper than elsewhere owing to the need for further excavation. The principles applying to the lining of this cavity are in all respects the same as in the case of the Class I cavity. The contour of the floor is established in the same way.

Enamel margins should be surveyed and smoothed with the enamel finishing bur, whilst at the same time establishing the 90-degree cavo-surface angle.

Management of the gingival margin

Up to this point note has been taken of the fact that these cavities are close to the gum and often extend below the free gingival margin. This is a most important circumstance and one which complicates preparation and filling procedures to a very significant degree. In extreme cases it may render restoration impossible.

In principle, a certain amount of damage to the periodontal tissues is acceptable since their reparative power is relatively good, *provided* that a highly effective restoration can thereby be achieved. A traumatized periodontium and a defective restoration is the least desirable result, for the latter will certainly continue to aggravate the former.

The difficulties presented by this problem are those of access visibility and the maintenance of a dry field; they can be met in many ways.

Wherever possible trauma to the gum should be completely avoided. With an intact gingival epithelium the maintenance of a dry field for a period long enough to place a satisfactory filling is possible in many situations.

Rubber dam may be adapted for the cervical cavity by the use of a cervical clamp. The clamp is designed to effect forcible retraction of the

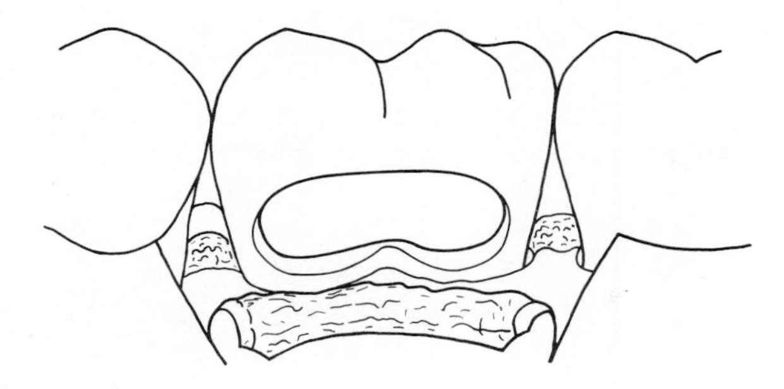

Fig. 6.17. Buccal gingival flap, reflected to allow preparation of the gingival margin encroaching on the root bifurcation and the insertion of a Class V restoration. Two sutures replace the flaps.

gingiva and the retention of the rubber clear of the working field. The jaws of a clamp may be modified by stoning and polishing, and the clamp stabilized in position by placing impression compound under the bows. Though effective in some situations it is limited in application; the degree of trauma inflicted depends upon the extent of the cavity and the amount of retraction required to expose it.

If, in the course of cavity preparation without rubber dam, a certain amount of unavoidable trauma is caused, the operator may well find the cavity obscured by haemorrhage. This can be controlled by gentle pressure with a pledget of cotton soaked in 1:1000 adrenaline, but seepage of serum may continue to preclude a dry field. If the laceration is limited, the application of trichloracetic acid (p. 54) may give adequate control for a period which allows the insertion of the filling. Failing this, it is better to insert a temporary zinc oxide dressing. After an interval of 7 to 14 days, during which the gum is allowed completely to heal, the dressing is removed with care to avoid further injury to the margin, and the permanent restoration placed.

A logical extension of this method, when trauma to the gum appears inevitable, is a gingivectomy localized to the tooth in question. The anaesthetized gum is resected at a level apical to the ultimate gingival outline of the cavity and haemostasis obtained, the cavity preparation can be completed without difficulty. If moisture is controlled an amalgam can be placed straight away. Otherwise a combined temporary filling and gingival pack can then be applied, the latter being taken firmly into the adjacent interdental spaces to gain adequate retention. After an interval for healing, the permanent filling is placed; the recovery of the normal gingival outline occurs surprisingly quickly.

For control of a deep cervical margin the reflection of a gingival flap is a very effective measure. Under adequate local analgesia axial incisions are made a few millimetres beyond the adjacent interdental papillae and a mucoperiosteal flap reflected. (Fig. 6.17). As soon as bleeding is controlled

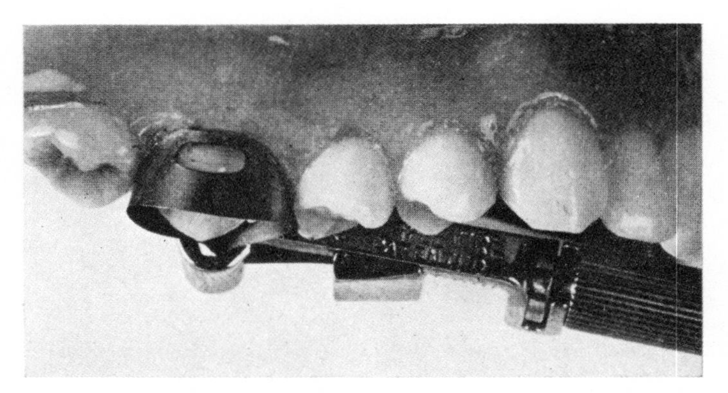

Fig. 6.18. *Tofflemire* retainer used with a perforated band suitable for a large Class V amalgam restoration.

the cavity is completed without difficulty. The wound is freely washed with warm saline and two strands of ribbon gauze are placed through the wound to collect excess amalgam.

The permanent restoration can now be placed, lining if necessary; the cavity can be kept dry and the cervical margin packed, trimmed, and lightly burnished under direct vision. This completed, after further irrigation, the replaced flap may be retained by two sutures, and if need be for comfort and protection, a gingival pack applied for three days or so.

Insertion of amalgam

The cavity toilet completed and a dry field obtained, the insertion of amalgam follows the same pattern as that described for Class I cavities. Condensation is carried out by stepping towards the cavity walls and the elimination of the mercury-rich excess. In the final stages, when an excess of amalgam must be built up to restore the convex surface of the ultimate restoration, packing must proceed with particular care lest the final layers should be inadequately condensed. At this stage the automatic condenser referred to on p. 123 and illustrated in Figs. 7.28 and 7.29 presents some advantage. With this condenser, a nib which is concave and oval in outline can be used in the final stages and it is probable that this method is capable of producing as good condensation as any of the available techniques.

The cavity which extends to the mesial and distal angles of the buccal surface may produce a problem in condensation owing to the obliquity of the mesial and buccal walls of the cavity, which no longer confine the amalgam to allow adequate packing. A method sometimes applicable to these extensive cavities involves the use of a perforated matrix band. This band encircles the tooth and is so placed that the perforation, which my be enlarged if necessary, lies over the centre of the cavity (Fig. 6.18).

Amalgam insertion starts in the normal manner. When the cavity is two-thirds filled the matrix band is placed and condensation proceeds through the aperture, previously made, packing towards the cavity margins. With the margins thus confined it is possible to obtain good condensation. The method is not, however, without its difficulties, but it is one which can be successfully used when dealing with the difficult case.

Carving and finishing

A smooth, highly polished filling is perhaps of greater importance in the Class V restoration than in any other *because of the inherent tendency of the area to stagnation and gingival irritation*. Care in carving and the elimination of fine irregularities after condensation is of primary importance. In particular, it is necessary to see that the cervical margin is trimmed quite clear and flush; the point of a sickle probe used flat upon the tooth surface is a convenient instrument for finishing this margin. A little extra time spent upon this is well repaid, for the trimming of the margin with burs when the filling is set is difficult, injurious to the gingival crevice and never so satisfactory in its final result.

Polishing proceeds as previously described except that a rubber cup is used in place of the cup-shaped brush. The former adapts well to the convexity of the surface, is less harsh upon the gingival margin, and can be insinuated into the gingival crevice with comparatively little harm.

Finally, it must be remembered that some parts of the cavity outline must necessarily remain in the stagnation areas. The operator fails in his duty if he does not explain the situation in simple language to the patient who is capable of intelligent co-operation. The use of disclosing agents and the correction of tooth-cleaning habit to ensure that plaque formation is reduced to a practical minimum, and the correction of dietary faults when these exist, are a logical concomitant to careful performance of a good restoration. With the cervical restoration this is particularly true and its effective life can be prolonged by attention to these details.

Treatment of sensitive dentine

Sensitive dentine occurs most frequently in areas subject to toothbrush abrasion, situated in the facio-cervical regions of the teeth in both upper and lower arches. They are the result of faulty toothbrushing habits, usually excessive cross-brushing and they are usually most marked in the canine and premolar regions, in the maxilla more than the mandible, and commonly more accentuated on the left side in right-handed subjects. Any tooth slightly buccally placed is more liable to this damage than the adjacent teeth, but abrasion can occur on any tooth. Cementum which normally covers radicular dentine is insensitive and being a very thin layer is very rapidly removed by abrasion, exposing dentine. Attrition can also expose sensitive dentine on occlusal and incisal surfaces and erosion of labial and lingual surfaces can have the same effect.

Exposed areas normally become insensitive simply by being exposed to the oral fluids and ingested substances, but occasionally this desensitization does not occur and small areas, sometimes minute in size, remain mostly around the periphery of the denuded area. However, the pain from several such areas may be very severe, completely preventing brushing or the ingestion of very cold, sweet, or sour food or drink.

Quite often the patient can locate the lesion fairly accurately and this is confirmed by gentle exploration with a probe. Further confirmation by the application of cool water or air should only be done when really necessary. Fortunately the treatment, once the area is located, could scarcely be simpler; it consists of applying an irritant to induce sclerosis of the dentinal tubules concerned. Many irritant substances have been used. The common ones are aqueous solutions of 2 per cent sodium fluoride: 8 per cent zinc chloride and formalin (40 per cent formaldehyde), all of which work well. There are also a number of proprietary pastes and varnishes available.

The area to be treated is isolated with cotton-wool rolls and dried with a pledget — even 'warm' air should be avoided for it can be excruciatingly painful—and a pledget moistened in the chosen drug is rubbed on the sensitive area for about a minute. This is normally no more than mildly uncomfortable. The area may be kept isolated for a further two minutes. Irritants of this potency could not be applied to areas of freshly cut dentine, as for example in a cavity, without serious risk of pulpal damage. Used in the manner described they are without clinical ill-effect. The patient should be warned that the effect of the treatment will not be noticed for a day or two and that a further application may be needed, though ony rarely does this prove to be necessary.

Summary

Restorations temporary and permanent.
Role of amalgam; failures in technique.

Class I amalgam. Preparation; initial entry, extension to sound dentine, margins in non-stagnation areas; extension for prevention, risks of over-extension.
Average cavity depth 2.5 mm; cavity floor made caries free.
Cavo-surface angle approximates 110 degrees in most areas: principles dictating this compromise.
Finishing enamel margins.
Larger cavity. Problems of deep caries approaching pulpal involvement. Access. Rationale of residual carious dentine; methods of treatment. Absolute requirement, caries-free periphery and hermetic marginal seal.
Cavity toilet; removal detritus, surface drying only.
Cavity lining. Non-irritant: mildly antiseptic, insulating cements; fortified zinc oxide, EBA, polycarboxylate.
Restoration of cavity floor.
Amalgam preparation. Proportioning constituents; 5 parts alloy 7–8 parts mercury, by weight. Weighing alloy; pellets, sachets, capsules. Spherical alloys; high copper alloys.
Methods of amalgamation, mechanical and hand.

Expression of mercury; avoidance of vapour hazard.
Insertion and condensation; instruments and use; removal of excess.
Carving, finishing, polishing.
Modification of Class I cavity to include lingual and buccal pits and fissures.

Class V amalgam. Factors controlling siting of margins. Gingival margin often sub-gingival. Cavity floor normally parallel to crown or root surface.
Retention by undercuts in occlusal and gingival walls.
Principles of lining similar to Class I.
Insertion and condensation of amalgam.
Use of mechanical condenser; matrix.
Importance of finishing and polishing.
Management of gingival margin; local gingivectomy or reflected flap to allow preparation and insertion.

Sensitive dentine. Causes and localization; simple treatment application of irritant.
Drugs used. Sodium fluoride, zinc chloride, formalin; proprietary pastes and varnishes.
Procedure, isolate. dry; apply drug on pledget.
Isolate for further two minutes.

7

The technique of compound amalgam restorations

The restoration of the carious cavity originating on the mesial or distal surface of a posterior tooth presents a number of characteristics which differ from those of the occlusal or cervical filling. These differences arise from three factors:

1. Because of its inaccessibility, the cavity must be approached from the occlusal surface, even though at the early stage when one would choose to treat this cavity, the occlusal surface has not yet been cariously attacked.
2. Since the carious attack, or its eradication, destroys the area of contact with the adjacent tooth, this must be restored.
3. Because the filling restores a substantial portion of the occlusal surface and of its margin, it is subjected to heavy functional stresses which tend to dislodge it. Special methods must be used to ensure adequate retention.

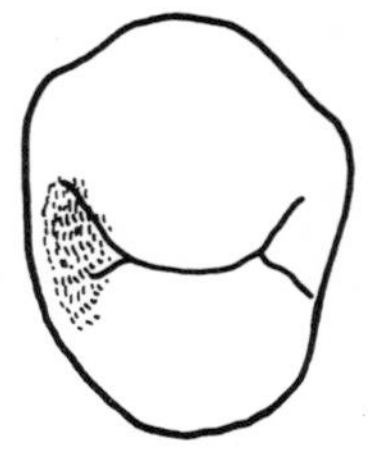
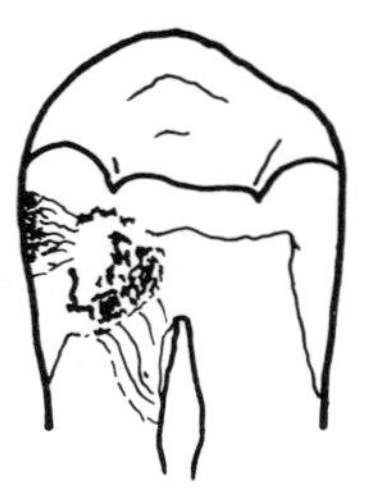

Fig. 7.1. Bluish-grey discoloration below enamel of marginal ridge, indicative of Class II cavity.

Reference has already been made to what has for brevity been called the *optimal cavity*, in which *carious involvement of dentine has gone far enough just to be visible with certainty on the bitewing radiograph*. If untreated at this stage the cavity extends in all directions until the enamel of the marginal ridge is attacked from its deep surface. At this stage a characteristic blue–grey discoloration is apparent on the inner aspect of this ridge (Fig. 7.1). This is a clear indication that the cavity is extensive and that immediate treatment is essential if the pulp is to be saved. The involvement of the marginal ridge is in due course followed by its collapse, and the central spread results in the invasion of the pulp in a matter of a few weeks or months (Fig. 3.7, p. 38).

Up to a certain size, this type of cavity is very suitable for restoration by amalgam, but the larger the cavity and the weaker the remaining tooth structure, the more the balance swings in favour of restoration by a gold inlay. The principles governing the choice of the latter type of filling will be discussed in Chapter 10.

Class II cavity preparation

In this description of the technique it will again be assumed that in the first instance we are dealing with the optimal cavity as previously defined and it is reasonable to assume that the actual size of the cavity is fairly accurately known to the operator from his study of the radiograph.

Access to the cavity is gained by using a high-speed handpiece and a No. 2 round or fissure bur, diamond or tungsten carbide. The bur is directed from the marginal fossa slightly towards the contact area of the tooth so that its direction of cutting is towards the area of affected dentine (Fig. 7.2). Using the lightest touch, intact enamel and then dentine are removed and the occlusal outline defined; proceeding in an axial direction the carious area is then removed (Fig. 7.4), taking care not to open the cavity too widely nor to allow the cutting head of the bur to pass even momentarily beyond the reach of the coolant spray.

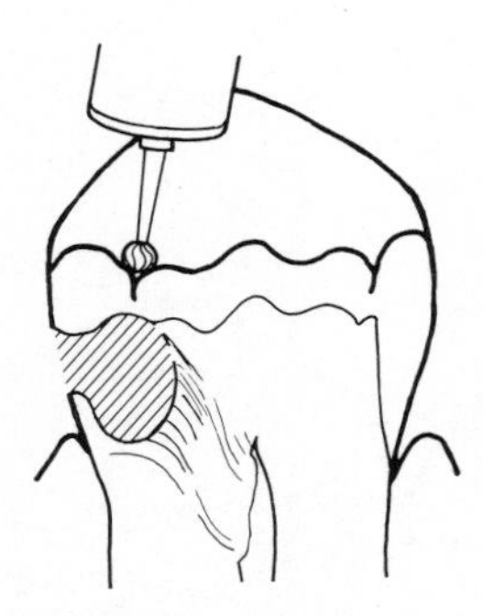

Fig. 7.2. Class II cavity undermining the marginal ridge. The cavity is entered through the marginal pit.

The risk of *damaging the contact area of the next tooth is very high* and the application of a steel matrix band to that tooth reduces the risk only slightly, because the band is so easily and rapidly perforated. The safest method is to reduce the marginal ridge to about 0.5 mm thickness and complete its removal with low speed, which may then be used for the final stages of the cavity preparation. An alternative method of removing the reduced enamel wall is to break it gently outwards with a chisel (Fig. 7.5).

Damage to the adjacent tooth surface is a serious failure of technique. Should the enamel be lightly damaged it should be possible to overcome this by taking down the surface very slightly with a medium grit sandpaper disc and then polishing the surface with one of very fine cuttle-fish grit. The ex-

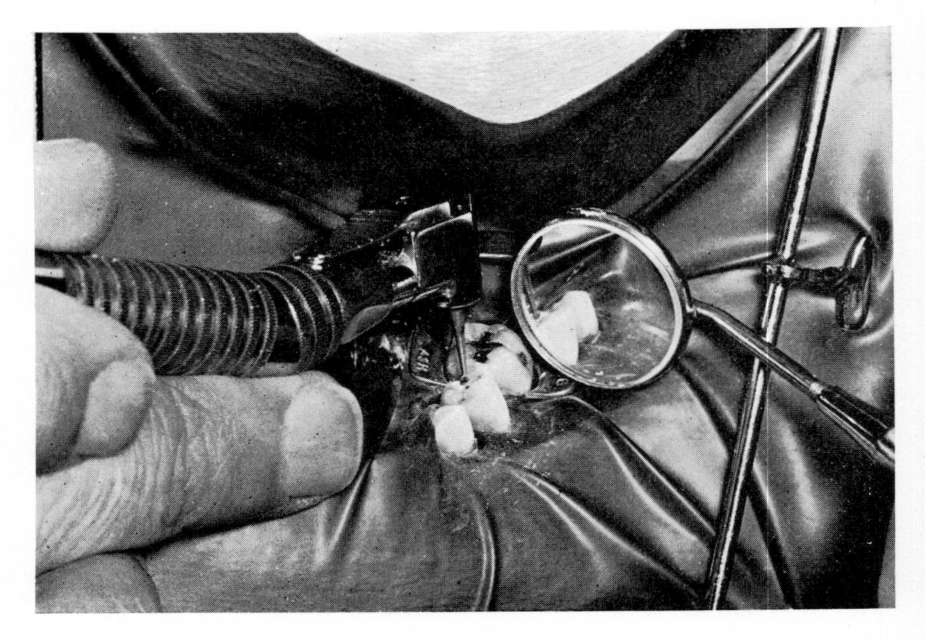

Fig. 7.3. Initial approach to a Class II cavity through the distal pit of a mandibular premolar, using a low-speed handpiece, without cooling spray. Note the finger support on the lower incisors.

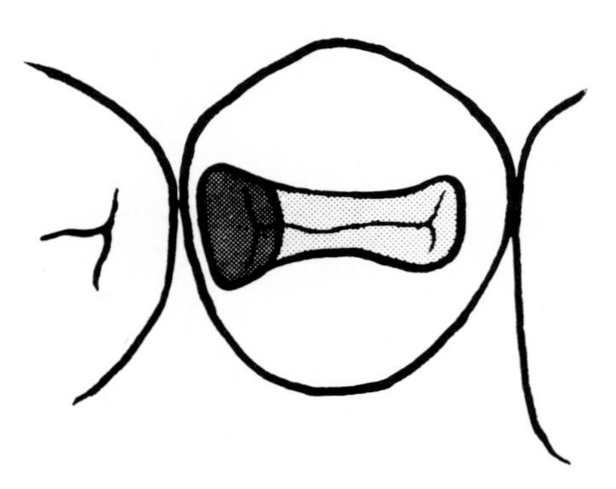

Fig. 7.4. Initial occlusal outline and entry into distal cavity, retaining interproximal enamel wall.

tent to which this is permissible is very limited, for the loss of enamel flattens the curve of the interstitial surface and may risk a defective contact. More severe damage than this, which could arise from lack of control, would require the insertion of a Class II restoration in the damaged tooth.

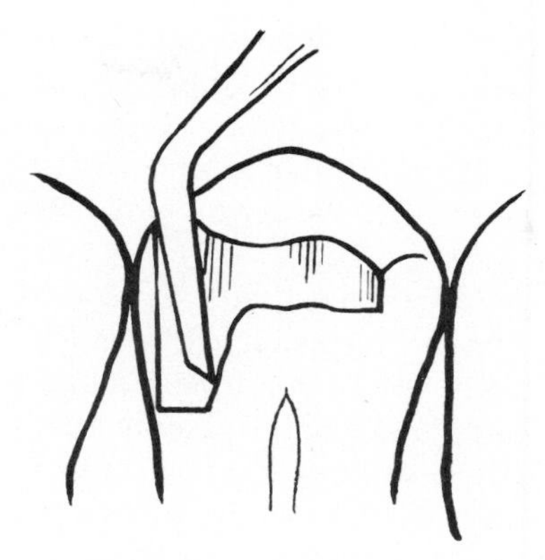

Fig. 7.5. Use of a chisel to break through final part of marginal ridge.

Hand instruments can also play a useful role in the completion of the interstitial part of the cavity, which will now be referred to as the box. Once the interproximal enamel has been very considerably reduced preparation can proceed by use of a sharp chisel, No. 85/82, or hatchet, No. 51 (Fig. 2.2 (a), p. 14). This is held by the grip which will allow firm pressure and control, and it is used to cleave small fragments away from the free axial edge, and the blade should travel the full length of the vertical margin at each stroke (Fig. 7.6).

Difficulty in the use of the enamel chisel may arise from several causes, all of them essentially simple:

1. A straight chisel should be chosen whenever access allows the use of that pattern (Fig. 7.6). The commonest cause of difficulty is bluntness of the cutting edge and this factor becomes all the more important when, as for example with a hatchet used upon a lower molar, the pen grip must be used and the pressure which can be applied is comparatively light; in these conditions a blunt chisel is quite ineffective.

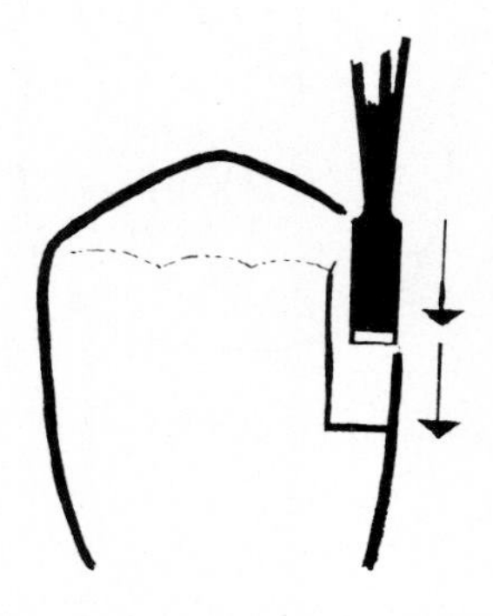

Fig. 7.6. Use of a chisel to define axial margins.

2. Only small decrements of enamel should be removed at each stroke. Whilst it is true that large portions of completely undermined and secondarily carious enamel may be easily removed, relatively sound tissue offers considerable resistance. As the resistance increases so should the decrement become small, to the extent that the action may become one of shaving or planing, rather than cleaving.

3. When difficulty is experienced with a wide blade, a narrower one should be tried. Since it makes a small line of contact it cleaves more easily, and the larger blade can be used for a final planing finish.

It was at one time maintained that the planing action of a sharp chisel was the most effective method of imparting a smooth finish to an enamel margin. With the increase in variety and efficiency of rotary instruments this appears no longer to be true. It is now known that fine plain cut steel burs and fine grit abrasive discs produce as smooth a surface but not all enamel margins are accessible to these rotary instruments. In practice both may be used in different parts of the outline. Present evidence indicates that very fine-bladed or blank tungsten carbide burs used at high speed produce a very smooth finish and this technique may prove to be the best available (see p. 167).

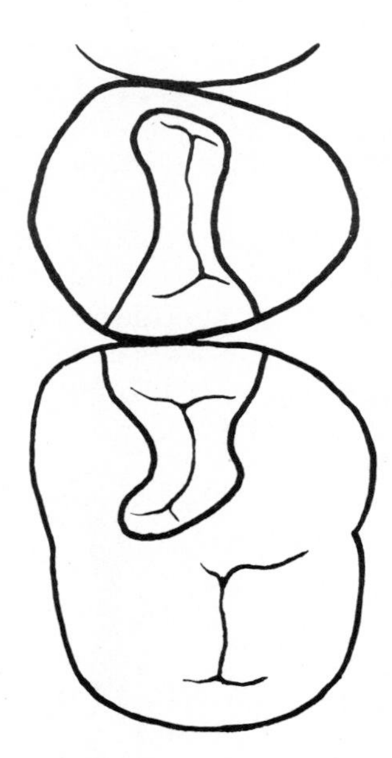

Fig. 7.7. Class II outline forms showing the position of axial margins in the interdental embrasures.

By the use of the chisel the axial margins on the buccal and lingual aspects are extended into the respective embrasures to the positions at which they

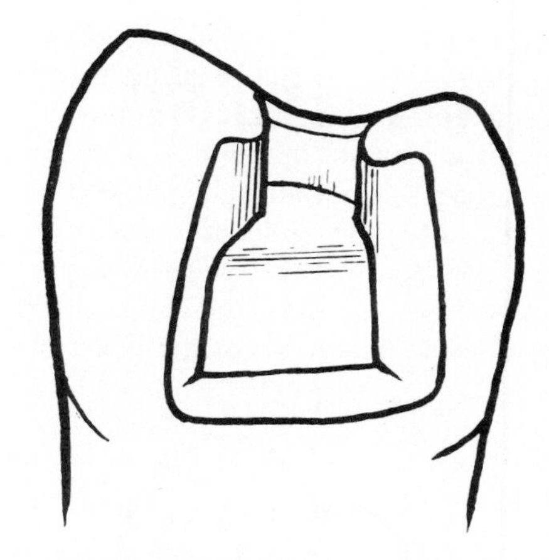

Fig. 7.8. Slight divergence of axial margins toward the gingival margin.

Fig. 7.9. In premolar regions the contact points lie nearer the buccal aspect than the lingual. Note the different shapes of lingual and buccal embrasures.

come to lie in self-cleansing areas (Fig. 7.7). As was mentioned in Chapter 5, a rough guide to this is that the margin should *just* be visible when viewed normally from the side. Viewed from the interdental space the axial margins may show a slight divergence as they pass from the occlusal to the gingival. This divergence should not exceed 5 degrees from the parallel (Fig. 7.8), and in many cases it is permissible that they should be parallel. They may not, however, be divergent towards the occlusal surface.

It should be observed, for it has a bearing upon this matter, that in the premolar region particularly, the buccal and lingual embrasures are of different shapes (Fig. 7.9). First, the contact point, or the centre of the contact area, is nearer to the buccal than the lingual aspect; also the buccal embrasure opens rapidly on to the buccal surface of the crowns. The lingual embrasure, however tends to be more acute and deeper. For this reason it is common to find that the linguo-axial margin comes to lie a little further from the contact point than does the bucco-axial margin.

On the other hand over-extension must be avoided. For reasons of appearance alone, it is desirable to bring the bucco-axial margin no further out than is essential for the purpose of avoiding the stagnation area. One of the most frequent causes of over-extension in these cases arises from the technique of cutting the occluso-bucco-axial margin with a fissure bur, starting centrally and moving along a radial path towards the embrasure. If the risk of over-extension is appreciated, it can of course be avoided, but a safer method of taking back the axial margin is first to undermine the enamel by removing the dentine under the margin to be removed. A groove, cut with a round bur, in a vertical direction along the amelodentinal junction (Fig. 7.10) will allow the enamel chisel to be used for the final placing of the axial margin. When completed and viewed from the occlusal, the axial margins of the box show some divergence towards the interproximal.

Fig. 7.10. Grooved axial margins of box prior to extension with chisel.

The gingival floor has now to be defined. The depth of the original cut will have determined the present depth of the box and, if the enamel chisel has been used to the full depth, it is probable that the gingival floor is already established at or near its correct position, which is normally recognized as the level at which a margin of sound tissue can be made. This can with advantage be supragingival, but in most cases must be at or below the gingival crest. Should it be necessary to deepen the floor, this should be done with an end-cutting fissure bur No. 958–960, the size chosen being such that the diameter of the cutting head equals the required width of the floor. Too small a diameter cuts an irregular furrow, too large damages the interdental papilla and causes bleeding which obscures the field.

A suitable end-cutting bur can be used at low or high speed; in either case it is carried lightly to and fro along the margin. Heavy pressure will produce stepping. Internal line angles can be sharpened by the use of a chisel or hatchet. In this way the outline form of the box is completed, but the occlusal outline has still to be created.

Retention

The retentive form which could be used in a box-like cavity can be obtained by several methods and any or all of them may be used in a particular case.

The most effective form of retention is the occlusal lock or dovetail (Fig. 7.11) in which the cavity is extended along an occlusal fissure and shaped to the form of a dovetail which is retentive in a horizontal direction. The fissure is often partly carious or likely to become so; in either case it is desirable that it should be included in the cavity outline. The factors determining this outline are those applicable to a Class I cavity with one important addition.

The efficacy of such a dovetail depends upon its strength at its narrowest portion. This depends upon the tensile strength and the shear strength of the amalgam, which are not high and the cross-sectional dimensions at its weakest point. An empirical rule which works well is: if the dovetail is cut to the elective depth of 2.5 mm, the width of the 'neck' of the lock should be approximately one-third of the distance between the axial margins of the box (Fig. 7.11). This is probably a fairly critical factor and *it can only be exceeded at the risk of weakening the adjacent cusps*. When it is desirable, the neck of the lock can be narrower but it should be deeper.

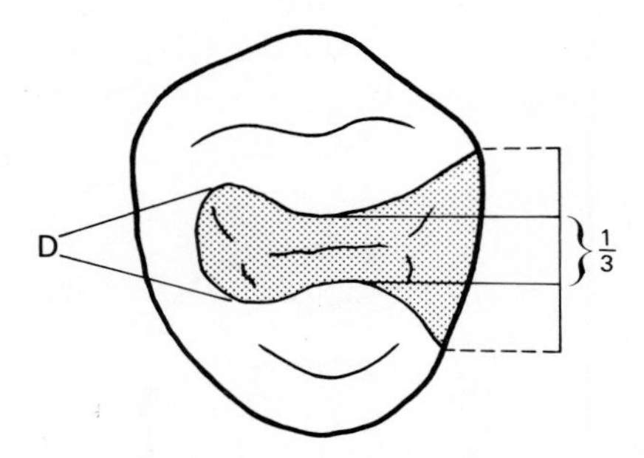

Fig. 7.11. Relation between width of neck of lock and that of box in Class II cavity preparation.

If the lock is deepened, remember that resistance to fracture of the amalgam under a load in the long axis of the tooth varies proportionately to the width of the neck, but to the *square* of its depth. This means that a small increase in depth confers considerably greater strength upon the amalgam. Against this, the deeper neck reduces the strength of the cusps: they are more susceptible to fracture under a similar load.

The use of the occlusal lock is open to criticism on several grounds in the case of premolars. If caries occurs later on the opposite surface of a premolar, that is to say, for example, a mesiocclusal restoration exists and caries then occurs on the distal, the preparation of the distal cavity damages and may destroy the greater part of the existing lock. In this case the pre-existing filling must be removed and a MOD restoration inserted. There appears to be no satisfactory alternative to this course.

It is sometimes claimed that such a sequence as this leads to unnecessary weakening of the residual tooth structure, with the result that the fracture of one or other of the cusps is a common sequel. This criticism is only justified when over-extension of the occlusal bridge has been committed. The extent of the mesial and distal cavities, as determined by caries, is more often the determining factor in the weakness of these cusps.

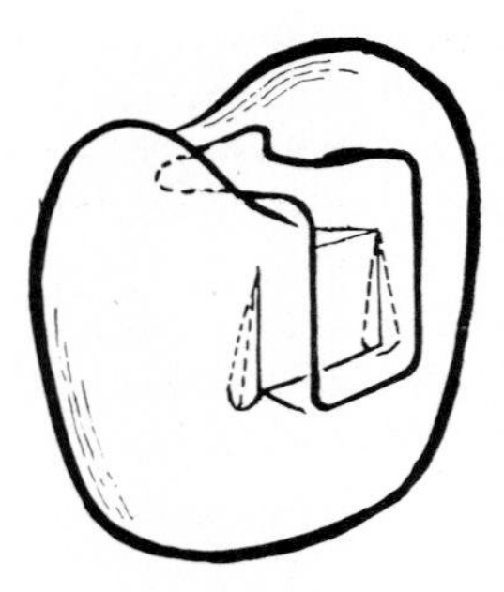

Fig. 7.12. Use of axial undercuts in box, in addition to occlusal lock.

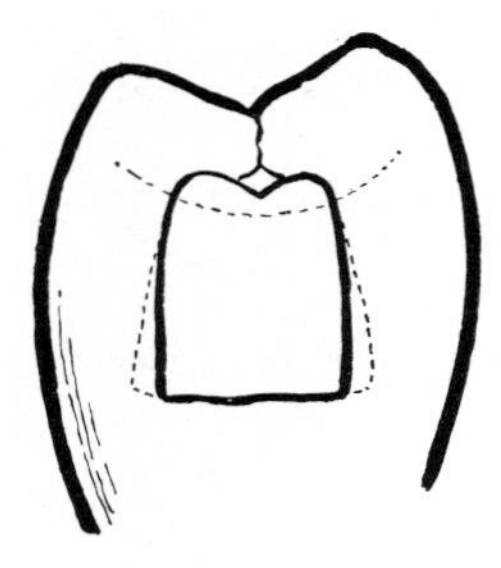

Fig. 7.13. Use of axial undercuts alone for retention of Class II restoration in lower premolar.

Another method of retaining this restoration where the axial surfaces of the box are not too widely divergent to the buccal and lingual, is the establishment of small opposing undercuts in the lower and inner parts of this surface (Fig. 7.12). The undercuts can be started with a No. 1 fissure bur and defined with chisels. They are more effective if they are joined by a groove along the cervical, to which reference will be made below. The lateral grooves should not be deeper than is required to provide secure retention and a balance must be held between the necessity for depth and the risk of weakening the wall. At all events, such retentive undercuts must be kept well away from the occlusal surface.

Retentive grooves in the axio-buccal and axio-lingual walls of the box may be used for all Class II cavities in addition to the occlusal lock. There are a few cases, of which the small Class II cavity in lower premolars is a fairly common example, in which the use of retentive grooves alone can be justified. In such cases the grooves must be rather more emphasized without weakening the substance (Fig. 7.13).

The gingival groove is a means of further retention and one which raises the question of the general design of the gingival floor. On the basis of theoretical considerations it should be of the outline shown in Fig. 7.14, basically flat, grooved along its inner margin, and its enamel edge trimmed parallel with the enamel prisms at this point. Defective prisms at the margin are as serious here as anywhere except that they are not subject to direct occlusal stress. In the absence of an adjacent tooth and on the mesial surface this shape is fairly simple to achieve. With an adjacent tooth and with other difficulties of access and vision, it becomes extremely difficult to achieve. Even with the full range of cervical trimmers it is sometimes impracticable, but the principles which determine the ideal form are logical and should be understood.

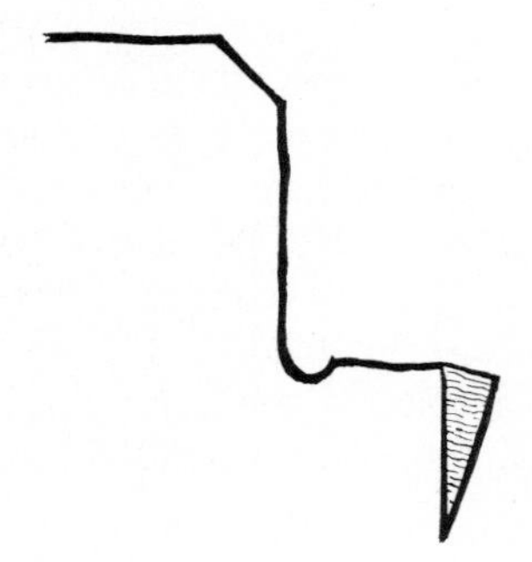

Fig. 7.14. Cross section of theoretical outline of gingival floor showing groove and cavo-surface angle.

In this context it is suitable to consider the use of cervical trimmers Nos. 77 to 79, for these are designed for the shaping of the cervical floor of Class II cavities. Cervical trimmers in general are difficult instruments both in use and maintenance, and a consideration of the following points will assist in their successful employment:

1. No excuse is offered for emphasis upon sharpness, for this quality is more necessary to this class of instrument than to any other in the whole range of hand instruments. The advent of tungsten carbide-tipped instruments has helped to solve this problem. In the process of sharpening the carbon steel blade, the oblique bevel must be established at the correct angle, which can be varied to suit the particular case. In the author's experience

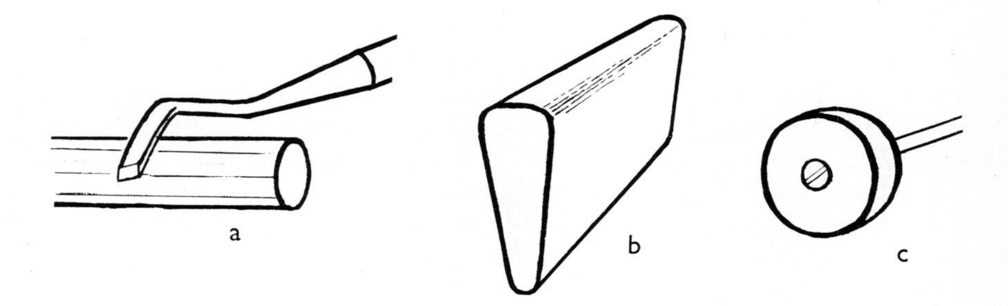

Fig. 7.15. Methods of completing honing of cervical trimmer blade: (a) cylindrical; (b) wedge-shaped Arkansas slips; (c) mounted stone.

the best result can only be achieved by the final use of cylindrical or wedge-shaped Arkansas slips, or a mounted Arkansas stone upon the concave aspect of the blade (Fig. 7.15). The width of the blade should be chosen to suit the width of the cervical floor.

2. The trimmer is not used as a hatchet; the stroke of the blade derives from the rotation of the handle on its long axis, so that a pattern with a thick handle is better than one with a thin handle. The action is one of planing and it follows that the pressure used should be very light in the first instance and only increased to moderate as the worked surface becomes smoother. The use of too heavy pressure results in the perpetuation of steps and notches rather than their elimination. The standard angulations of the blade are intended to allow the trimming of the margin to the flat outline (Fig. 7.16). The dimensions of the box limit this application. This limitation can to some extent be overcome by adapting the bevel on the sharpening stone as referred to above. The reverse-bladed instrument may be used to plane enamel prisms along the margin.

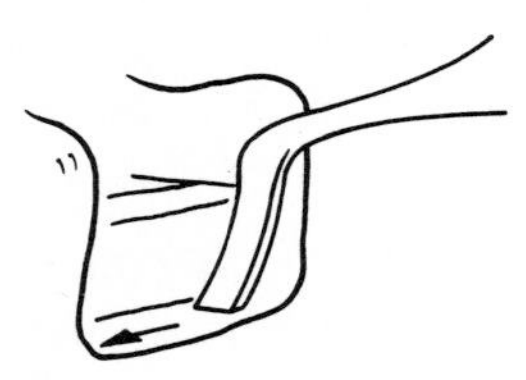

Fig. 7.16. Application of trimmer to flat cervical floor.

The gingival floor must be correctly sited and must, in common with all other margins, be in *unquestionably sound tissue*. A flat margin subserves its purpose very well and if the retentive groove is required it should be added. It must also be mentioned that these grooves can be formed by hand instruments alone, or, after the use of a bur, can be sharpened with these.

That all the internal surfaces should be as plane as possible and that line angles should be sharp is of importance, for these contribute significantly to stability and retention. On the other hand it is also true that sharp line angles imply the removal of dentine and give rise to concentrations of stress in restoration and tooth tissue when a load is applied. In the design of a restoration a balance must be sought between these conflicting factors. As applied to the cervical, this leads to the conclusion that, theoretically at least, the axio-gingival angles should be as shown in (Fig. 7.17 A), with sharp line angles but a rounded outline form. This is a little easier to achieve.

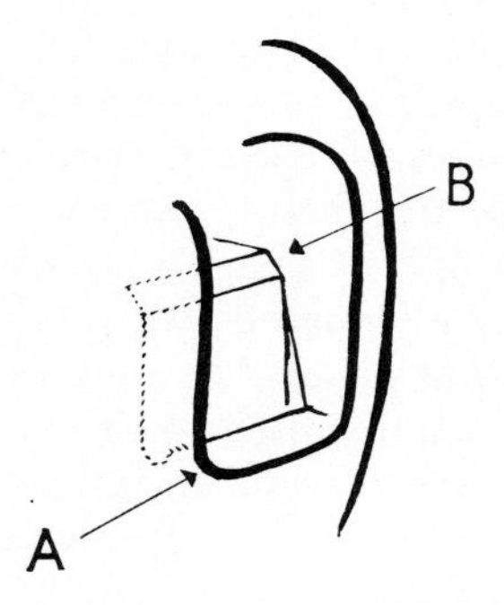

Fig. 7.17. A: Axio-gingival outline rounded, with sharp internal line angles. B: Bevelled angle between occlusal floor and axio-pulpal wall.

The occlusal floor of the lock is normally flat and would meet the axio-pulpal wall of the box at a solid line angle. This should be bevelled (Fig. 7.17 B) or rounded in order to give greater strength to the overlying amalgam at this point.

Removal of caries

In the optimal cavity it is not unusual to find that when outline, resistance, and retention forms are completed there remains no carious tissue; more often a small residue is seen. This is removed with sharp excavators until dentine of normal consistency is reached. In this size of cavity there should, by definition, be no risk to the pulp and the cavity can therefore be rendered clinically caries free.

A review of the enamel margins must confirm the smooth sinuous outline, the absence of all unsupported enamel and the correct cavo-surface angle throughout. The final finish is applied with the fine-cut or blank enamel finishing bur.

The cavity toilet and the isolation of the tooth, prior to lining and filling, then follows.

The extensive cavity

The optimal cavity referred to above is *determined in all its main aspects by the theoretical requirements of the outline, resistance, and retention forms.* It is, again, the smallest Class II restoration conforming to these principles which can be placed. In the larger carious cavity (Fig. 7.18) the marginal ridge has collapsed and caries of dentine has extended much farther. In cases such as this the *outline form is determined very largely by the extent of the carious process* (Fig. 7.19).

The main modifications of the larger cavity arise directly from the necessity for a larger box, since the margins must lie in sound tissue. This implies that the axial walls diverge more widely and their margins lie farther out into the buccal and lingual embrasures (Fig. 7.20 A, B) and its cervical margin nearer, and perhaps below, the amelocemental junction. It should, however, be noted that there is no intrinsic merit in placing the axio-cervical angle (Fig. 7.20 C) farther out than is required to render it self-cleansing.

The box is also deeper and the pulpo-axial wall is no longer a plane surface but is concave and may approach the pulp quite closely. The excavation of a large cavity should proceed by establishing the outline with high speed in areas of sound tissue without over-extension, but because of the loss of a sense of touch, these instruments are not best suited to the removal of caries. This should be done with a sharp excavator or with a No. 5 round bur at lowest speed using, if necessary, a speed-reducing contra-angled handpiece. **A caries-free periphery should first be established before excavation in a pulpal direction**. It is important to establish the ultimate size of the box before the occlusal retention is shaped, for the size and shape of the latter depend upon the former.

The rule governing the width of the neck of the lock (Fig. 7.11) has been stated. The width of the distal end of the lock (Fig. 7.11 D) need not vary proportionately but it must remain sufficiently wide to preclude lateral displacement.

Lining the cavity

The general rule laid down in connection with Class I cavities apply. Complete control of moisture is essential. In the minimal cavity there may be a slight concavity of the axio-pulpal wall. This can be built up by a lining cement strong enough to withstand condensation of amalgam. The fortified zinc oxide EBA, polycarboxylate, and phosphate cements all have their place; the rationale and the method of using them are described on p. 73 *et seq*. By this means the cavity is re-formed and the cement acts as an insulation against thermal shock. It is possible that the box may be so shallow that a cement lining cannot be placed on this wall without reducing the thickness of amalgam to a degree where it is no longer thick enough to bear the occlusal force. When this is so, it is generally accepted that no lining is required. The alternative, to cut the box deeper in order to place a lining, would be unjustifiable.

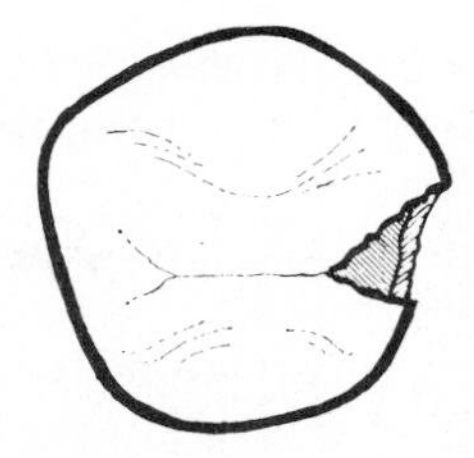

Fig. 7.18. Extensive Class II cavity with destruction of marginal ridge.

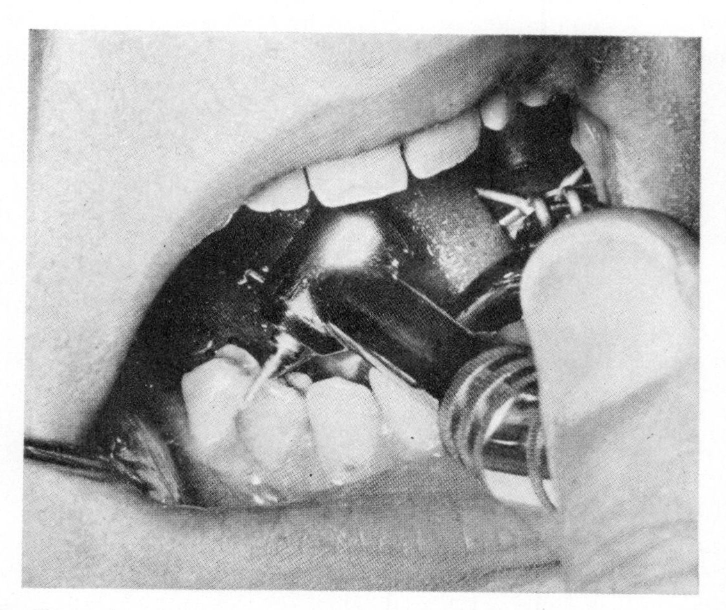

Fig. 7.19. Preparation of an extensive Class II cavity in a mandibular premolar using an air-turbine handpiece with a water jet.

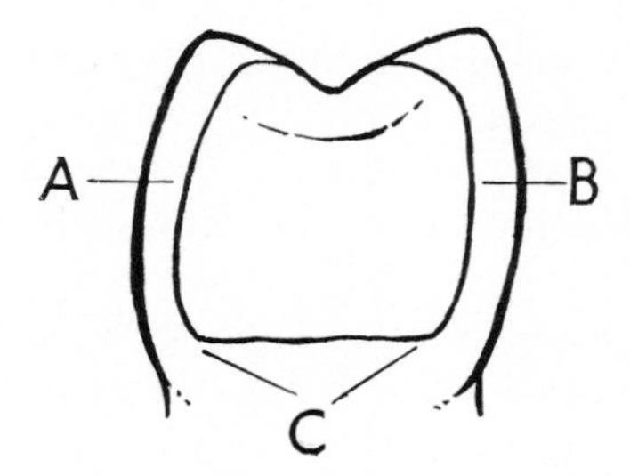

Fig. 7.20. Interproximal view of larger Class II restortion. A and B: lingual and bucco-axial margins extended. C: axio-cervical angles.

The maxim which applies with fair accuracy to this situation is that **the axio-pulpal wall should always be lined if this can be done without undue loss of strength in the amalgam**.

If the occlusal lock is cut at the elective depth of 2.5 mm it is not necessary to line the floor and, by definition, it would be impossible to do so without reducing the strength of the amalgam below the permissible minimum. When by reason of caries the lock is deeper than this, the lining should be placed. The occlusal and pulpo-axial lining become one mass (Fig. 7.21) and the line angle between these facets is again bevelled or rounded. This is an example of a lining with both protective and structural functions (see p. 73).

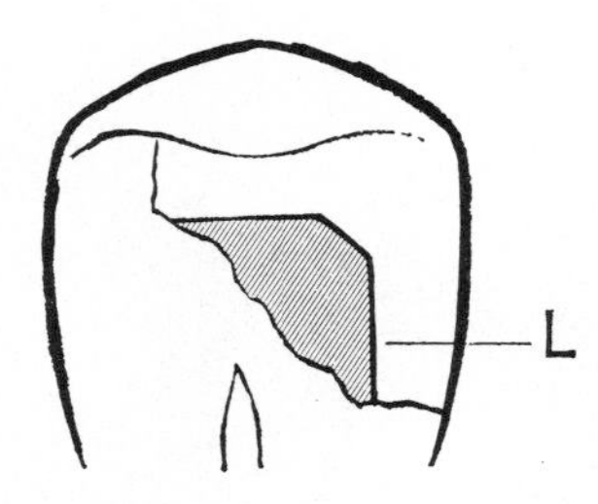

Fig. 7.21. Restoration of occlusal floor and axio-pulpal wall by cement, L, which also acts as insulation.

The matrix and its manipulation

The Class II cavity, prepared as described, is incompletely enclosed; it lies within four walls and a floor. A fifth wall must be provided if amalgam is to be contained and firmly condensed. The fifth wall, in the form of a metal strip, is called a matrix band, and, in most types, the device which retains the band during use is called a matrix band holder, or retainer.

The primary functions of the matrix band are as follows:

1. To retain the amalgam within the cavity during condensation.

2. To permit the close adaptation of the amalgam to cervical and axial margins.

3. To allow the restoration of the contact area and the external contour of the crown.

Matrices made of substances other than metal are not sufficiently firm and resistant for this purpose. Those types of matrices which preclude the placing of a cervical wedge are only of very limited value.

In general matrices fall into three types:

1. In which the band encircles the tooth and is secured by a retainer on the buccal or, in some cases, the lingual aspect (Fig. 7.22 A). This is the commonest form and there are many patterns of this type of retainer in use (Fig. 7.23 (1) (2)).

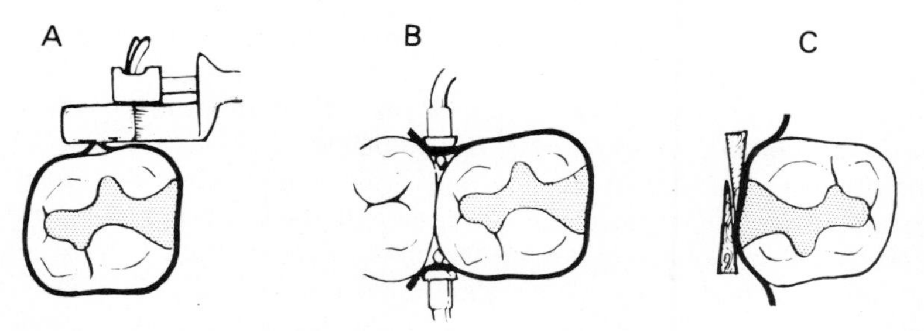

Fig. 7.22. Types of matrix bands: (A) encircling the tooth; (B) covering three lateral surfaces; (C) covering mesial or distal aspect only.

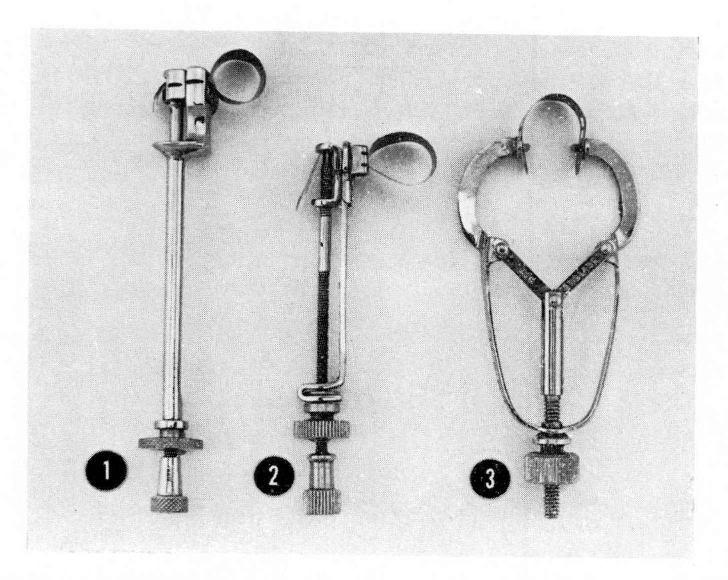

Fig. 7.23. Matrix bands and retainers: (1) *Bonnalie*; (2) *Ivory No. 8* patterns, both of which encircle the crown. (3) *Ivory No. 1* which adapts to either mesial or distal aspects of the crown.

The chief advantage of the type is that it can be very firmly adapted to the tooth and it allows free access from both buccal and lingual to the distal interdental space, and from the lingual to the mesial interdental space. There is one encircling pattern, the T-band, which can be used without a separate retainer, but, it cannot be applied tightly and is not in common use.

2. In which the band encircles three-quarters of the crown and is retained by jaws impinging into the free embrasures (Figs. 7.22 B and 7.23 (3)).

Its advantages are that it leaves the operative embrasure completely unobstructed and it only passes through one contact point.

3. This type covers a number of variations in which a small band covering the mesial or distal aspect is retained either by ligature or by wedges or

coil springs of various types (Fig. 7.22 (c)). It is not frequently used but has some variants which are extremely effective in some cases.

The most important features of a good matrix are, that the band shall be thin, about 0.05 mm, smooth and strong; that it shall be capable of close adaptation, in particular to the cervical margin; and, finally, that it can be made to restore contact with the adjacent tooth. To achieve the last, it may on occasion be necessary to stretch the band in the contact area by gentle burnishing. The width of the band should allow it to extend about 1.5 mm above the marginal ridge to allow for over-packing of the occlusal surface.

The adaptation of the cervical margin is a matter of primary importance. If amalgam can escape between the band and the cervical margin, it is certain that defective condensation will result. Further, the filling will overhang the cavity margin and must be trimmed back in the plastic stage. Such an overhang, if left until set, can only be corrected at the cost of much time and effort, trauma to the inter-dental papilla, and a generally inferior result. If left permanently untrimmed chronic periodontal irritation results until the filling is replaced.

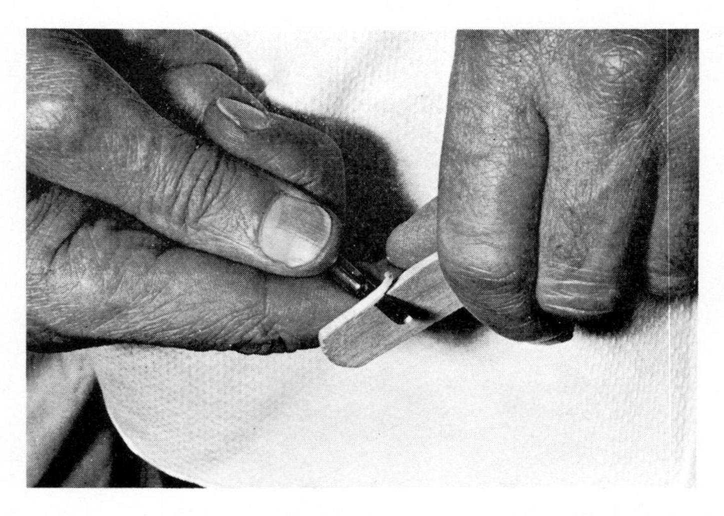

Fig. 7.24. A sliver of wood, cut from the edge of a tongue spatula, may be shaped to form a cervical wedge.

In order to preclude marginal failure of this sort it is generally necessary to use a cervical wedge. Ready-made wedges of wood, celluloid, and silver are also available and serve their various purposes very effectively. A satisfactory way of making a wooden wedge is by cutting a sliver from the edge of a wooden tongue spatula (Fig. 7.24). By angulation of the knife a wedge which is trapezoid in cross-section can be produced; this is the most effective shape for the purpose (Fig. 7.25 A). The matrix band is arcuate in

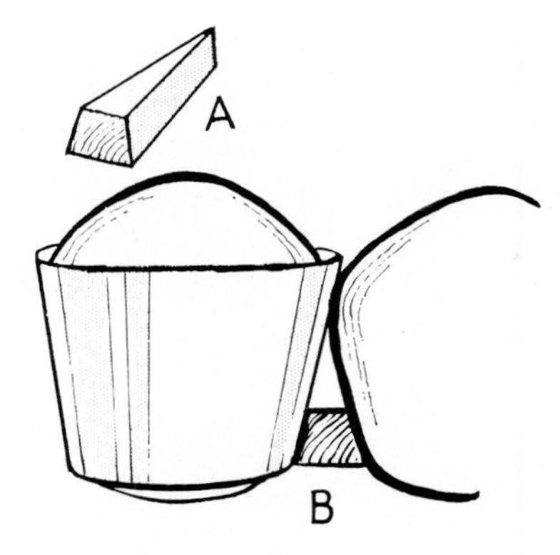

Fig. 7.25. (A) Wooden wedge of trapezoidal section (B) wedge in position against matrix at cervical margin.

form and so, when encircling the crown, adapts fairly closely to the cervix. When adjusted, the band must overlap the cervical margin of the cavity, and a wedge inserted interdentally, compressing the papilla, supports the band in very firm contact with the margin (Fig. 7.25 B). When properly applied it should almost be impossible to insinuate a fine probe between cervical margin and band.

The accurate adjustment of matrix and wedge is a matter calling for experience, ingenuity, and knowledge of many minor variations of patterns and their applications, so that when one method fails, the end may be achieved by another means.

Insertion of amalgam

When the prepared cavity is cleared of detritus, the access of moisture is controlled whenever necessary by the application of a rubber dam. The cavity is lined according to the principles previously discussed and the matrix band applied. With the more commonly used Type 1 matrix, of which representative examples are the Bonnalie and the Ivory No. 8 patterns, a clean, polished, and undistorted band, held in a retainer, is gently passed between the contact points and adjusted lightly to enclose the greatest diameter of the crown. The band is then carefully pressed towards the cervix, *closing it slightly as it passes down* so that its lower edge comes to lie between the papilla and the cervical margin. It is essential to see that no part of the papilla becomes trapped between the band and the cervical margin. The band can then be fully tightened and a wedge inserted from the lingual or the buccal aspect of the interdental space (Fig. 7.25 B). The fit of the band at the margin must then be checked.

The first load of amalgam is directed into the deepest part of the box, at the cervical margin, and the object of the first stage of condensation is close adaptation of amalgam to that margin and to the axio-cervical angles. The

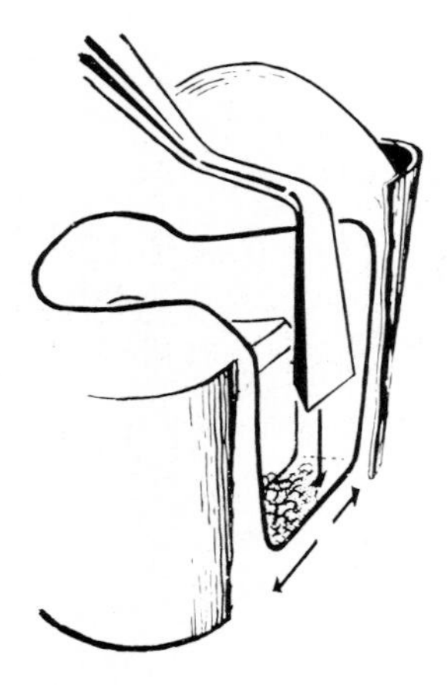

Fig. 7.26. Condensation of amalgam at cervical margin.

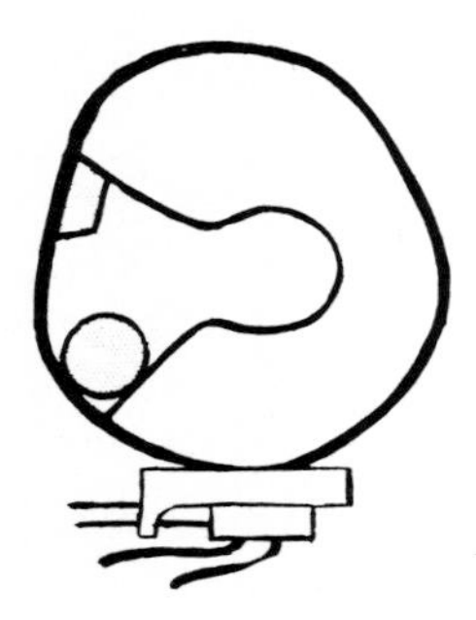

Fig. 7.27. Comparison of round and trapezoidal condenser nibs applied to axial margins.

nib, which must be small enough to fit the cervical floor, starts in the centre of that floor and steps first to the bucco-axial (Fig. 7.26) and then the linguo-axial angles. It is obvious that a circular end to the condensing instrument is not the shape most suited to good condensation of the axial margin (Fig. 7.27) and a trapezoidal form such as that shown is much more effective. Such instruments are available both in hand and mechanical forms.

When using spherical alloy larger loads of amalgam and, where possible, broader condensing instruments can be used to advantage. This favours quicker condensation, better control of the mass and reduces the tendency to penetrate rather than consolidate the more mobile mass which is characteristic of this type of amalgam.

Condensation proceeds as above, by progressive filling of the box and systematic removal of excess mercury until the floor of the lock is reached, when the whole now resembles a Class I cavity. Thorough condensation is

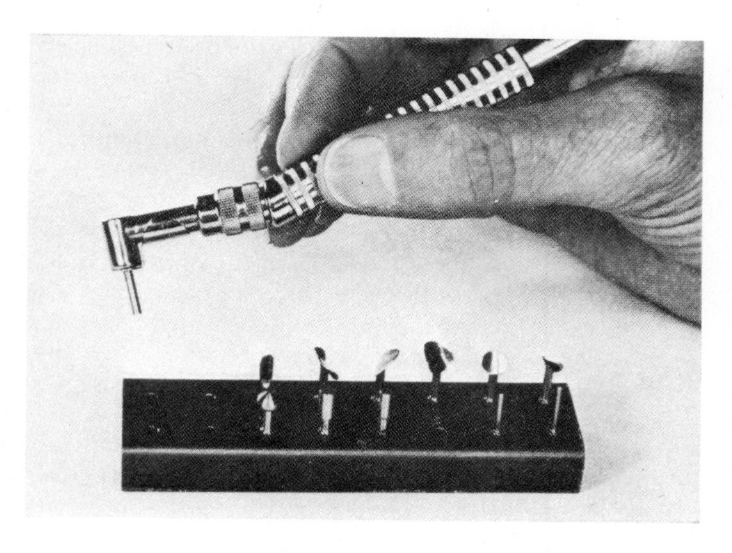

Fig. 7.28. A *Dentatus* mechanical condenser with a set of condensing nibs suitable for use in various shapes of cavity.

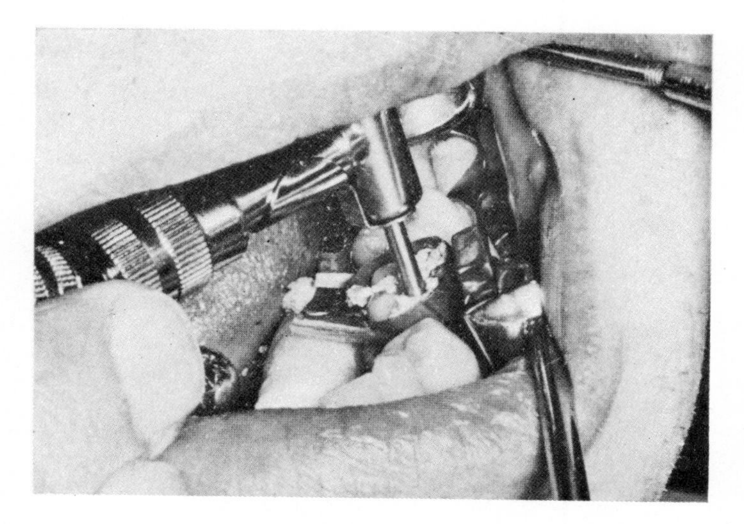

Fig. 7.29. Final stages of mechanical condensation of an amalgam restoration in a MOD cavity.

necessary throughout, **but particularly along the area next to the matrix band, with extra attention to the axio-occlusal angles, and in the neck of the lock.**

It is appropriate at this point to consider the role of the mechanical condenser (Fig. 7.28, 7.29). It is derived from the mechanical condenser used in cohesive gold technique and most designs depend upon a plunger actuated by a cam driven in a low-speed handpiece. It delivers a series of rapid blows of short amplitude through a bit, the end of which is suitably shaped for amalgam work; a variety of shapes are available. Run at 3000 r.p.m., the condenser can be used more as a vibrator; opinion seems divided as to whether this is an advantage.

This condenser speeds the process considerably and produces a well-condensed filling with less effort. Most operators like to use both hand and mechanical methods at different stages. There are some Class V cavities in which the process of packing is greatly facilitated by the mechanical condenser. There is one possible danger; care must be taken to see that the condensing bit does not come into direct contact with enamel margins; injury to these could lead to failure at a later date.

In the Class II restoration overpacking of final margins must also be assured and trimming back to these proceeds in the manner as described on p. 93. As much as possible of the occlusal carving must be completed before the removal of the matrix. The outline of the lock and axio-occlusal angles must be defined and there are also two other most important features which must be established.

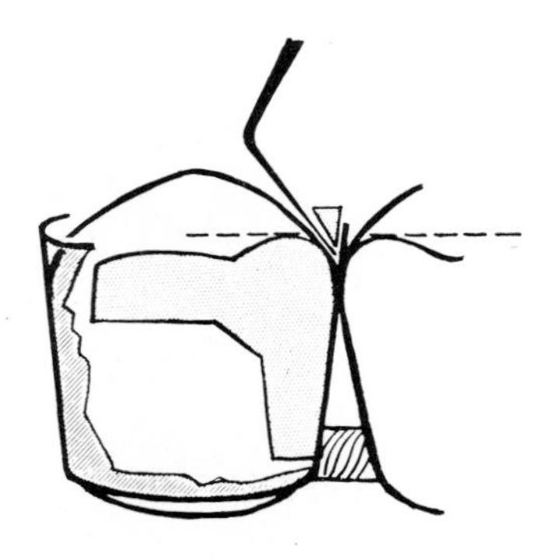

Fig. 7.30. Adjusting shape and height of marginal ridge; removal of fillet of amalgam. NB adjacent marginal ridges are at the *same level.*

First, the *height of the marginal ridge* must be adjusted exactly to that of the next tooth (Fig. 7.30). This is done by removing, with a sharp probe, a fillet of amalgam from against the matrix, then reducing the height of the ridge with a carver. Second, *the marginal pit* should be excavated in its correct position relative to the ridge. The axial margins may also be trimmed as far as they are accessible.

The retainer is then loosened and removed. The band is first removed from between the contact points not involved by the restoration. It is then wrapped back against the tooth adjoining the new restoration. From this position it may be lifted in an axial or oblique direction free of the filling;

alternatively, one end of the band can be cut off short near the embrasure and the remaining end pulled through laterally to clear the contact area.

If the former method is chosen, it sometimes happens that the marginal ridge, or part of it, becomes detached and is lifted off with the band, an occurrence which may spoil the occlusal form to a serious extent. This mishap may be due to one or both of two errors in technique. **Either the marginal ridge has not been sufficiently condensed or it has not been adequately trimmed in the manner described**. Attention to these points eliminates the risk.

With the removal of the band the axial and cervical margins are now accessible. The axial margins are trimmed towards the gingival with a downward stroke of a sharp carver such as a Ward's No. 1 or 2. If correct adaptation of the band was achieved, the gingivo-axial angles and the gingival margin should require only the lightest of trimming with a sickle probe inserted into the embrasure. Should more than this be required, a fine-bladed trihedral scaler (Fig. 7.31) can be used very effectively for sharp trimming and it must be remembered that correction of slight overhang at this plastic stage is essential if a good result is to be obtained.

The rubber dam may now be removed, avoiding taking the rubber through the new contact, by cutting the isthmus of rubber passing through the interdental space. The occlusion may then be checked. It is inadvisable simply to ask the patient to bite, for in some cases this may result in the

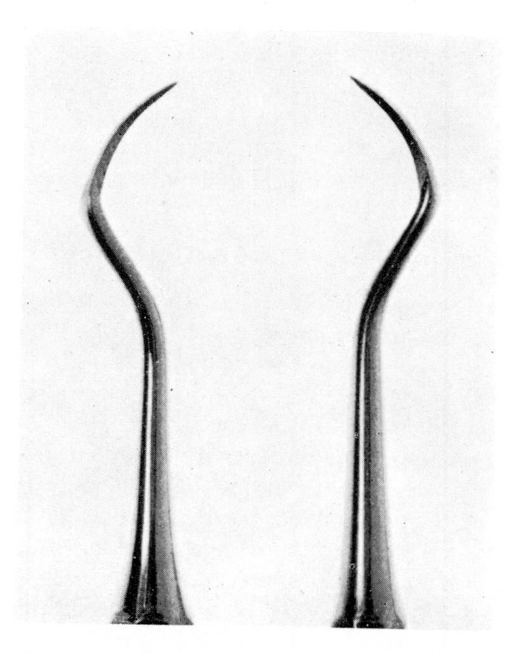

Fig. 7.31. Long-bladed trihedral scalers suitable for cervical trimming of amalgam and wax.

crushing of the marginal ridge. He should be asked to put his teeth gently together and if no abnormality is noted, a light 'rubbing' movement is allowed before inspection of the filling for the small burnished spot which shows high contact. If this is present it is carved down with a sharp instrument until the bite is cleared. Failure to adjust the occlusion at this stage may result in a broken restoration, or in the development, during the next twelve to twenty-four hours, of traumatic periodontitis of the restored tooth and its opponent. Light burnishing and smoothing with a cotton pledget completes the finish at this stage and care is taken to see that all residual fragments of amalgam are removed from the interdental space by dislodging them with a probe and the use of the atomized spray. The patient should be instructed not to chew on the restoration for a period of three hours.

The finishing of the restoration after an interval of twenty-four hours follows much the same sequence as described for Class I restorations. The occlusion is again checked and the occlusal surface lightly worked over with a small finishing bur. Fine cuttle-paper discs are useful for polishing both interstitial and occlusal surfaces and sulcus discs are also of value. The contact area, having been condensed against a highly polished band, should need only the lightest of polishing with the finest cuttle linen strip. If the contact is unduly tight it may be necessary to use a separator for this purpose. If this is done great care must be used to see that only the lightest polishing is effected, otherwise the contact may be destroyed and food packing will result.

As regards occlusal morphology, it should be recognized that the features of greatest functional importance are the *marginal ridge and pit*. This should be reviewed and checked as to its formation and finish in order to ensure its correct function, which is to deflect food from the contact area and so to reduce the tendency to food packing.

MESIO-OCCLUSO-DISTAL (MOD) RESTORATIONS

When cavities exist on mesial and distal aspects of a premolar or molar it is often necessary to prepare one cavity which involves the mesial, occlusal, and distal aspects of the crown, the mesio-occluso-distal or MOD restoration (Fig. 7.32).

A fairly common exception to this is seen in maxillary molars where the mesial and distal occlusal fossae are separated by an intact oblique ridge (Fig. 7.32 (b)). In this case, if the oblique ridge is sound, it may well be unjustifiable to join up the occlusal locks of the mesial and distal restorations. These are better treated as separate mesio-occlusal and disto-occlusal restorations, although it may be convenient to prepare and fill them simultaneously.

The technique for the preparation and filling of MOD cavities does not vary in any significant manner from that already described. The principles

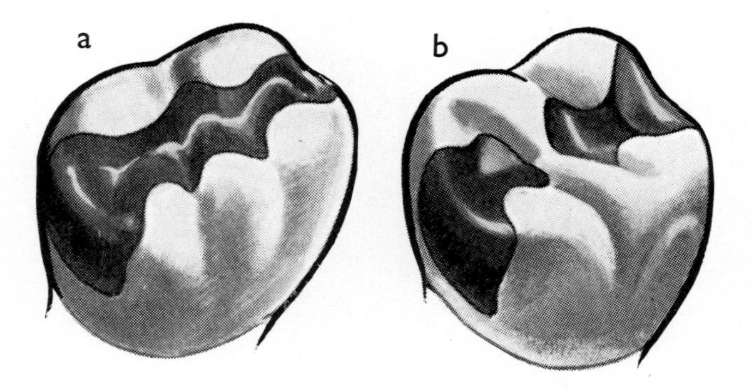

Fig. 7.32. (a) MOD amalgam restoration of mandibular molar; (b) mesial and distal restorations separated by intact oblique ridge, in maxillary molar.

and their application are those of the Class I and Class II restorations. The matrix band must be so shaped that it can be adapted to both mesial and distal boxes. With a large restoration, assistance in mixing and loading amalgam is a great advantage. Two mixes are frequently required if adequate condensation is to be achieved during the correct working time of the amalgam. The occlusion must be adjusted with greater care and the patient should be warned to avoid the filling for four to six hours since the risk of damage due to overloading is rather greater.

OTHER TYPES OF AMALGAM RESTORATIONS

The Class II and MOD amalgam restorations are common in occurrence and durable in practice. It not infrequently happens that extension beyond these forms becomes necessary, either as the result of untreated caries or because repeated replacement of restorations has, over a period of years, undermined the strength of the occlusal surface.

The destruction of one or two cusps is a defect which can be restored by amalgam. It is also possible to replace a complete molar or premolar crown in this way, though it is arguable whether the difficulties of technique warrant the use of a material which is unsuited to this purpose, when stronger and more satisfactory restorations can be constructed in gold.

The replacement of a single cusp, as for example the disto-lingual cusp of a lower molar (Fig. 7.33), involves an extension of the general principles discussed in the preceding chapters. It must be emphasized that an adequate bulk of amalgam must be used to allow strength, and any attempt to replace less than one-half of the vertical height of a cusp should be considered inadequate. It is wrong to consider the amalgam as a thin protective layer to an underlying cuspal outline; it is not strong enough for this purpose; it must be used as a replacement.

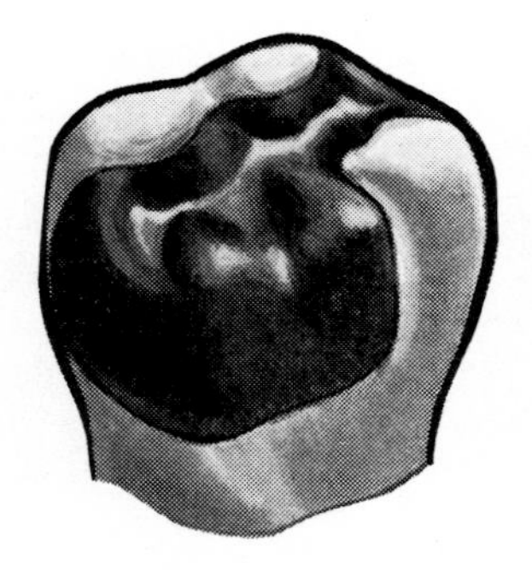

Fig. 7.33. MOD restoration of mandibular molar replacing one lingual cusp.

The restoration shown in Fig. 7.33 develops from the MOD by the loss of a lingual cusp. In this case the outline passes from the distal gingival, rising slightly around the disto-lingual before rising steeply to regain the occlusal midway along its lingual border. The line should be smooth, devoid of sharp angles, and the cavo-surface angle is 90 degrees. The internal aspect of the cavity should present a well-established resistance form in one or more horizontal surfaces. Figure 7.34 shows the theoretical requirements of such a design.

Usually, the retention in this design presents little problem for, as in the simple MOD cavity, one aspect serves to key the other in position and it is only necessary to see that the restoration is strong enough in the occlusal.

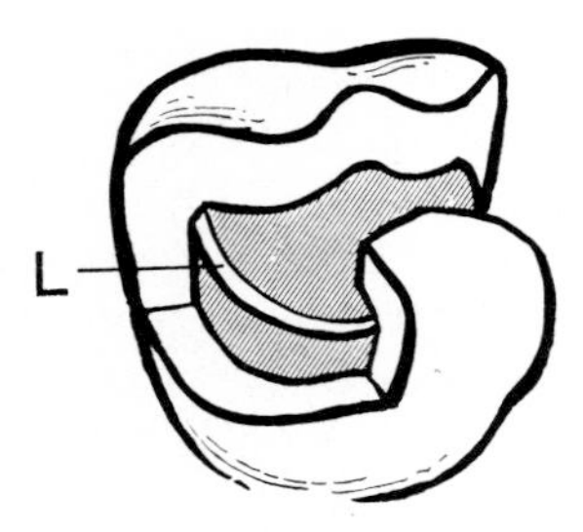

Fig. 7.34. Cavity preparation for restoration shown in Fig. 7.33. L: cement lining.

A metal restoration of this size requires a substantial insulating lining (Fig. 7.34 L). This is built with phosphate cement or fortified zinc oxide cement strong enough to withstand condensation of amalgam. The lining restores the theoretical resistance form, but in fact it only contributes a small mechanical factor to resistance by virtue of its limited adhesion to the base of the cavity.

The matrix band used is the type normally used for an MOD restoration. It must, of course, encircle the crown; it must be carefully contoured to the gingival crevice and must stand high enough above the occlusal level to permit considerable overpacking. At the mesial and distal gingival margins, wedges may be used in the manner described, and care is taken to see that the band, stretched by burnishing if necessary, allows the restoration of both contact points. Where a cusp is missing and the gingival margin passes on to the lingual surface, close apposition of the band depends upon its tightness as a whole. A slight amalgam overhang in this area, though undesirable, is accessible to trimming during the plastic stage.

The help of the surgery assistant and use of mechanical amalgamation and condensation are of particular value in packing a large restoration since the degree of overpacking required to get a good condensation is considerable. The cusp must be overbuilt in height against the matrix band, and at all margins, to allow carving back to fully condensed material. As much carving as possible should be completed before removal of the matrix band; by this time the amalgam is in the advanced stages of its carving phase.

On occasions it is possible to use a standard copper or stainless-steel ring, suitably trimmed, as a matrix. It must fit the gingival contour with great accuracy and, in the area of the contact point, the ring may be thinned by stoning its outer surface almost to penetration. This type of matrix has the advantage that, when relieved clear of the occlusion, it may be left in position for twenty-four hours and removed when the amalgam is completely set, by cutting it on lingual and buccal aspects.

In adjusting occlusion anything but the lightest contact should be avoided lest an undue premature strain be applied, and the final load carried by the restored cusp must be less than normal and distributed over a wider area by rounding any opposing cusp which may be impinging on it.

The loss of two lingual cusps on the tooth under consideration presents a new problem of retention, for the distal and mesial portions no longer retain one another and a new method of retention must be sought.

Pin retention

This introduces the use of pins and we should stop briefly to consider the bases of their use as means of retention. There are three general types:

1. A threaded pin which cuts a thread in a slightly smaller hole in dentine. This is best distinguished as a screw-pin.
2. A roughened pin which is forced into a slightly undersize hole and is retained by the elasticity of the dentine; we could call it a friction pin.
3. A roughed pin which is cemented into a slightly oversize hole. The cemented pin is not much used except that, when using a screw or friction type pin it sometimes happens that one finishes with a hole which is slightly oversize, rather than undersize. It is then useful to know that the screw or friction pin can be cemented in

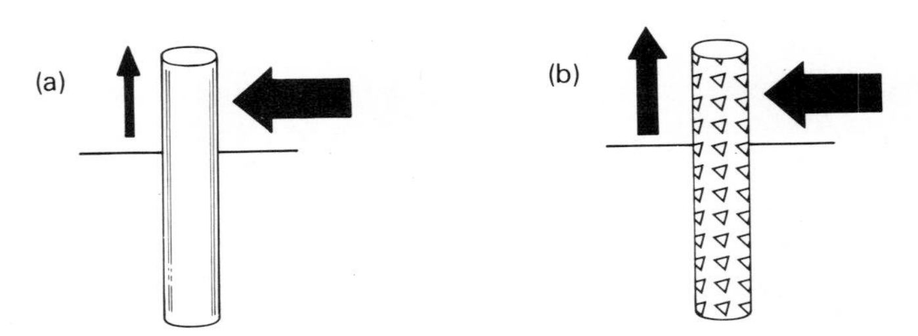

Fig. 7.35. (a) A smooth polish pin in hole of the same diameter. A very small axial force will remove it. A transverse force meets great resistance. (b) Roughening or threading the pin greatly increases axial resistance. Transverse resistance remains unchanged.

Pins and screws can resist forces in two axes at right angles one to the other. To take the simplest example (Fig. 7.35). A single smooth-sided dentine pin embedded in solid dentine and amalgam will resist lateral displacement up to the shear stress of the pin. Axial displacement, however, has only to overcome the frictional forces between the pin and the walls of the holes in which it is contained; the magnitude of these forces could be quite small but roughening the pin and using an undersize hole will greatly increase them. In the case of the dentine screw-pin of equal dimensions, it again strongly resists lateral displacement and the thread will greatly enhance frictional resistance to axial forces.

This is an over-simplification of the problem. Theoretically, the force transmitted in any known loading can be analysed into axial and transverse components. In most cases, pins or screws will bend before they shear and they may cause dentine or amalgam to break under the forces of occlusion which vary rapidly in magnitude and direction. It is clear that in general terms the retentive power of pins and screws cannot be greater than the strength of the dentine and amalgam in which they are embedded. Drilling one or more holes in dentine weakens it; similarly pins in amalgam weaken the amalgam mass, particularly if their presence makes adequate condensation difficult.

The use of two *non-parallel* screws (Fig. 7.36) introduces another retentive factor. The axial withdrawal of one pin is opposed by the other as long as they both remain rigid. In theoretical configurations greatest resistance is offered when the pins are at right angles, but this condition could not be met in amalgam restorations. In fact, in normal use it seems enough to have the pins inclined to one another by about 20 degrees.

From these considerations certain general rules can be deduced for the practical use of pins and screws in amalgam restorations.

1. Never use pins or screws unless the retentive form of the cavity is truly insufficient. In many cases they are used unnecessarily.

2. When two pins are used they should be widely spaced to distribute stress and allow condensation, and inclined at about 20 degrees, though a smaller inclination is certainly allowable.

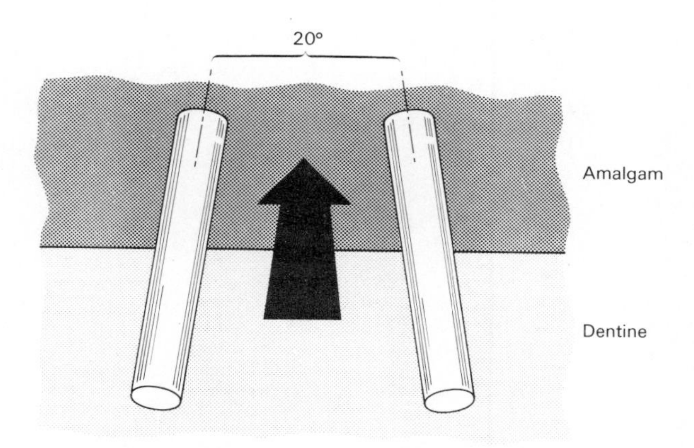

Fig. 7.36 20° angulation of the pins gives considerable increased resistance to withdrawal. This may be further increased by roughening or threading the pins.

3. The pin hole should be placed in the greatest bulk of dentine available, never in enamel *and never at the amelodentinal junction.* If possible the dentine should be 1.5 mm thick in all directions—not always possible but a minimum of 1.0 mm on one aspect is acceptable. The depth of the hole should not be less than 2.0 mm, and 2.5 mm is preferred where this can safely be done.

4. The pin or screw must not encroach on the pulp or the periodontium. The risk of penetrating the periodontal ligament is usually greater than that of entering the pulp, because the area of the periodontum is much larger and because the operator does not sufficiently allow for the taper of the root surface.

5. However firm the hold of pin or screw in dentine, its hold in amalgam will be defective unless thorough condensation is achieved. The regions between the pins, and between the pins and the matrix band are particularly prone to underpacking.

6. As in all operations involving the use of small objects and instruments in awkward positions, the application of rubber dam is a wise precaution.

Figure 7.37 shows a very practical form of screw-pin designed for use in a right-angled handpiece. Two diameters of special drill are available, colour coded, 0.53 and 0.68 mm which accept screws of diameters 0.60 and 0.75 mm respectively. The proximal end of the screw is a shank to fit a latch type handpiece. As to method, it is best to mark the point of entry for the drill by a pit made with the smallest round bur. A flat-bladed plastic instrument may be placed in the adjacent gingival crevice and aligned with the root surface. The drill, in a speed-reducing handpiece, is lined up with this and the cavity and with the handpiece at moderate speed, say 200 r.p.m., sufficient pressure is used to allow a single plunge cut to the elected depth, and out again.

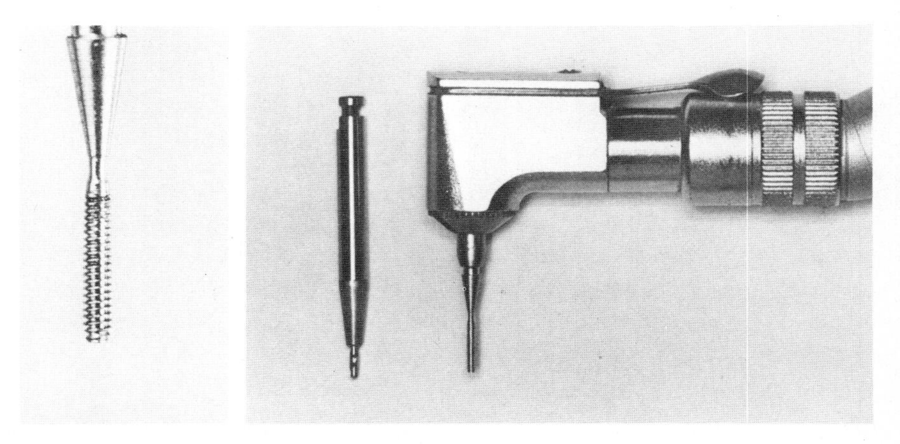

Fig. 7.37. A drill and self-shearing screw pin for use in a RA handpiece; *inset*: detail of threaded pin.

Two practical points; make sure the drill is turning forwards, that is clockwise when viewed from behind; do not stall the drill whilst cutting, as this leads to breaking the drill in the hole.

The appropriate size of screw is now placed in the handpiece, correctly aligned with the hole, run at moderate speed until it reaches full depth, when it shears automatically at its narrowed neck. The screw may be

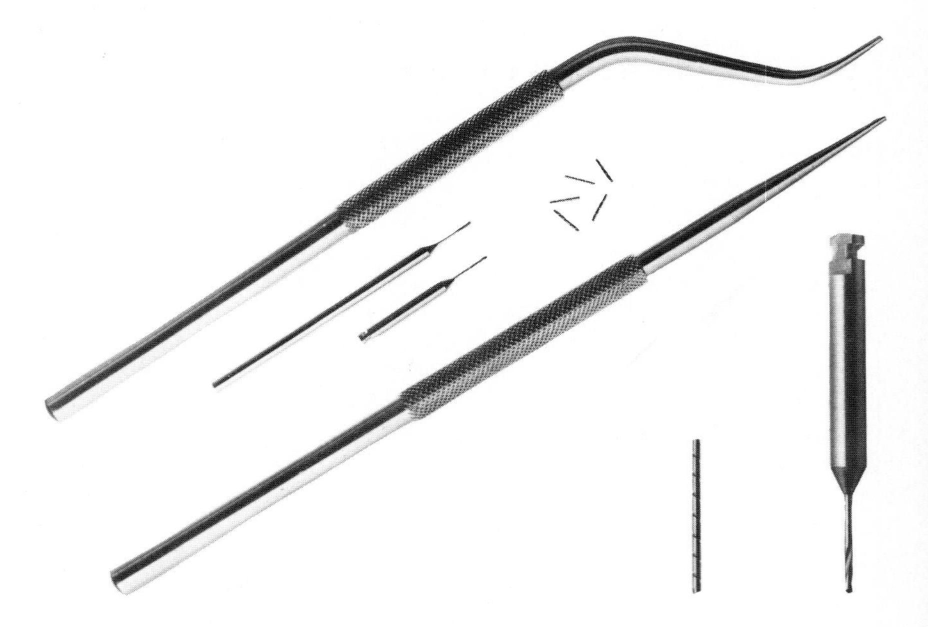

Fig. 7.38. Drills, friction pins, and pin setters; *inset*, detail of drill and friction pin.

Fig. 7.39. Use of screw pins for retention of extensive amalgam restoration.

shortened by using a small high-speed bur with very light touch. It can also be bent to locate the caval portion more advantageously. Directed towards the greatest bulk of amalgam the screw should not reach the surface of the restoration, or be uncovered by future preparation, for example, to accept a full crown.

Figure 7.38 shows an example of the friction-fit pin system. In the manner previously described a hole 0.53 mm diameter is drilled to receive a pin 0.56 diameter. Three lengths of pin are available, 4.8, 7.1, and 9.5 mm. The correct length selected, the pin is placed in a pin setter, a simple instrument which works better if the tip is dipped into water so that the pin is easily retained in the loose socket by surface tension. The pin is set by firm pressure to the full depth of the hole where it is firmly retained by the elasticity of the dentine. This pin can also be bent, but cutting it *in situ* is likely to loosen it; the choice of a shorter pin is preferred.

It is important in both techniques that the drill hole be kept free of water or saliva, for the hydraulic and bacteriological effect of a driven screw or pin upon recently cut dentine, possibly close to the pulp, could cause avoidable damage. Both of these systems are very effective in use and each has its adherents. There are also modifications to the patterns described.

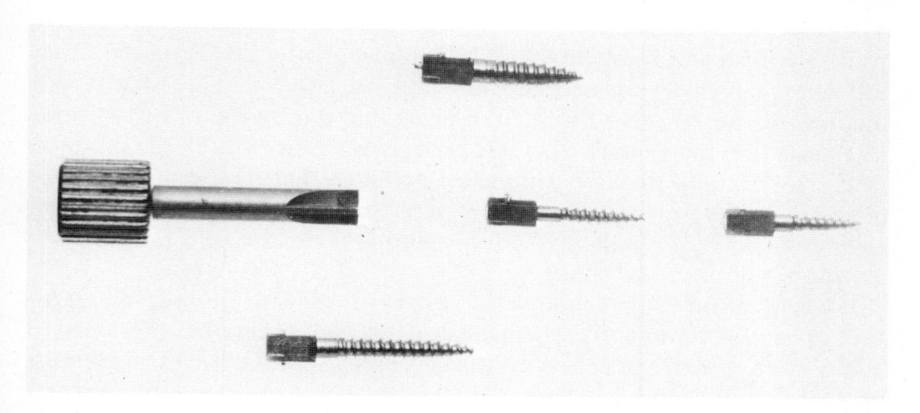

Fig. 7.40. Square-headed conical screws for use in root canals.

Used with precision and understanding of the principles involved there may be little to choose in mechanical retention, though the axial retention of the screws might seem to be greater than that of the roughened pins.

Lining the cavity for thermal insulation, adaptation of the matrix and packing of amalgam proceeds in the manner described, to ensure complete uniform condensation. With all extensive restorations in amalgam the carving of lateral contours and occlusal anatomy calls for skilled and rapid working. It may often be an advantage to use a slower-setting amalgam to allow more time for carving.

There is also a larger type of screw (Fig. 7.40) with a conical thread and square head. This is screwed into the pulp canal of a root-filled tooth to give anchorage to amalgam. Its method of use is easily apparent and it certainly gives good retention in suitable cases of non-vital teeth to which its use is confined.

THE BONDED AMALGAM RESTORATION

In the early days of amalgam many attempts were made to give it adhesion by the use of cement. Of many methods, probably the best known was that described by Harry Baldwin in 1896 and much used for many years. Today, with improved techniques and a more systematized approach, the method, though not completely in disuse, is reserved for the occasional patchwork repair, and in these circumstances it can be surprisingly successful.

To us, its main interest is that it has given rise to a technique which represents an attempt to bond, or cement, an amalgam restoration into a cavity. This is a recognition that *the marginal seal is all important, that it is a frequent cause of failure*, and that a method of bonding amalgam to dentine might help to solve the problem.

From research upon leakage at the margin of amalgam restorations it appears that all those done under normal clinical conditions probably leak to a small extent within a few days of insertion, or a few weeks at most. These fine leakage pathways exist at the dentine–amalgam interface and probably in small interstices left in the amalgam after condensation. They are fine channels, very irregular in shape and are no more than a few micrometres in effective width. It appears that during the first few weeks after insertion the channels become progressively blocked by the products of corrosion, until they are completely occluded. From an examination of prepared cavity surfaces (Fig. 2.13) it is easy to understand that such cavities could exist, and a method of ensuring their complete and early occlusion would be an advantage.

It has been shown that coating the cavity surface with a thin layer of varnish or cement eliminates early micro-leakage and is therefore presumed to give a more effective seal. Some clinicians apply copal–ether varnish to the prepared cavity before inserting the amalgam. This varnish is a solution of a natural resin and would be expected to block leakage pathways, but there is some doubt whether in long-term clinical conditions this seal remains impermeable. The use of phosphate cement is justified by the fact that not only is the leakage eliminated but the cement provides a significant mechanical bonding effect to the dentine.

The technique now to be described is in no sense an attempt to retain a filling in a cavity by means of adhesion due to cement; it is a means of reducing the risk of marginal leakage.

In the application of this technique to a Class II restoration, steps 1 to 6 in the cavity preparation proceed as described; full advantage is taken of retention form. The cavity is lined in the normal manner to give full insulation against thermal change. The matrix band is tried on, carefully trimmed and adjusted, the wedge prepared; it is then loosened just enough to allow its removal, and it is placed aside.

The requisite quantity of amalgam is now prepared simultaneously with a slow-setting, medium creamy mix of phosphate cement so that both are ready for use at the same time. The cement is worked into the cavity on a blunt probe or small plastic instrument, to cover the whole of the prepared cavity surface and immediately one, or in the large cavity, two, carriers full of soft amalgam are discharged into the cavity. With a small burnisher this is worked into the cement and on to the surface until only a very thin film of amalgam and cement is uniformly spread over the cavity.

The cavity margins are *not* cleared of this layer but care is taken to see that no large portion of un-incorporated cement remains on or near the margin.

The matrix is now rapidly re-applied and, being already closely adjusted, it discards the excess along the interstitial margins. The wedge is placed and packing of the cavity, carving, and finishing of the filling now proceed to completion. The cavity margins should be closely inspected to preclude the possibility of a visible cement margin; if the technique is performed with care, this is a very rare occurrence. There should be no visible difference between fillings completed by the bonded and normal techniques but it is easy to prove that the former is fairly firmly adherent to the cavity walls. The clinical impression is that with the bonded technique marginal failure is delayed. It is reasonable to suppose that the cement has some physical keying effect and experimentally these restortions certainly prove to be less liable to leakage. This technique has long been used by many conscientious clinicians and probably deserves more careful evaluation as a restorative measure.

CLASS III AMALGAM RESTORATION

The Class III cavity is normally restored with composite resin when appearance is paramount, or with gold when durability is more important. There is, however, a very limited role for amalgam to play. The distal cavity in canine teeth, more particularly in the mandible, is one which can be effectively restored with an amalgam filling. This can be durable and not unduly unsightly provided that steps are taken to avoid discoloration of the labial surface. The composite resin has largely supplanted this type of amalgam restoration.

Cavity preparation

Access to the cavity (Fig. 7.41 (a)) is gained with a No. 3 round bur or a diamond instrument at a point opposite the distal contact area; the surrounding enamel is removed and the rounded outline of the lingual aspect established (Fig. 7.41 (b)) avoiding weakening of the distal incisal slope.

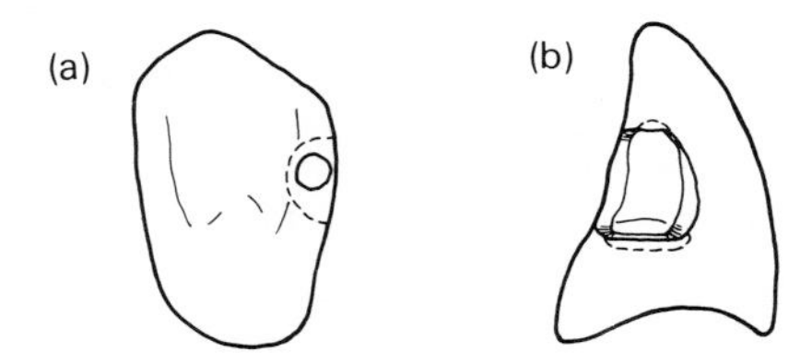

Fig. 7.41. (a) Point of entry to a distal cavity in a mandibular canine; (b) small preparation with cervical and incisal retention.

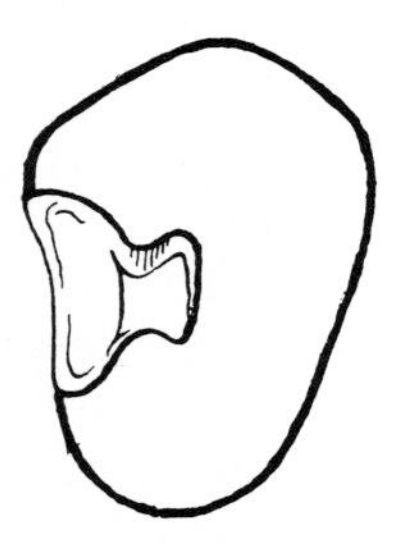

Fig. 7.42. Larger preparation with lingual lock retention.

The accessible lingual outline can be finished with a fine-cut fissure bur, while the gingival and labial enamel margins are defined with chisels and the smallest finishing bur. The labial should be taken only just on to the labial surface to allow cleansing; its visibility from this aspect must be kept at a minimum.

In an optimal cavity this form can be made self-retentive by a gingival groove and a shallow incisal pit (Fig. 7.41 (b)), and where this is possible it should be done. In the larger cavity it may be impossible to form the incisal undercut without risk of weakness here. It then is necessary to obtain retention by the use of lingual lock (Fig. 7.42). This should be cut to the depth of 2 mm and with due regard to the strength of its narrowest point. The walls should show slight inward divergence, but when doubt exists the use of a small inverted cone bur along the floor wall angle is permissible.

Lining of the main cavity is often necessary and more generally desirable since the extension of the cement on to the labial aspect of the cavity will help to reduce the discoloration of the dentine and so maintain the appearance of the tooth.

The insertion of amalgam presents problems because no completely satisfactory matrix band exists, though most patterns can be adapted. The Bonnalie or Ivory No. 8 can be applied as they would be for a distal premolar cavity. The palatal aspect of the band is scribed with a sharp point, and cut away when the band is removed from the tooth, until access to the palatal is free enough to allow condensation with the band in position and an effective wedge in place (Fig. 7.43).

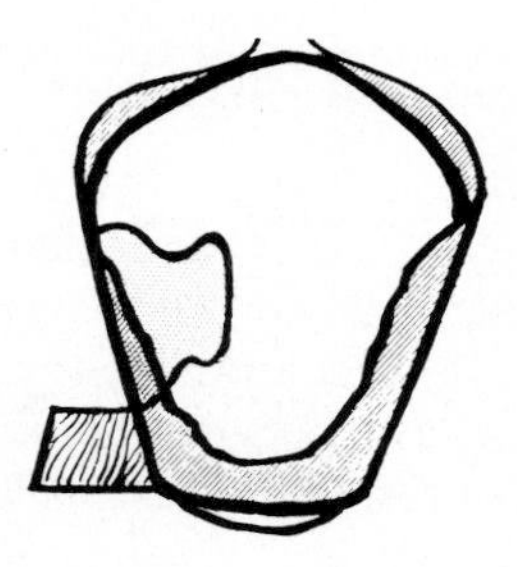

Fig. 7.43. Matrix band in position for Class III cavity.

Small-ended contra-angled condensers must be used for access to the distal box and thorough condensation along the labial and cervical assured before proceeding further. The restoration is carved and finished in the normal manner. Such a restoration, well finished and polished, in a mouth with good hygiene, will give many years of good service without significant detraction from good appearance.

Removal and repair of amalgam restorations

The removal of a defective amalgam restoration is a relatively simple task. What is not so simple is to avoid unintentional removal of sound tissue or damage to an adjacent tooth or restoration. Using a small or medium round or cylindrical bur, copious water spray and suction, an entry can be made in the neck of the lock of a Class II amalgam which is removed to full depth. Then continue mesio-distally. (Fig. 7.44), removing the lock from centre to periphery, stopping just short of enamel. The practice of cutting all round the periphery and dislodging the centre intact may be slightly quicker on occasion, but carries a greater risk of enlarging the occlusal outline unnecessarily. The removal of all the occlusal portion can either precede or follow removal of the approximal portion. In most positions the removal of the occlusal first allows a better view of the approximal box and is worth doing for this reason. As the cut enters the amalgam filling the box it can be

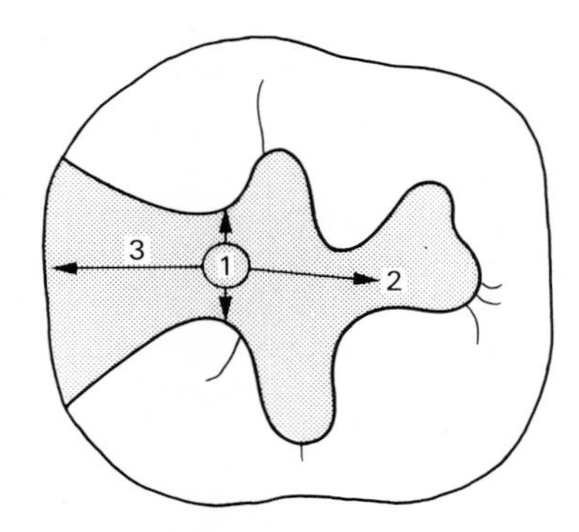

Fig. 7.44. Removal of defective restoration by (1) a plunge cut to floor of box; (2) removal of the occlusal lock; (3) bisection of the approximal portion.

deepened, and approaching the embrasure, hold the bur half a millimetre short of the surface until the wall disintegrates. Enlarge and deepen the cut until the two separated portions can be tipped towards one another.

The extent and position of recurrent caries will determine the detail of the removal. Certainly the experienced operator will complete some, if not all, the new cavity preparation in passing, but remember that *over-extension is always easy to commit and is irretrievable.*

For the same reason repair of an amalgam should always be considered in suitable circumstances. The great majority of amalgams fail by marginal leakage and carious recurrence. It follows that at an early stage it will be possible to cut out only the defective margin if it is accessible; this emphasizes the importance of regular inspection. Recurrence at the cervical margin, however early, always calls for complete replacement when the adjacent tooth is standing, because of its inaccessibility. A recurrent lesion on the occlusal surface can usually be repaired, if treated early enough. In between are the defects of axial margins, much more problematic usually on account of possible cervical involvement, but each is to be judged on its own merits.

The procedure is simple. The marginal defect is cut out and treated as a Class I cavity. The absolute requirements are that *the cavity must extend to sound margins and that the amalgam–cavity interface must be macroscopically caries free.* It is at this stage that the decision to remove in entirety may be made. Only slight undercuts are needed for retention; the cavity is filled with well-condensed amalgam. Repairs of this type will last as long as the restoration, and in their economy of tissue and operative time must clearly be in the patient's best interests.

Summary

Characteristics of Class II cavity: inaccessibility, contact area involved, occlusal approach; optimal cavity described.
Collapse of marginal ridge.

Cavity preparation. Occlusal entry, removal approximal wall, siting axial margins. Retention; occlusal lock shape, dimensions; axial and cervical undercuts. Internal line angles.
Removal of caries. Review enamel margins. Extensive cavity, modifications required. Lining; thermal insulation.
Factors concerning shallow and deep cavities.

Matrices. Three primary functions; three types. Commonest encircles crown; screwed retainers. Placing band. Use of wedges; various types.
Amalgam insertion. Condensation procedure; choice of instruments.
Mechanical condensation. Important areas. Occlusal carving.
Matrix band removal. Carving approximal margins. Occlusal check, adjustment. Marginal pit and ridge.
Finishing and polishing.
Extensive restorations. MOD based upon Class II.
Separate M and D restorations.
Replacement of cusp; cavity design; resistance, retention.
Matrices. Completion.

Pin and screw retention. General principles of retention. Axial and transverse stresses; non-parallel pins. Three types, screw, friction, cemented. Rules applied to use of pins.
Sizes of drills and pins. Detailed methods, drilling and insertion, screw and friction pins. Conical screws, use in root canals.

Bonded amalgam. Characteristic of early marginal leakage. Methods of reducing. Use of phosphate cement, marginal seal, mechanical bonding. Technique.

Class III amalgam. Occasional use, special sites. Palatal entry, self-retentive distal cavity. Palatal lock. Lining. Modified matrix. Insertion and finishing.

Removal and repair. High-speed instrument, copious spray. Removal neck of lock, lock, bisection of box.
Avoidance of damage and over-extension.
Repair of early marginal recurrence. Cervical, never, occlusal often, axial on merits. Treat as Class I cavity. Economy of tissue and time.

8

Restorations in tooth-coloured materials

The ideal restorative material does not exist, for the properties required of such a material are very exacting. With all existing materials we must accept the advantages we seek together with the less desirable properties that go with them. A balance must be struck between the desirable and undesirable in every restoration undertaken.

Class III and IV cavities, and many Class V cavities, arise in positions in which they are very easily seen. Because of this, the importance of appearance is greater than with fillings which are less visible. Tooth-coloured materials can be made to match the colour of the tooth with great accuracy. Translucency and refractive index are not so easily matched but they can, nevertheless, be fairly closely approached. The result should normally be that any anterior restoration should be virtually undetectable when viewed from the front of the mouth and under normal lighting — at least when first inserted. In time the restoration may become discoloured by staining of the surface, by 'ditching' at the margin, or by physical change within its substance. Any of these, as well as loss of surface due to wear, and marginal leakage with or without recurrent caries, may be a reason for replacing these restorations. Their effective life may vary between three and ten years, chiefly because of errors in clinical technique. So replacement must be done fairly frequently and on each occasion a little more enamel and dentine are lost, and reference will be made to this again on p. 143 when considering extension of the cavity.

All tooth-coloured materials described in this chapter are compounds which are mixed to form a plastic conformable mass which sets initially about two or three minutes after insertion. They fall into the following categories:

1. Auto-polymerizing composites, based upon the reaction product of bisphenol A and glycidyl methacrylate (often shortened to BIS-GMA), with the addition of a finely divided siliceous filler. The physical properties of these composites show distinct improvement in most respects of clinical importance over those of methyl methacrylate.
2. Silicate cements. These have been in use since the early decades of this century and are still used, in much improved form, though they are gradually declining in popularity.
3. Auto-polymerizing methyl methacrylate, which is still in use though largely supplanted by the composites.
4. Glass ionomer cement. This is the product of polyacrylic acid acting upon finely divided alumino-silicate glass. It is often referred to by the

acronym ASPA, alumino-silicate polyacrylic acid. Although in clinical use for several years it has limitations due to its opacity but it is probably capable of further improvement by development.

The only remaining tooth-coloured material used for restorations is fused porcelain. With this a perfect match can be achieved; it is durable and unchanging. It is used almost entirely for full crown restorations and lies outside the immediate scope of this book.

Though all these materials have differing chemical properties the fact is that the plastic materials *when set* are physically alike. This is not surprising, for they are all trying to reproduce the combined properties of enamel and dentine. It follows that the cavities prepared for the reception of these materials are in most cases identical, whether silicate or composite is to be used. Cavities must be internally retentive and there is need to conserve as much sound tissue as possible, because the smaller the filling the less conspicuous it will be, and to allow repeated replacement should it be necessary over a long period of time.

Cavity preparation

In the absence of an adjacent tooth the Class III cavity can be approached directly from the lateral aspect. More often the incisal arch is intact and an approximal cavity must be entered through the lingual or labial surface. When the crowns are in normal relationship, the factors for consideration are these:

1. Access through the lingual surface involves indirect vision and indirect access during preparation, but this is normal procedure. It preserves the labial wall and therefore avoids the majority of the aesthetic disadvantages mentioned. The filling is exposed on the lingual surface to such wear as may take place by the oblique masticatory impact. This does not often seem to be an important factor.

2. Access through the labial surface is less common, but allows the lingual aspect to be preserved; this is said to reduce lingual wear, and therefore to increase durability. It allows direct access and visibility and is, therefore, easier to achieve and more certain in result. It produces a filling which is entirely visible on the labial and in view of the inevitable discoloration which follows in due course, this is a disadvantage. As the cavity increases in size and additional retention is required this consideration of visibility becomes even more serious.

3. On balance, in most cases the lingual approach is preferable.

4. When the crowns are not in normal relationship, other decisive factors may come into play. If the crowns are imbricated, a cavity on the lingually placed surface is best attacked from the lingual aspect (Fig. 8.1); similarly the labially placed cavity is best reached from the labial. This circumstance provides the commonest exception to the general rule of access through the lingual aspect.

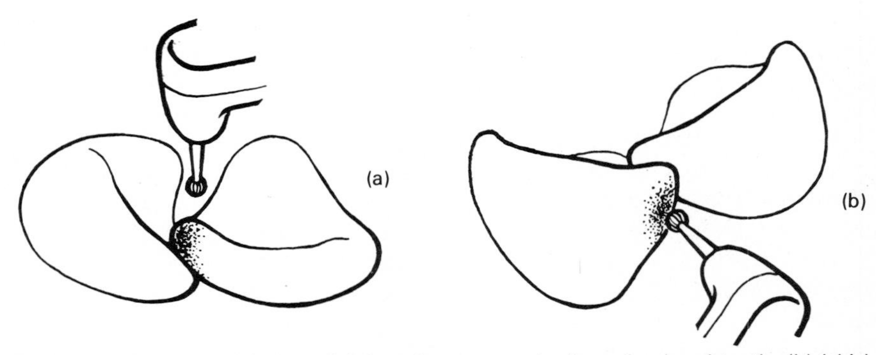

Fig. 8.1. Imbrication of incisors (a) lingual access to the lingually placed tooth; (b) labial access.

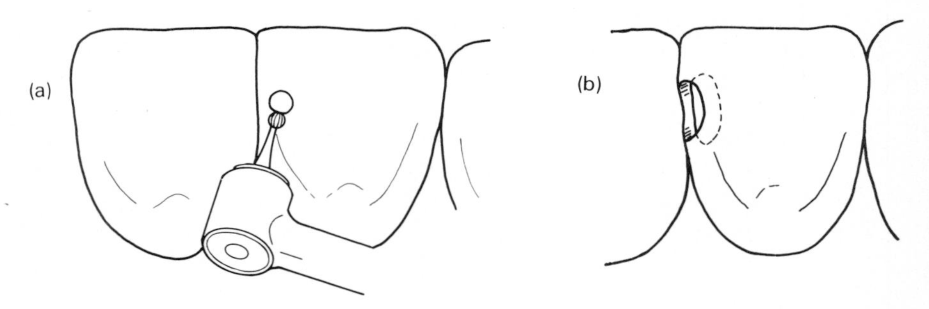

Fig. 8.2. (a) Site of initial access to small early Class III cavity; (b) outline of cavity enlarged to final proportions.

Initial access is attained through the intact enamel overlying the mesial or distal margin of the lingual surface (Fig. 8.2 (a)) and definition of the cavity is achieved with a small round bur or diamond instrument, and any extension for retention on the lingual surface can be best done with a small fissure instrument. Risk of damage to the adjacent tooth is still present, especially as visibility in the mirror is limited by the spray. For this reason some part of the preparation is done under direct vision when possible, though the scope for this approach is obviously limited. Hand instruments may be used to break down and shape the final portion of the interproximal enamel.

A common solution is to establish the form of the cavity at high speed and to resort to lower speed for completion of detail. This is a compromise which may be acceptable in the present state of development of these instruments.

The interstitial outline may be defined using a narrow hatchet No. 10 and 22 to form a curved labial margin connecting the incisal margin to the gingival margin, with very little incursion on the labial aspect (Fig. 8.2 (b)).

In Chapter 3 the general principle was deduced that, to avoid recurrence of caries, the outline form must not lie within a stagnation area such as is provided by the area of contact and of close approach. In the early Class III cavity this principle must, when possible, be modified in its application, for these reasons:

1. The Class III restoration whether of composite resin, silicate cement, or acrylic, however well performed, is essentially a short-lived restoration and it may be replaced three to six times over a period of years, before resorting to a more extensive design of restoration such as a crown. It is therefore necessary to conserve tissue as much as possible.

2. To place the outline in relatively immune areas involves bringing it close to the incisal margin, just on to the labial surface and possibly into the gingival crevice. This means the destruction of sound tissue which is difficult to justify, at this stage, with this type of restoration.

3. The relatively caries-immune outline described above (Fig. 8.2 (b)) is that which is achieved after, say, the second replacement of the smaller filling, and this may well cover a period of ten to fifteen years. For all patients except those with the highest caries experience, it is wise to design the outline form conservatively.

The resistance form of this cavity consists essentially of the labial wall of the cavity and the retention is derived from a slight general undercut which can be more pronounced in the cervical and incisal aspects than elsewhere. Thus the axial internal diameter of the cavity is a little greater than the labio-lingual (Fig. 8.3). There can be no question of squaring walls and internal angles, for this is destructive of sound dentine and contributes nothing. This cavity must be kept small and the masticatory stresses upon it are very low.

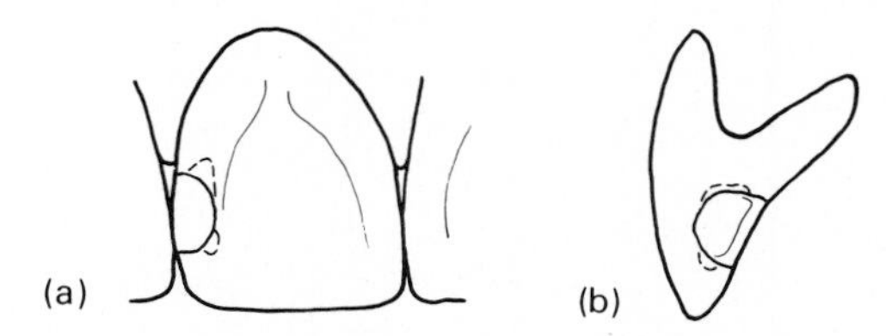

Fig. 8.3. Outline of small Class III cavity, retained by cervical and incisal undercuts (a) from lingual (b) from mesial aspects.

The enamel edges are as important as ever. All enamel must be supported by sound dentine and the margins can be finished with the smallest pear-shaped finishing bur in order to smooth the outline, remove loose prisms, and to effect a 90-degree cavo-surface angle. The cavity is then ready for lining, which in this case presents some difficulty owing to its small size; this will be referred to later.

The larger cavity (Fig. 8.4) is one which may be attained as the result of several replacements of successive restorations, or it may arise initially by

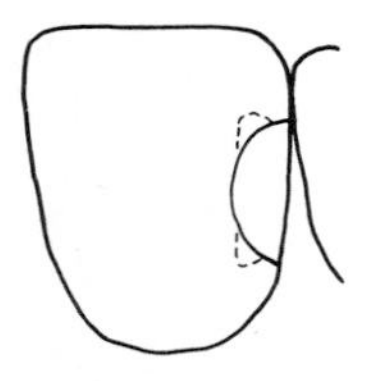

Fig. 8.4. Larger Class III cavity using internal retention.

preparation of a cavity which is the result of untreated caries. In the latter case access is gained as before and the more extensive outline form is defined by the high-speed bur or diamond instrument for further extension. Many operators prefer to use the rotary instruments where possible, for heavy chisel pressure and the recoil that follows, exerted in a line transverse to the long axis of the tooth, can be very uncomfortable.

At this stage of larger development it is easier to place most of the outline in relatively immune areas as detailed above without undue loss of tissue. Eradication of stained and carious dentine, by using a No. 3 round bur or a contra-angled spoon excavator No. 243, proceeds *first around the periphery of the cavity*, defining the sound amelodentinal margin and, if necessary, cutting back small areas of lingual enamel to ensure adequate *access and visibility* to every part from the lingual aspect.

Retention may still be obtained by undercuts in the cervical and incisal extremities of the cavity. In the cervical, this is more of a groove cut with a small round bur, running along the cervical floor a short distance away from the cervical margin. In the incisal, the form is that of a rounded pit cut in the dentine, and it is obvious that the risk of weakening the incisal angle could develop; this risk increases as the size of the cavity grows and the residual tissue at the incisal angle decreases.

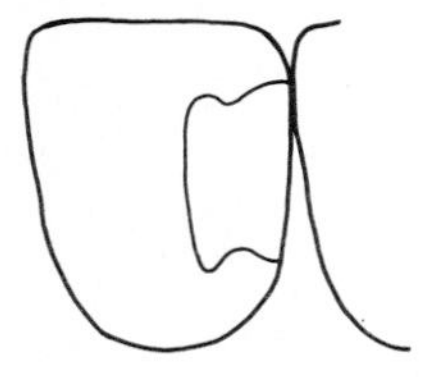

Fig. 8.5. Use of modified palatal lock as retention in large Class III cavity.

To avoid this complication in the larger cavity, the position of the retention can be modified (Fig. 8.5). This modification is in effect the cutting of a lock on the palatal surface, a device which is suitable to the use of silicate cement, but not used with composites.

For composite restorations, acid etching as referred to on p. 154 is a less destructive and quicker method of increasing retention. Large silicate restorations are now uncommon, so there is little call for the use of the lingual lock in practice.

The similarity between this lock and the form of the internal undercuts will be noticed, and the very wide 'neck' appropriate to the low strength of the material is another essential feature of its design. The lock is cut by using a cylindrical fissure bur in a contra-angled handpiece. The bur is held normal to the lingual surface and the cut should enter dentine to the depth of 0.5 mm. The area of the lock must be reduced to a minimum consistent with adequate retention, for this part of the preparation is almost wholly within sound dentine and the degree of pulpal injury is a matter for consideration here. The walls of the lock should be divergent inwards, or lightly undercut along the floor–wall angle, and in this way a very effective retention can be obtained which serves until such time as the collapse of the proximo-incisal angle occurs, as eventually it must, as the result of repeated filling and incisal wear. The main disadvantage of the lingual lock is of course the destruction of coronal dentine, particularly as a cavity on the opposite aspect of the tooth is likely. The eventual need for a full crown must always be kept in mind.

The removal of caries from the pulpal aspect of the cavity again calls for care and discrimination. Its complete removal is not difficult when no threat to the pulp exists, but in young teeth with large pulp chambers and persistent cornua the position needs careful evaluation.

The retention of a small amount of frankly carious dentine is in this case a much more dubious procedure for two reasons. First, there exists a much higher risk of leakage at the margin, and then there is also the added disadvantage of staining, which in due course can become unsightly. The degree of staining would be directly related to the amount of carious dentine left and the permeability of the normal dentine.

There are occasions when, in a large tooth, a minimum of stained dentine may be left in the absence, or presumed absence, of any pulpal infection. This area is covered with a non-irritant lining such as calcium hydroxide cement.

In cases of greater uncertainty, particularly in young patients, it is preferable to insert a dressing of zinc oxide and eugenol and to put the tooth on probation for a period of six months. At the end of that time a vitality test, and replacement of the dressing if required, gives information as to the condition of the pulp which, if vital, will have laid down a barrier of secondary dentine. This delay sometimes gives the opportunity for virtually complete removal of all stained dentine from the base of the cavity or from the labial, where it could cause visible discoloration. If necessary two dressings may be placed over a period of nine to twelve months before the permanent restoration is inserted.

In many cases, however, where the integrity of the pulp is in doubt due to serious risk of exposure by excavation of carious dentine, or by early bacterial invasion from such dentine, *it is better treatment to elect to extirpate the pulp at that stage* and proceed to root canal treatment. The reason for this is that in this early condition when, at the worst, the pulp is only lightly infected in a circumscribed area, the prognosis for endodontic treatment is very much better than at any subsequent time. Later, infection is certain to spread and will almost certainly have involved the apical canals and periapical tissues. In these conditions treatment is more complicated and prolonged, and a good result more difficult to attain.

To summarize the situation, it may be said that the arguments in favour of complete carious eradication are very strong. The occasions for the retention of carious dentine are few. That dressings and reliance upon the natural reaction of dentine can sometimes be resorted to, and finally, when the pulp is at risk early recourse to endodontic treatment is always preferable to delayed treatment.

In regularly conducted practice, problems of this sort are more easily resolved than their discussion might suggest. The cavity is seen at its earliest development and filled for the first time when of optimal size. Thereafter, when six-monthly inspection may reveal marginal failure, replacement is undertaken long before carious destruction of dentine becomes established. In these conditions invasion of the pulp rarely enters the clinical picture and successive replacements may be carried through a period of two or three decades before progressive weakening of the crown calls for a different type of restoration. Indeed it may be said that many operators err in continuing with cement or plastic restorations for too long before resorting to porcelain or bonded crowns. Repeated replacement of mesial and distal restorations, particularly when the lingual surface has been involved for retention, may led to extensive loss of coronal dentine. From the point of view of appearance and function it is often better to resort to a jacket crown earlier rather than later.

Treatment of carious dentine

The removal of carious dentine follows the principles previously discussed. In establishing the outline form, the periphery of the cavity is cleared of softened and stained dentine and can fairly be said to be composed of normal tissue.

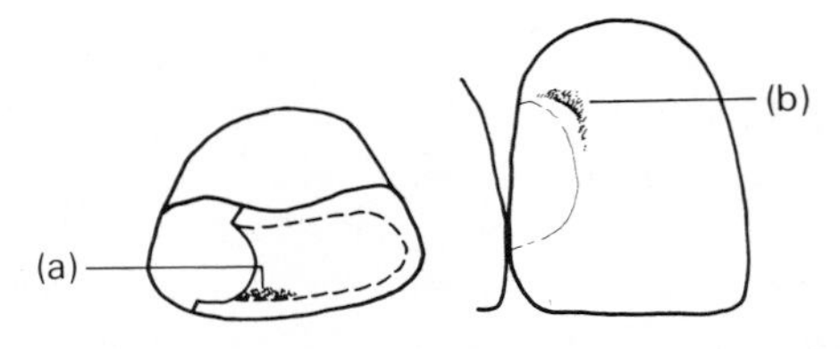

Fig. 8.6. (a) The persistence of stain at the amelo-dentinal junction; (b) the appearance of stain at labio-cervical margin deep into enamel.

It is occasionally noticed that there has occurred in the gingival region of the labial aspect of the cavity, a very fine hair-line extension of stain along the amelo-dentinal junction (Fig. 8.6 (a)). This occurs more often on the replacement of a filling. The location of this stain in the cavity is difficult to see and it is not often accompanied by softening. From the labial aspect, when the cavity is filled, the stain is apparent as a very unsightly greyish discoloration, deep to enamel, and conforming in shape to the inner surface of the filling mass (Fig. 8.6 (b)). The last traces, difficult to see in the transmitted light incident on the labial enamel, may escape notice during cavity preparation. When the cavity is filled the persistence of the shadow is more easily seen.

Enamel margins of all Class III cavities must be finished with a 90-degree cavo-surface angle. This and the elimination of fine irregularities is done with a fine-cut multi-bladed or blank finishing bur. The lingual outline, for example, is trimmed with a cylindrical bur, since it is readily accessible to this form. The remainder, that is to say incisal, gingival, and often the labial, must be reached with a round or pear-shaped bur of suitable size (Fig. 8.7). It is used with an even brush-like motion along the cut enamel surface. Where the labial margin is easily accessible from the labial aspect, the fissure bur is preferable.

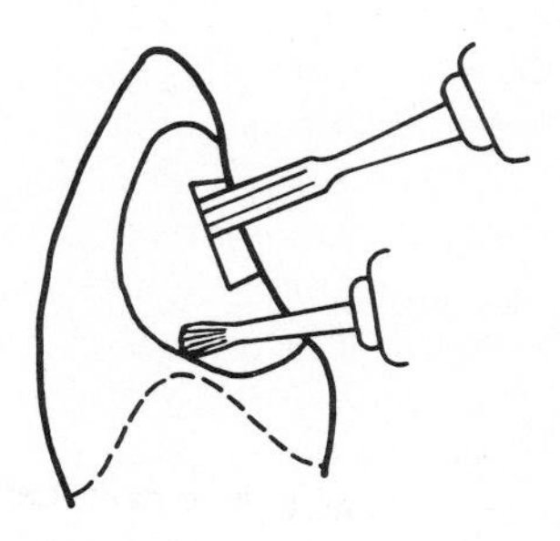

Fig. 8.7. Use of finishing burs on enamel margins.

Cavity lining

Composite resins and silicate cement are examples of filling materials with irritant properties which must be mitigated because of their other relatively desirable characteristics. The irritancy of the silicate cement has been variously attributed to a number of factors, but there seems little doubt that it is directly related to the acid content and it probably varies with the hydrogen ion concentration of the mass at the time of insertion into the cavity and for a period afterwards. In the case of resins the monomer is toxic, and can exert its action during polymerization and subsequently if an excess of free monomer remains for some time.

For practical purposes it may be said that all Class III cavities must carry a protective lining of some description. In the majority of cases the insertion of an unlined restoration would certainly result in the death of the pulp. The incidence of non-vital incisors from this cause—the maxillary lateral is particularly susceptible to toxic attack—is all too high. It is important for this reason to understand the factors which control the reaction of the pulp.

On the assumption that the mixing of the material is standardized to a minimal free monomer or acid content, its irritancy upon the tooth will be greater:

1. The larger the cavity and the closer the approach to the pulp.
2. The smaller the tooth and, therefore, the closer the approach to the pulp.
3. The younger the tooth and the more permeable its dentine.
4. The greater the extent of fresh dentine exposed by recent cavity preparation.

All these factors are variable; some are apparent on examination, some are controlled by the operator, and the remainder are a matter of clinical judgement.

In general the following rules may be accepted as giving a reasonable guide:

1. Below the age of sixteen, an arbitrary figure, it is safest to insert a temporary filling in the fully prepared cavity for a period of six months prior to the insertion of a Class III restoration. If, when reopened the cavity needs further preparation which involves fresh dentine, then the dressing should be repeated.
2. All restorations must be lined, having regard to protection not only of the deepest part of the cavity but also of the freshly opened dentine near the periphery.

The types of linings commonly used are calcium hydroxide cement and polycarboxylate cement. The principles governing the use of these linings are explained on p. 73 *et seq.*

The technique for lining a Class III cavity varies slightly according to the material used. There are for example several proprietary varnishes which claim to protect dentine against chemical irritation, presumably by blocking the ends of recently cut tubules. Although the evidence produced suggests that some of these varnishes are at least partly protective there remains some doubt as to the degree of their effectiveness. A fully impermeable and adherent surface film would be of the greatest value for it would give protection to recently cut dentine around the periphery, where it most often lies, and where lining cements cannot be placed without risk of losing retentive undercuts.

The linings suitable for composites and silicate cements are applied in a viscous state, *using a blunted probe*—a very useful instrument for placing cement accurately and cleanly. The cement is teased across the floor of the cavity and on to the walls, covering as much of the recently cut dentine as possible without reducing the retention of undercut areas. The thickness of the cement need not exceed 0.5 to 1.0 mm. *It is important to place the correct amount in the right place first time*, just a small amount, followed by small additions, avoiding the enamel margins. Excess cement can be removed when partly set, with a small sharp excavator, but this is wasteful of time and effort.

Restorations in composite resin

The techniques for preparing and inserting composite and acrylic resins are in general similar, though in detail they vary. They employ several catalytic systems and the technical details provided by each manufacturer should be closely followed.

The physical properties of composites allow them to be used for retentive Class III cavities, Class IV cavities when acid-etching and pin retention are used, and Class V cavities. Clinical opinions agree that composite resin is not suitable for Class II stress-bearing restorations except in special conditions.

In general the handling of composites is simpler than that of silicates. Proportioning and mixing is less critical, and colour-matching is easier. The harmful effect of moisture on the setting material is less but it must be kept dry, so the use of rubber dam where practical is good technique. The ability to polymerize a new mass on to a pre-existing one, with complete bonding is a useful property which can legitimately be used on some occasions with great advantage in economy of tooth tissue and operating time.

Colour-matching

The selection of the correct colour and shade of material is obviously very important. For the whole purpose of using 'tooth-coloured materials' is to obtain as nearly as possible an exact match with the tooth. Sometimes a compromise must be accepted because the shade near the cervical is markedly different from that near the incisal. Colour-matching should be done with a shade-guide, in a good light and whilst the tooth is still moist. The accuracy of the match is less critical with composites for their optical properties allow them to some extent to take on the colour of the surrounding tooth.

If a deviation from an absolute match is allowable it is probably better to have a restoration slightly darker, in spite of the fact that surface staining may darken it still further. The reasons are these. First, all teeth darken with age and there is no natural process which causes dental tissues to lighten in shade; so too light a colour always *looks* wrong. This is a precept which applies generally in all shade- and colour-matching in restorative dentistry. A darker tooth could always be natural — a lighter one never, unless the adjacent teeth had all darkened abnormally, which is uncommon. Then, a *slightly* darker Class III restoration can mimic the natural interproximal shadow. Finally, surface staining can be removed if necessary. A little experience with any particular material will soon allow the operator to match colour and shade very accurately.

Preparation and insertion of composites

These materials, having various polymerization systems, are supplied in several forms, two pastes, a paste and a liquid, and, in the case of acrylic, a polymer powder and monomer liquid. The two-paste system is commonest and the proportioning not very critical. The extrusion of a measured length of paste from a tube or syringe is a convenient method. The pastes are thoroughly mixed with a plastic spatula on a disposable paper pad. For exact details the operator should closely follow the manufacturer's directions in order to standardize the working and setting characteristics. Accuracy and uniformity of mixing are facilitated by the encapsulated materials now available for mechanical mixing.

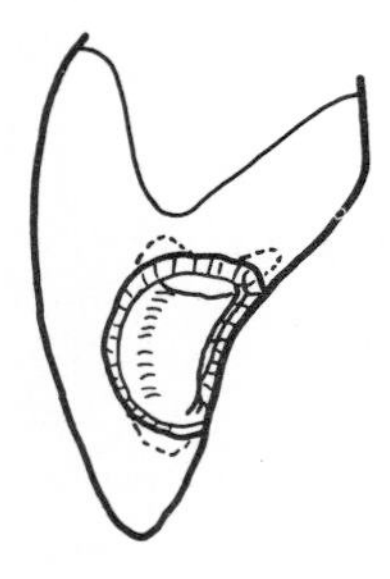

Fig. 8.8. Class III cavity showing incisal and cervical retention areas suitable for a composite restoration.

In the medium-sized Class III cavity the internal retention must be well pronounced (Fig. 8.8) and inspection of the cervical and incisal undercuts should be made after lining to ensure that they have not been blocked out by cement. Moisture is controlled, if necessary by rubber dam, and restorations in adjacent teeth protected by vaseline or silicone grease. The matrix commonly used for Class III cavities is a strip of transparent plastics about 8.0 mm wide and 0.08 mm thick. Most matrices are made of cellulose acetate, polyester, or polythene which do not react with composite or acrylic monomer; they separate cleanly, macroscopically at least, from the surface of the set material. They can be moulded slightly to fit the contour of the tooth better by stretching over a warmed rounded instrument. The matrix band is eased between the contact area and located so that all margins of the cavity are covered; in fact this is only possible in the case of the smallest cavities. The matrix band is a flat strip but the surface of the crown has varying convexities on the labial and approximal—and the lingual is concave! There is, furthermore, a limit to the extent to which the elasticity of the strip under tension can adapt to these margins without the surface of the composite becoming concave. So it is quite clear that the strip cannot occlude all margins equally well and this means that some margins will carry a heavier excess of filling material and others less. The margin which will be most difficult to trim will be, in most cases, the cervical so it is to the adaptation of the matrix to the cervical margin that we must pay most attention at this stage, whilst it is still easily visible.

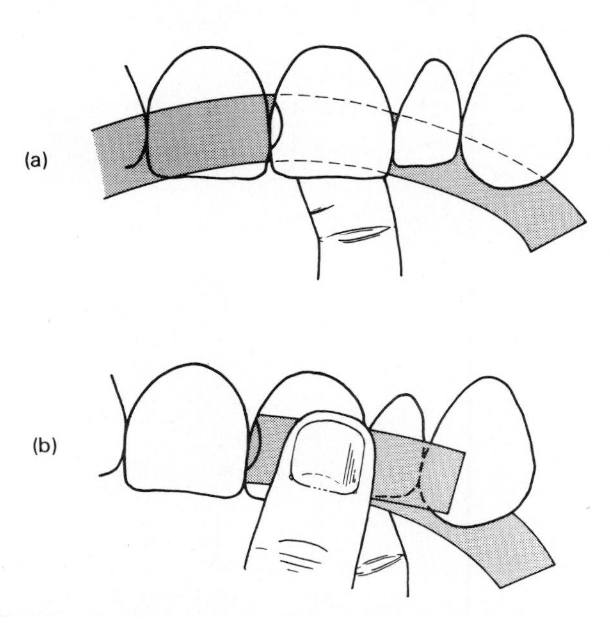

Fig. 8.9. (a) The matrix strip supported by a fingertip in a palatal fossa. (b) When the cavity is filled this strip is drawn across the labial and held in position.

The mixed material may be semi-fluid or a soft dough consistency. With the cavity dried and the matrix strip in position, supported by a fingertip on the palatal (Fig. 8.9 (a)), the mix is first placed into undercuts with care to displace any air which might be trapped. Then more material is worked into the cavity until it is judged to be *slightly* overfilled. This achieved, the matrix strip is brought across the labial and, whilst maintaining contact along the cervical margin, is adapted wherever possible to the proximo-incisal, labial, and lingual margins. When correctly positioned the strip is firmly held between the forefinger on the palatal and thumb on the labial (Fig. 8.9 (b)) for the period of initial set, about two and a half to three minutes. When the interdental space and the level of the cervical margin permit, a cervical wedge can be used as shown in Fig. 8.12, p. 154.

The occlusal adjustment of the palatal and surface finishing of composite materials is difficult because of the extremely hard silica particles embedded in a relatively soft matrix. Many manufacturers now produce fine-grained hard abrasive stones, wheels, and discs for trimming and shaping the set resin. Fine-grained diamond instruments, fine bladed tungsten carbide burs, and smooth tungsten carbide finishing burs can all be used, the last of these, used at high speed probably gives the most controlled cutting effect for final finishing of margins without too much loss and roughening of enamel. All abrasive instruments should be lubricated with vaseline or silicone grease to reduce the heat of friction.

The time and care needed to produce a well contoured, smoothly finished composite restoration with accurately trimmed margins and little damage to adjoining enamel will surely re-emphasize the importance of a well shaped and closely fitting matrix band.

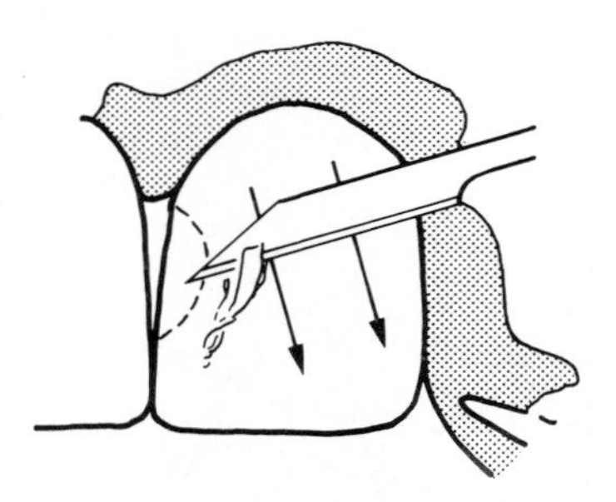

Fig. 8.10. Trimming labial excess from Class III restoration.

With acrylic restorations correct occlusion must be checked at the time of insertion, but finishing is better left until the next visit if possible. As the material is softer, excess in the form of a flash can be trimmed with a sharp blade cutting in the plane of the enamel surface (Fig. 8.10), from filling to enamel. Steel finishing burs and fine abrasive strips and discs, the last at moderate speed and lubricated. Polishing with fine pumice followed by stannous oxide powder imparts a smooth surface to the restoration.

Class IV restorations

The restoration of the mesial or distal angle has always presented difficulties due to the exposure of these restorations to frequent and heavy occlusal loads, and the need for good appearance. Silicate cement has never been satisfactory, for though its crushing strength is good, its tensile strength is poor. The Class IV gold inlay described on p. 204 *et seq.* is a good restoration but unsightly if it involves the labial surfaces more than marginally, unless it is faced with silicate cement or composite. It is a complicated and expensive restoration.

The improved properties of composites have made more practical the rebuilding of the incisal angle with pin reinforcement, and the development of resins originally used as fissure sealants, has opened up the 'acid-etch' technique for achieving greater retention.

Pin retention

This is most suitable when a Class III cavity, having been progressively enlarged over the years, becomes or has to be converted into a Class IV cavity by removal of the incisal edge. Before describing the method, a note of warning must be sounded. For reasons already indicated Class IV restorations have a limited life. **Excessive weakening and loss of coronal dentine due to repeated Class IV cavity preparation may eventually prevent the effectual construction of a full crown on that tooth, as a vital tooth.** It is not unusual to have to proceed to extirpation of the vital pulp and the construction of a post-crown, only because the amount of residual dentine cannot provide sufficient retention.

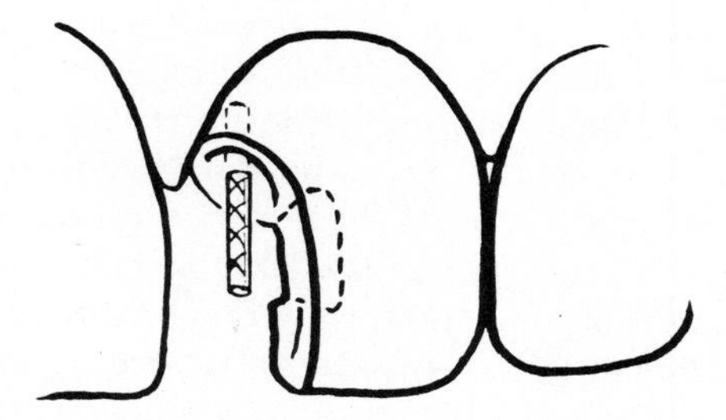

Fig. 8.11. Siting of a pin to give added retention to a Class IV composite restoration.

In this type of cavity it is possible that some part of the original retention form may remain and if a palatal lock has previously been cut, this would be used (Fig. 8.11). The main increase of retention comes from a cervically placed pin. The pin hole is drilled in the cervical floor of the cavity, well within dentine, at least 2.0 mm deep and parallel to the root surface (see p. 130). The pin, when firmly in place, may be slightly bent by gentle pressure until its free end, about two-thirds of the length of the preparation, lies centrally in the cavity (Fig. 8.11).

As the cavity is long-established, much of the floor and walls will consist of dentine which is sclerotic and already isolated from the pulp. Any freshly cut dentine must be protected by a non-irritant cement.

The use of a special cellulose acetate or polystyrene crown-form as a matrix or container is an important part of the method. This is made by cutting the appropriate angle from a transparent crown form of suitable size. The part retained for use must be large enough to cover, most importantly, the cervical margin of the crown and all the incisal edge. It should be shaped to cover all cavity margins by about 2.0 mm and carefully checked for cervical and incisal location; in other words it must be easy to place it accurately and definitely in position.

If required the acid-etch technique described in the next section may be used to provide further retention.

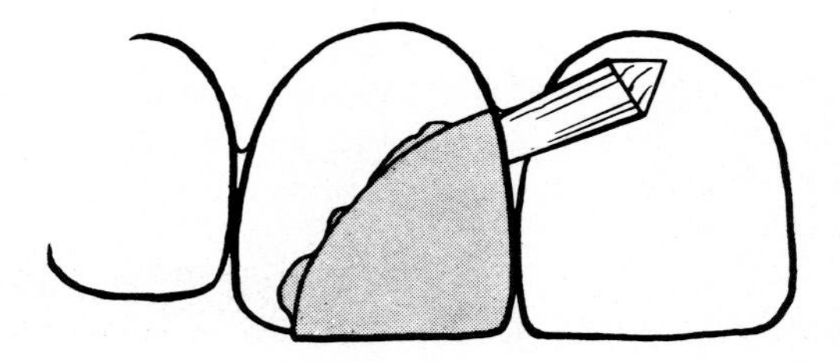

Fig. 8.12. Restoration of incisal angle using a reduced crown-form wedged at the cervical margin.

The prepared composite is vibrated or wiped into all the retentive features of the cavity, using the plastic instruments provided by the manufacturers of the material. Care is taken to see that no air bubbles are included and the pin is completely enveloped before the transparent matrix filled with composite is carried to place. At the cervix the matrix can be retained by an interdental wedge; any excess beyond the matrix can be removed whilst soft and the restoration left to set (Fig. 8.12). To remove the matrix it can be cut along the interproximal margin and the matter of shaping and finishing can then be assessed. The margins of this cavity are accessible to rotary instruments. With the interdental papilla retracted, pointed instruments can often be used even at the cervical, where, if control of the matrix has succeeded, the need for trimming should be least. Finishing and polishing proceed as described. It is of course very important to see that the occlusion is cleared before final polishing of the palatal surface and dismissal of the patient.

There are available some composite systems in which polymerization is activated by exposure to ultraviolet and visible light. They may of course be used for any anterior restoration but are particularly suited to the larger ones. The thickness of the restoration and the penetration of the light must be considered; polymerization to a depth of 2.0 mm in 20 seconds has been claimed. Their great advantage is that filling one or more cavities and careful adjustment of the matrix can be completed without hurry, before application of the light stimulus starts the setting reaction.

Acid-etch technique

The development of plastics for use as fissure sealants for prevention of Class I cavities has introduced a technique which uses the ability of these resins to adhere to roughened enamel. The enamel surface is etched with buffered 30–50 per cent solution of phosphoric acid which dissolves the enamel prisms more readily than the interprismatic substance, producing a microscopic roughness which is intimately penetrated by the resin applied in liquid form. The method can be used to increase the retention of any type of Class IV restoration. In its simplest form it is particularly suited to the early repair of a fractured incisor in a child. The loss of a mesial angle of a central incisor is a common injury caused by a direct blow; the fracture often involves enamel and dentine without exposure of the pulp. The exposed dentine may be protected by a complete cover of non-irritant cement, usually calcium hydroxide. Buffered phosphoric acid is then applied with a brush, not only to the fractured enamel surface, but to a margin of about 1.5 to 2.0 mm of normal enamel surrounding the fracture. After a period of 2 minutes' contact the acid is washed off thoroughly with warm water and the tooth dried, leaving the etched area whitish and opaque.

The layer of liquid resin is now applied with a brush to the fractured surface and the surrounding etched area. This is followed by a composite of suitable shade contained in a trimmed crown-form as previously described. Excess composite can be trimmed, whilst soft, from the edge of the crown form (Fig. 8.13). When set the restoration can be trimmed but a margin of the surface flash must be retained to assist retention.

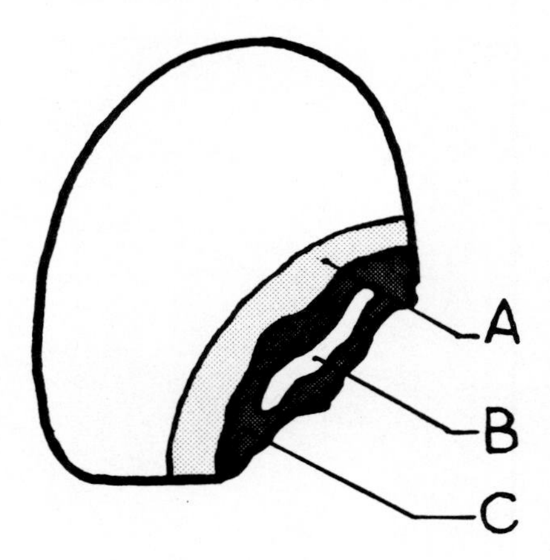

Fig. 8.13. Restoration of incisal angle by acid-etch technique. A: etched surface enamel. B: cement protecting dentine. C: etched fracture enamel.

This technique requires no mechanical cavity preparation, which would need local anaesthetic and could still further damage dentine. It is quick, painless, it protects the pulp; it produces an aesthetically acceptable result which can be easily repeated when necessary until a full crown is indicated.

Another less common but very valuable application of this technique concerns the treatment of defective and unsightly enamel on the labial surface of incisors affected by hypoplasia, staining, erosion, or abrasion, when conditions contraindicate the use of a full veneer crown (Fig. 8.14). The labial enamel, where still existent, is reduced to about a quarter of its normal thickness, etched and covered with a veneer of composite in a crown-form adapted to the labial surface. Proprietary labial crown-forms of suitably shaded plastics are also available for this and similar techniques. Though no more durable than other composite restorations their appearance can be

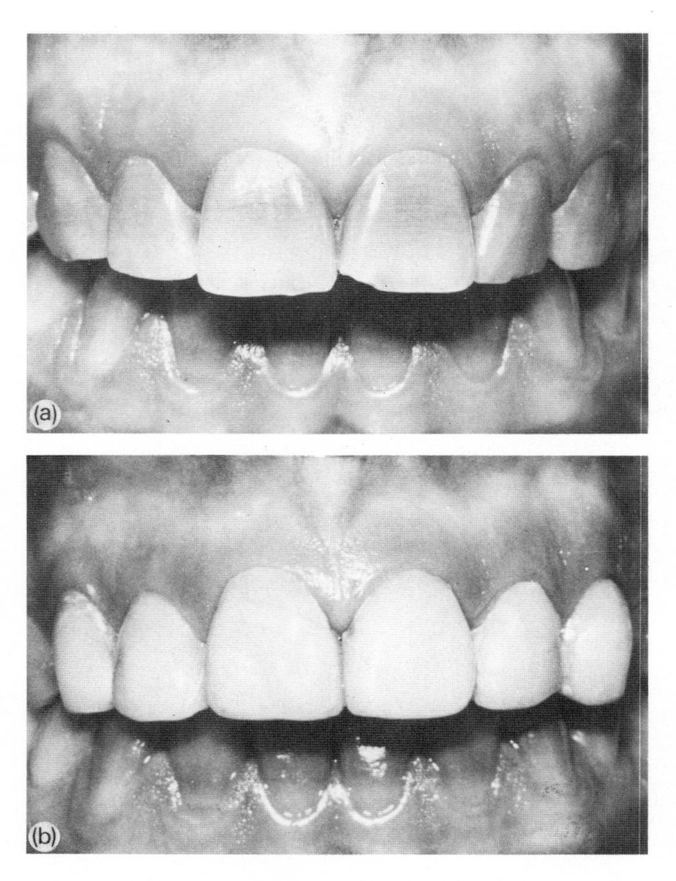

Fig. 8.14. (a) Tetracycline staining of incisors; (b) treated by labial composite veneers using acid-etch technique. (Courtesy Dr G. Roberts.)

very good. This type of veneer therefore provides a quick and simple solution to what is otherwise a difficult clinical problem.

Class V restorations

The preparation of a Class V cavity for composite resin restoration is similar to that described for amalgam on p. 96. The restoration bears no direct occlusal load and the forces tending to dislodge it are small. Once the outline and depth are determined, all that is necessary is to form undercut areas along the occlusal and cervical walls. This can be done with an inverted cone bur, but it can equally, and perhaps better, be done with a small round bur, number 2 or 3. There is no advantage in having a sharp floor–wall line angle in this cavity and a rounded cavity allows a little more latitude in lining.

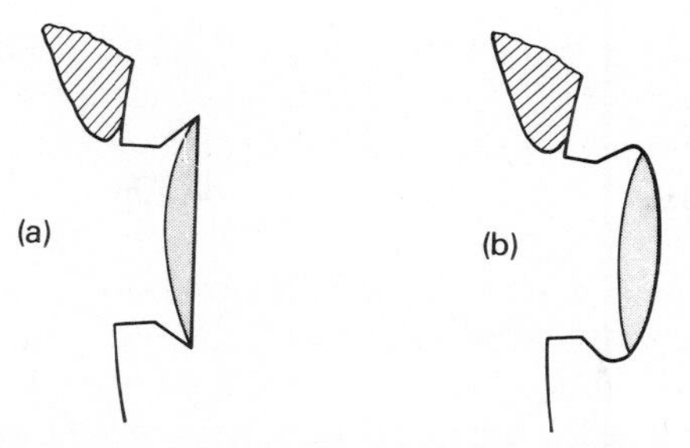

Fig. 8.15. Diagrammatic comparison between, (a) the sharp, and (b) the rounded undercuts on occlusal and cervical walls of a Class V cavity.

The lining of the cavity covers the floor, or pulpal surface, with a thin layer of calcium hydroxide cement—a thin layer because the cavity is often shallow, and because the undercuts must not be filled (Fig. 8.15). The problem of protecting the pulp in this simple case is one which presents to some extent in all cavities. In a prepared Class V cavity the centre of the floor will be underlaid by a tract of sclerosed dentine; the freshly cut dentine will lie chiefly on the periphery of the cavity. It is the protection of this peripheral zone without obliterating undercuts which calls for care and precision in placing cement.

When mixed the composite is worked into the undercuts and the cavity filled with sufficient material to reach all cavity margins and to reproduce the correct convexity of the cervical area restored. At this stage of setting when the mass is passing from a semi-fluid to a doughy consistency this convexity can be controlled. The mass may be allowed to set without a matrix, but it must be kept dry. Some operators, however, prefer to use a form of simple matrix to cover the surface. These matrices are flakes of cellulose or coated tin foil, slightly convex and of the same general shape as Class IV cavities, in a variety of sizes (Fig. 8.16). Select one slightly larger than the

Fig. 8.16. Metal matrices for use on Class V plastic restorations.

cavity and settle it on the surface of the setting composite so that the margins are covered. This may be left in position until the next visit, if necessary, and removed for the minimal finishing which the method allows.

Glass ionomer cement

This material, which in its structure has much in common with silicate cement, has introduced two properties which have been much sought after. It is relatively non-irritant to dental tissues, therefore no lining is needed except in the case of a near exposure, and it adheres to clean enamel and dentine without special treatment of the surface. This means very little cavity preparation, no local analgesia in many cases and a commensurate saving of patient discomfort and operating time. As with silicate cement, the fluoride content is said to confer local resistance to adjacent enamel. Its colour stability and abrasion resistance are good. It is not as translucent as the composites, and though this has recently been improved it remains unsuitable for the restoration of large areas of easily visible enamel. It is therefore particularly suitable for Class V cavities, and among these especially for the shallow abrasion cavity.

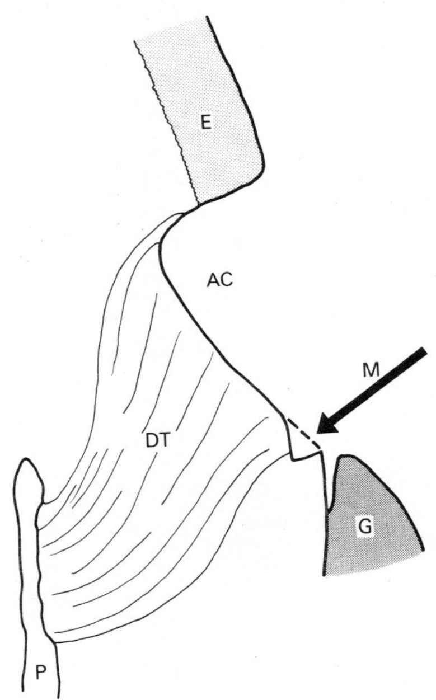

Fig. 8.17. The base of an abrasion cavity, AC, is underlaid by sclerosed dentine of a dead tract, DT. E, enamel. P, pulp. G, gingiva. A shallow margin may be required at M.

When cavity preparation is required it essentially involves the removal of carious dentine and enamel; where retentive undercuts exist they can certainly be kept but the formation of undercuts is not necessary. Where an abrasion cavity margin fades imperceptibly on to the surface of crown or root, a finishing line not less than 0.5 mm deep must be formed (Fig. 8.17). As in all cavity preparation the pain involved in this operation is precisely related to the invasion of vital dentine. To the extent that cutting is confined to areas of dead tract it will be painless.

The surface of the prepared cavity must be chemically clean. An abrasion cavity in which *no fresh dentine has been cut* may be cleaned with a 50 per cent aqueous solution of citric acid to remove protein debris, but in the presence of freshly opened dentine 20 volumes hydrogen peroxide should be used. After cleaning recontamination must be avoided.

The difficulties posed by the proximinity of gingival margins to Class V cavities are discussed on p. 98. The same principles apply but because these cavities are more visible they are, on the whole, treated at an earlier stage, preparation is less and control is generally that much easier.

Glass ionomer cement consists of a powder which is finely divided glass and a liquid, polyacrylic acid, which are mixed together in the ratio 3 to 1; the proportioning is fairly critical. The glass mixing slab and the powder may be refrigerated to slow the reaction, but not the liquid. Using a non-corrodible spatula powder and liquid are mixed to a putty-like consistency which retains its gloss and wets the cavity surface. Like silicate, glass ionomer cement is sensitive to water contamination and must be kept dry during and after insertion and setting. A matrix (Fig. 8.16) should be used to compress the material during a setting time of five minutes, then varnished and left for a further five minutes before trimming. The matrix carefully removed, excess cement can be trimmed away with a sharp scalpel, excavator or scaler rather than with rotary instruments at this stage. Marginal carving should run parallel to the edge. Over-drying must be avoided and the surface is finally protected with copal–ether varnish.

It is suggested that these cements can be used for fissure sealing, for Class II cavities in deciduous teeth, where minimal cavity preparation helps, and in the temporary repair of broken cusps where adhesion is an advantage.

Restorations in silicate cement

Silicate cement is still in use though it seems likely that it will finally be displaced by newer materials. It has a higher crushing strength than the composites and is capable of accurate matching but is subject to surface staining. It is irritant and must always be lined. The powder has a small fluoride content which has been shown to leach out during the life of the restoration and this, it has been claimed, confers some mild, local caries immunity. The setting shrinkage of this cement is very small, thus micro-leakage is not common if the mixing techniques are carefully followed. The proportioning and mixing must be more precise than with the composites and moisture must be excluded until at least half an hour after the initial set.

These conditions are most easily met by:

1. Using the capsulated form of the cement with mechanical mixing.
2. Operating under rubber dam.
3. Application of varnish to the finished restoration.

The capsule consists of two separated chambers containing liquid and powder which are forced into contact under pressure. The capsule is then vibrated in a machine similar to the mechanical amalgamator. The method is simple, quick, and uniform in its results, but carefully performed slab mixing can be equally effective and needs no complicated apparatus; details of this method are provided by the manufacturer.

The insertion of the cement is in all respects similar to the procedures described for composites. Excess material must be avoided and the restoration is trimmed and finished with white alpine points and fine abrasive strips and discs which must be lubricated in use.

All the tooth-coloured restorative materials used in the front of the mouth have a short life when compared with the durability of restorations such as cohesive gold, a gold inlay, or full crown. It is claimed for silicate and possibly glass-ionomer cement that its fluoride content has a caries-inhibitory effect; no such advantage can be claimed for the resins. All these materials are subject to a considerable degree of discoloration, either on the surface or in the substance. They are also subject to 'ditching'—the formation of a small, discoloured linear defect running along the cavo-surface outline and to marginal leakage. These, particularly the last, are serious defects and it is necessary to watch all anterior restorations very closely so that they may be replaced at the proper time, and especially before recurrent caries becomes established. Although mechanical and chemical bonding of these materials can be demonstrated, these properties have to be improved. Research and the development of new materials continues and there is no doubt that the next few decades will see the introduction of improved and possibly new tooth-coloured materials.

Summary

Composites commonest material; silicates rarely; glass ionomers developing.

Class III preparation. Lingual approach preferred.

Round bur, high speed to define minimal cavity; complete with medium speed and hand instruments. Generally undercut cavity; 90 degree cavo-surface angle. Factors in outline form; replacement, visibility. Larger cavity; modifications, palatal lock. Conservation of coronal dentine. Carious dentine; complete removal; temporary dressings; early extirpation of pulp preferable to delayed pulpitis.

Cavity lining. Irritancy demands lining.

Four factors on irritancy. Two guiding rules. Calcium hydroxide and polycarboxylate cements.

Composite restorations. Class III. Properties of material. Colour matching; slightly darker better than lighter. Two-paste systems. Moisture control; strip matrix; adaptation, manipulation. Trimming, occlusal adjustment, finishing.

Class IV. Pin retention. Lining. Reduced crown-form matrix. Acid etching; use in incisal fractures; protection of pulp; minimal preparation. Labial veneers for defects of labial enamel.
Class V. Preparation and undercut retention. Lining periphery.
Insertion, cervical matrix.
Glass ionomer cement. Non-irritant adhesive, moisture senstive. Suitable Class V, particularly abrasion cavities. Cavity preparation nil or minimal. Protection for near exposure; chemical cleaning cavity surface. Proportioning, mixing critical; slow setting; cervical matrix. Trim with sharp blades; protect with varnish.
Silicate cement. Use declining. Irritant; lining essential. Mixing by capsule or slab methods. Moisture sensitive. Trimming, fine lubricated abrasives. Protect with varnish.

9

Direct gold inlay technique Class I and Class V cavities

A gold inlay is a restoration of cast gold, made accurately to fit a prepared cavity, into which it is cemented. The cavity is so designed that the inlay is in a high degree self-retentive even before it is cemented. Although the cement aids retention, its primary purpose is to seal the crevice between gold and tooth against the ingress of fluids.

In conformity with the general principles of bonding one solid to another, the thinner the layer of cement between the inlay and the tooth, the stronger the bond. Moreover, the thinner the line of cement exposed at the inlay margins, the slower will be the rate of solution of the cement in the oral fluids. It will therefore be apparent that a very close fit between the inlay and the cavity is one of the most important requirements of the satisfactory inlay.

The particular advantages of gold over amalgam are its relative inertness in the mouth, its greater strength, and its malleability. The yellow, lustrous appearance of gold, though generally undesirable in easily visible parts of the mouth, is more acceptable than the colour of amalgam. The colour of gold is permanent and it does not discolour the tooth tissue in contact with it. If gold alloys of suitable physical properties are used, the amount of wear and deformation occurring under normal occlusal stress is insignificant and the restoration is less liable to marginal failure.

The advantages of gold over amalgam are most easily demonstrated in the medium- or large-sized compound restorations. *The greater the extent of coronal destruction, the more important does it become to restore that part of the crown which is missing and to protect the remaining tooth substance by the use of gold.*

In the Class I and Class V types of restoration, with which this chapter deals, the differences between gold and amalgam are not so outstanding. It is nevertheless true to say that on balance the gold restorations should be more pleasing in appearance and more durable in function than amalgam. These inlays are not widely used in practice and their inclusion here is justified simply to illustrate the basic principles of theory and technique of inlay work, which will be further elaborated in following chapters. The greatly increased cost of gold in recent years has, however, forced upon us a reassessment of its use in all forms of restorative dentistry. No doubt cheaper materials and techniques now being evolved will come to fruition in due course.

Gold inlays may be constructed by the direct or the indirect technique. In the direct technique a wax pattern is constructed in the cavity prepared in

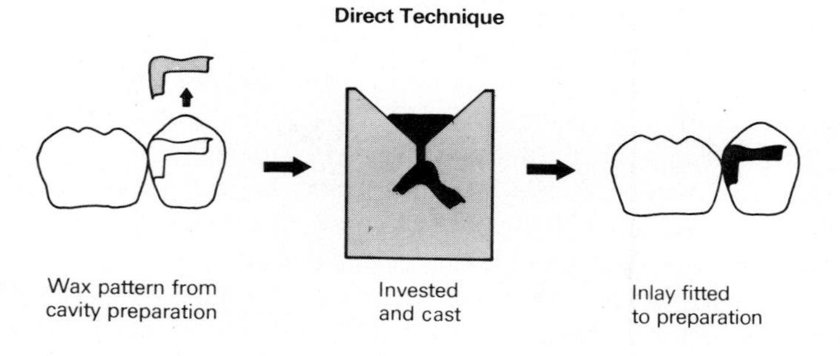

Fig. 9.1.

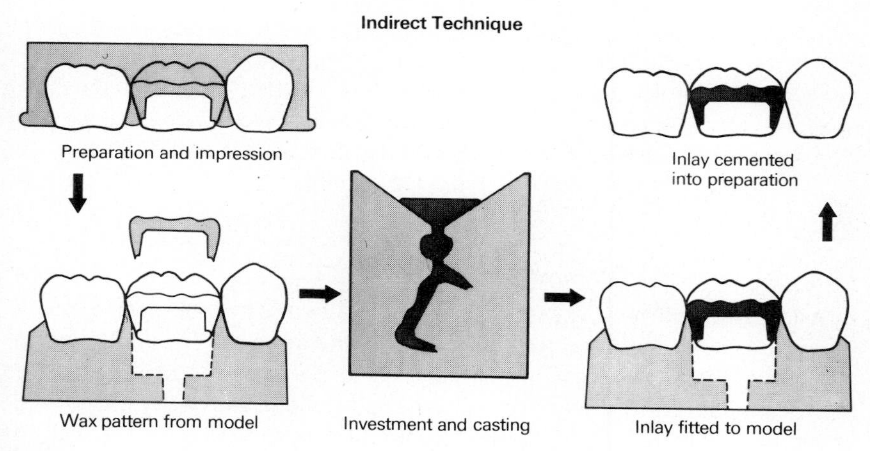

Fig. 9.2.

the tooth (Fig. 9.1). In the indirect technique an accurate model of the prepared cavity is constructed and an inlay made to fit the model; the inlay is subsequently cemented into the tooth (Fig. 9.2). This involves two further processes, of impression and model production, compared to the direct method. On account of the possibilities of error which these involve, it is sometimes claimed that the indirect method results in a less accurate inlay than does the direct method, and it is not difficult to see how this could be so. In the direct method volumetric changes and loss of detail may take place in the wax pattern, the investment, and the casting process. In the indirect method there are the added changes in the impression and model materials.

To summarize the situation, it is possible to make durable restorations of excellent function and good appearance by both methods. The simpler the type of restoration, the more are the indications in favour of direct technique. The more complicated the design and the more inaccessible the tooth, the more are the indications in favour of indirect techniques, because the preparation of a wax pattern is more easily and better carried out on a model. In addition, much of the indirect procedures can be deputed to a skilled technician, thus providing a saving of chairside time.

The direct inlay techniques described in this and the next chapter show the essentials of all inlay techniques and are in themselves suitable for comparatively small and uncomplicated cavities. They are, by any standard, valuable restorative work, and from them may be learnt the truth that, in the sequence of operations from cavity preparation to completed inlay, each stage must be correctly performed if the final result is to be satisfactory. It may be said that this statement applies to all operative work. This also is true, but the results of failure in inlay procedures are usually much more obvious than in plastic techniques. When a fault develops in inlay production it is essential to trace and remedy the defect before proceeding. For this reason, inlay work in general requires more meticulous application to detail and better understanding and control of the materials.

Furthermore, not all dental surgeons have skilled technical assistance available, in fact in many parts of the world no such facilities exist. In these circumstances the direct technique is of the greatest value.

Class I cavity

Principles of cavity preparation

Consideration of the Class I cavity shows the two principal differences between the cavity designed for an inlay and one intended for amalgam.

The first main difference is that, since a wax pattern has to be constructed in the cavity and withdrawn from it to be cast in gold, the cavity must not be undercut in the line of withdrawal. In practice, this means that the walls of the ideal cavity must taper; they must diverge occlusally and converge apically. The angle between the converging walls, the angle of taper, should be about 5 degrees, but it could vary 2 degrees on either side of that value, giving a permissible range of 3 to 7 degrees. Figure 9.3 shows what tapers in this range actually look like. It is helpful to know that tapered fissure burs are in the range 5 to 7 degrees and most long diamond instruments have a taper of 3 degrees.

Transferring this now to the theoretical cavity in a molar (Fig. 9.4), the straight walls form an angle of about 92.5 degrees with the floor which, in this case, is roughly normal to the line of withdrawal. This represents the optimal taper for a simple cavity. Referring again to Fig. 9.3, it will be clear

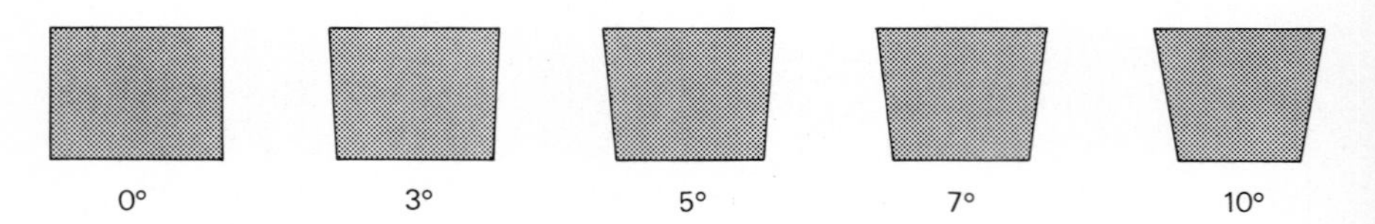

Fig. 9.3. Diagram illustrating tapers from 0 to 10 degrees.

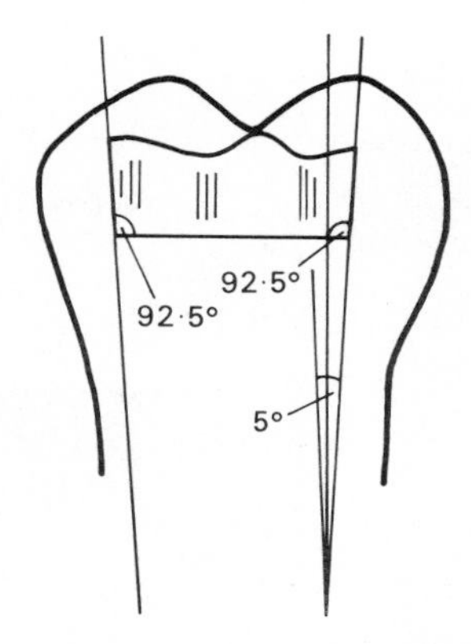

Fig. 9.4. Sectional molar Cl. I cavity showing angles 92.5 degrees and 5 degrees ± 2 degrees.

that working freehand as we do with burs and diamond instruments, to reduce the taper below 5 degrees though greatly increasing retention would risk an undercut, perhaps undetected until removal of the wax pattern proved difficult or impossible. To increase it beyond 7 degrees would seriously reduce retention. It is within this small range that the most retentive restorations must lie.

The term 'taper', borrowed from a more exact discipline, may imply precision and uniformity to which our freehand cavity preparations cannot lay claim. The alternative term 'near-parallel' is less precise, perhaps more descriptive, and is a useful reminder that the nearer to parallel, the greater the retention. Both concepts serve a purpose in the consideration of these principles and their practical applications.

Associated with the angulation of the walls, the depth of the cavity obviously has a direct bearing upon retention. The deeper the cavity, the greater the retention. These two properties may be brought together by stating that in cavities of equal depth the most retentive shape is the one which has the longest single line of insertion (Rosenstiel 1957).

In Fig. 9.5 overleaf the first diagram shows a cavity with widely divergent walls. An inlay cast to this shape could be inserted along a variety of axes lying between the lines AB and CD. In the second diagram part of the cavity walls, AX and CY, are now parallel. Over a considerable part of its travel the inlay can still be inserted along a number of axes between AB, and CD, but from the points A and C its axis of insertion must be a single one, parallel to the walls AX and CY. The third diagram shows an extension of this principle, in which the inlay must have a single line of insertion for nearly the whole of its travel AX. This is clearly the most retentive form. If divergent walls must be accepted, all the more is the *length* of the insertion important. The fourth diagram makes the point that the shallow cavity ABDC is less retentive than the deeper XBDY, though the taper of the walls is the same in both cases.

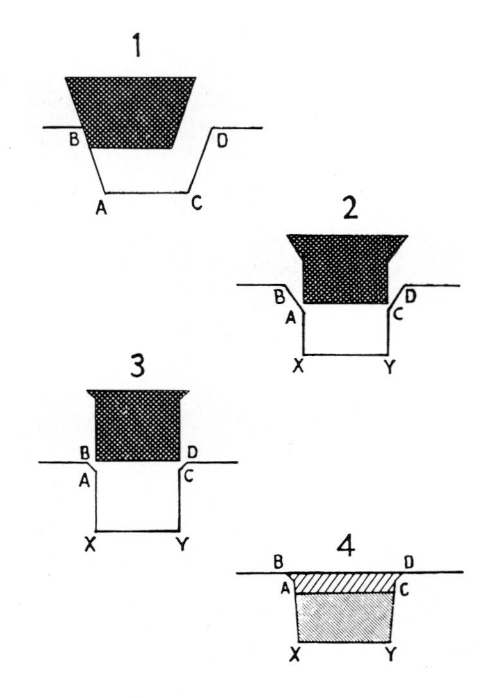

Fig. 9.5. Diagrams to demonstrate the longest single line of insertion; *see* text (adapted from Rosenstiel).

These principles, simple and obvious though they may be, apply to the retention of all fixed restorations, from the smallest direct inlay, through all types of crowns, to the most extensive bridge.

The second main difference between the cavity for amalgam and that for an inlay relates to the cavosurface angle of the occlusal margin which is bevelled. Taking advantage of the ductility of gold the most effective margin is one thin enough to be burnished, spun, or swaged into contact with a robust enamel margin. On the other hand the gold must not be too thin otherwise, over years of use it will wear and break away. It has been found empirically that to be most effective the gold margin should have an angle between 30 and 45 degrees (Fig. 9.5 (a)), which corresponds with a cavosurface angle of 150 to 135 degrees. We have seen, however, as in the Class I amalgam cavity, that the cavosurface angle rarely *is* 90 degrees. If due to the slope of the cusp the cavosurface angle is already 135 degrees (Fig. 9.6 (b)), the angle of the gold margin will be 45 degrees, and any further attempt at bevelling will produce an ill-defined, friable margin, difficult to reproduce in wax, difficult to finish in gold and vulnerable to wear.

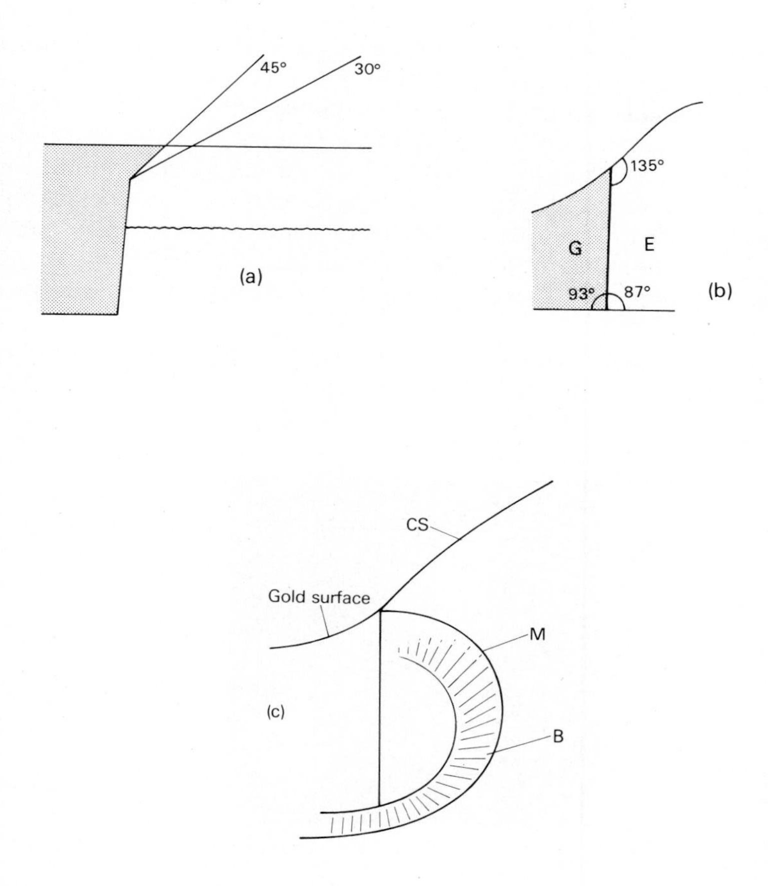

Fig. 9.6. (a) Diagram showing 30 degrees and 45 degrees marginal bevels on a flat surface. (b) If the cuspal slope is 135 degrees the gold margin of 45 degrees requires no bevel. (c) Diagram of 45 degrees cavo-surface bevel merging with the cavity wall where the cuspal slope is steep. CS Cuspal slope; G eventual level of gold surface; B bevel; M inlay margin.

It follows that in forming the occlusal bevel the cuspal slope must be taken into account. There may well be places where the bevel runs into the cavity surface as the cavosurface angle exceeds 135 degrees (Fig. 9.5).

The preparation of the occlusal cavity proceeds in all respects as described in Chapter 6, p. 80, with regard to outline, resistance form, and removal of caries and lining. In the inlay cavity the floor should be level and smooth. The floor–wall line angle should be sharp and clearly defined; this is clearly important for it directly affects the taper of the walls and the retentive depth of the cavity. The two factors which increase retention are the near-parallelism of the walls and the greater depth of the cavity. The best instrument for defining the walls is tapered plain-cut fissure bur of tungsten carbide used at high speed.

As with finishing enamel margins in general, so with bevelling, a number of instruments can be used, but rotary instruments are most favoured where margins are accessible to them. With the application of the scanning electron microscope to the study of margins, of restorations as well as cavities, much has been learnt, but the clinical significance of all that can be seen at high magnification is not by any means clear. The general concensus at present, so far as the use of rotary instruments is concerned, seems to be as follows.

1. Two instruments commonly used for many decades, the steel finishing bur and the green carborundum point are unsatisfactory, the bur because it

Fig. 9.7. Multi-bladed tungsten carbide finishing bur. (Courtesy of Dr I. E. Barnes.)

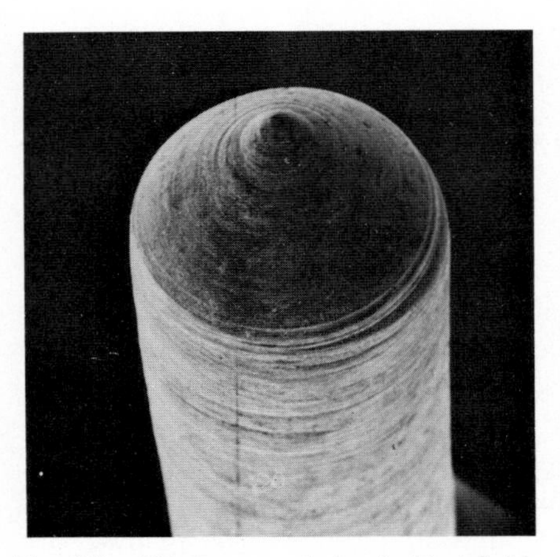

Fig. 9.8. 135 degree-ended cylindrical finishing bur of the Baker–Curson type (by permission of *Br. dent. J.*) see also Fig. 10.9.

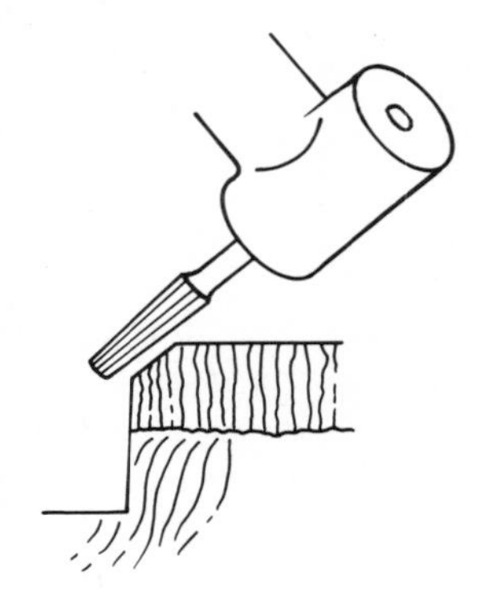

Fig. 9.9. Finishing bur held at a correct angle to form the appropriate bevel of 30 degrees to 45 degrees.

blunts so rapidly and then ceases to cut efficiently, the green point because it leaves a very rough surface.

2. The finest diamond instruments also leave a roughly scored surface.

3. A plain cut tapered tungsten carbide bur such as would be used at medium or high speed to cut the cavity, leaves an acceptable margin but of course it cuts very rapidly and is therefore difficult to control for bevelling.

4. A multibladed tungsten carbide bur (Fig. 9.7) used at high speed gives a good margin but cuts fairly rapidly.

5. A blank tungsten carbide bur (the Baker–Curson bur previously referred to), cuts very slowly at high speed and so can be effectively controlled for bevelling (Fig. 9.8).

All high-speed cutting must be water cooled and the eventual choice of instrument will probably be determined by the preference of the operator and the equipment available. The bur, of whatever type, is held at an angle to the cavity margin (Fig. 9.9) in such a manner that the gold resulting will contain an angle of between 30 and 45 degrees, bearing in mind the variations in the cuspal slope as previously described. The cavity form and outline, and the smooth finish of all surfaces are finally reviewed before proceeding to the wax pattern to ensure that they conform to the principles stated.

The wax pattern

For the purpose of taking an impression of this cavity a small cone of inlay wax is required, about 10 to 20 mm long and 6 mm wide at its base. This is formed by gently warming the end of a stick of inlay wax above a spirit or gas flame. It is important that the wax should not be over-heated to the extent of melting it till it drips. The end of the stick is moulded to pointed form between finger and thumb and a portion of the required length is then cut off and allowed to cool.

To prevent the wax sticking to the prepared surfaces a thin film of separating fluid, usually a weak aqueous solution of glycerol and a detergent, is applied to the cavity. Alternatively, a thin smear of liquid paraffin may be used; this must be carefully washed off the pattern with the detergent used before investment. Moisture alone is an unreliable separating medium.

The wax cone, held at the base by the fingers, is now gently warmed above the flame and this proceeds until the point of the cone is almost molten, whilst the base remains firm. The cone is then rapidly transferred to the mouth and inserted straight into the cavity along the axial line of the tooth (Fig. 9.10). Firm finger pressure carries the softened wax into all parts of the cavity and the patient is asked to bite until his teeth meet in centric occlusion. This exerts even greater pressure upon the hardening wax and, if the patient 'grinds' a little from side to side, helps to define the occlusal form of the wax.

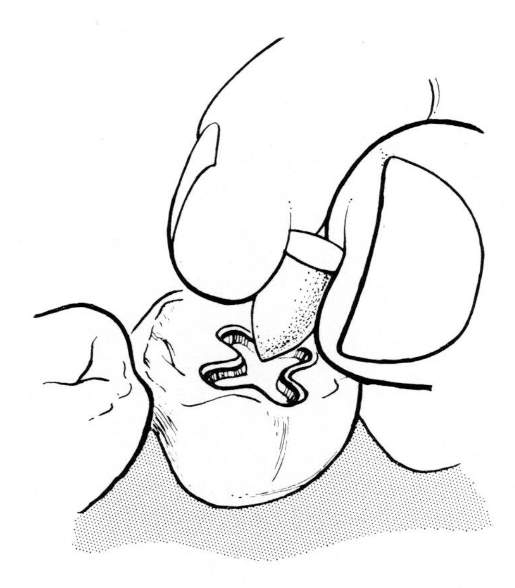

Fig. 9.10. The insertion of the wax cone into an occlusal cavity.

When the wax has cooled to mouth temperature, carving of the occlusal surface may start. Gross excess may be removed with a warmed flat plastic instrument, and the finer carving proceeds with a Ward's carver No. 1 or 2. Great care should be taken of these carving instruments; the edges of the blades should be even, smooth, and sharp. They should never be heated or used for adding wax, for which their shape is unsuitable. For cleaning they should only be lightly warmed.

As described in the carving of amalgam, the blade of the instrument as it

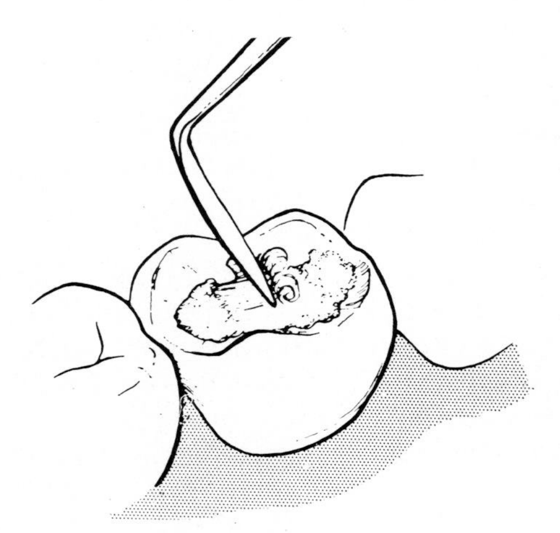

Fig. 9.11. The use of the Ward's carver to define the occlusal outline of the wax pattern.

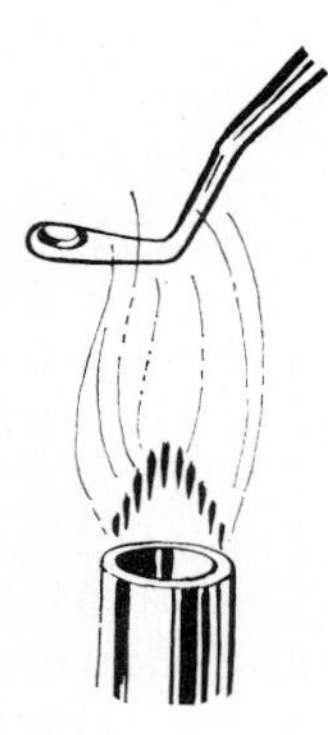

Fig. 9.12. When melting wax on a blade, the junction of the blade and shank should be the hottest point.

moves along the cavity outline lies upon and is guided by the occlusal enamel surface (Fig. 9.11). As the cavo-surface bevel is reached the transparent margin of the wax is seen to thin to a feather edge, which must be carefully preserved. If the surface is carved beyond this point, a little molten wax must be flowed on to the deficiency. To do this, the wax and tooth must be dry and, in the upper jaw, some difficulty may be found due to the gravitational flow of the wax away from the blade towards the shank (Fig. 9.12). It is therefore important to avoid over-carving and to reduce as far as possible the need to replace with molten wax. It is held that addition of molten wax, particularly to an unrestrained pattern, can lead to distortion of the remainder due to the formation of internal stresses, and for this reason the additon of molten wax should be kept to a minimum. Any added wax must chill to mouth temperature before carving proceeds; for wax may be carved when hard set, or with due precaution flowed on in the liquid state. *It should not be worked in the intermediate plastic condition.*

The outline form being clearly and accurately defined, the fissure pattern can now be carved with the round end of carver No. 2, with a small flat plastic Ash No. 79, or with a pear-shaped excavator No. 220/221. The bottom of the fissures and pits remain rounded and cuspal slopes are brought to a smooth finish.

The final smoothing of the pattern may be done in either of two ways. A small tightly rolled pledget of cotton wool, held in tweezers, is lightly moistened in eucalyptus oil. Any excess must be removed by touching it on a gauze square. This is used with a light, circular movement on the wax surface, *avoiding the actual margins*, until a smooth and even surface is obtained. This method has two minor disadvantages: first, it leaves the surface slightly softened and, second, it can remove wax from the feather edges unless care is taken.

The other method is to moisten a pledget with water and heat it to near boiling over the flame. This can be used with a similar circular movement to smooth out fine irregularities. It only softens the surface temporarily and does not remove any wax. As the pledget cools rapidly, several applications may be required.

Spruing the pattern

The 'sprue', which should correctly be called the sprue former, is used in this type of inlay for the removal of the pattern from the cavity. The sprue is a piece of straight round wire of 1 mm diameter, reduced to a blunt point at the end which is to enter the wax. It should be about 15 mm in length.

In order to reduce the heat transferred to the pattern in the process of spruing, and so to reduce the risk of warpage and deformation of the carved surface, a sprue of 1 mm stainless-steel tubing may be used. This reduces the heat applied and some of the molten wax displaced flows up the lumen of the tube.

The sprue is warmed in the flame and a very small quantity of inlay wax melted on to it. The behaviour of this wax gives a visible indication of the heat of the sprue, which is now held between finger and thumb at such an angle that it can be inserted axially into the pattern. It is gently warmed over the flame and inserted into the centre of the pattern. If it is too hot, undue melting of the pattern results, with loss of detail of the carving. *It must not be inserted deep enough to touch the cavity floor.* It is held motionless in position till it cools and becomes firmly embedded. The pattern and sprue are chilled with room-temperature water and, holding the sprue, preferably with the fingers, the pattern is gently withdrawn from the cavity along the axial line (Fig. 9.13).

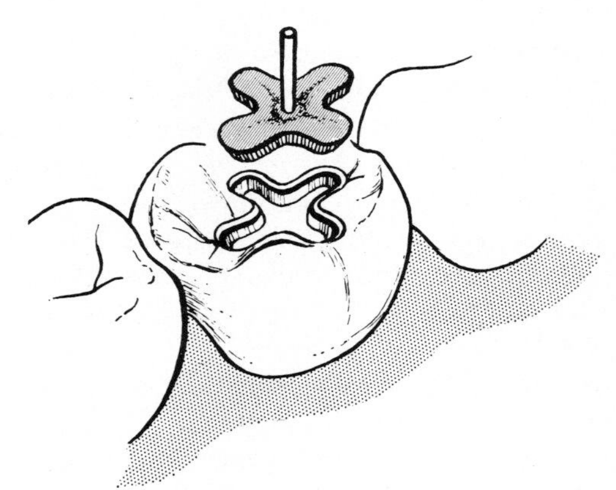

Fig. 9.13. The removal of the sprued wax pattern from an occlusal cavity.

Removed from the mouth, the pattern should now be inspected for detail; the floor–wall line angle should be sharply reproduced throughout and it should be possible to trace the cavo-surface feather edge around the whole periphery, relating this to the bevel imparted to the outline.

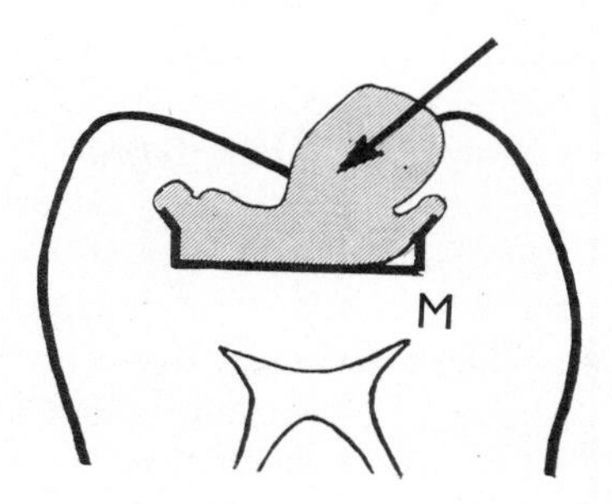

Fig. 9.14. The insertion of the wax cone with a mesial inclination can result in the rounding of the pattern at M the mesial floor–wall angle.

If the impressed surface does not show full detail, then either the wax was not warm enough, or it was not forced home whilst still plastic, or the *direction* of the pressure was incorrect. For example, if the cone were to be inserted with a mesial inclination (Fig. 9.14), this might result in the rounding of the wax in the mesial floor–wall angle. Axial insertion and correct pressure obviate this fault.

Should difficulty arise in the withdrawal of the sprue and pattern, with the result that the former comes away without the latter, either the sprue was not hot enough on insertion, or the cavity is undercut somewhere along its walls. If an undercut is present there is no alternative to removing the wax with an excavator, locating the fault, and correcting it by taking the wall back with a suitable tapered bur before repeating the pattern.

Investing the pattern

In preparation for investing, the sprued pattern is now placed in the crucible former and the length of the sprue adjusted so that the pattern will come to the centre of the investment mass (Fig. 9.15). It is firmly fixed in this position.

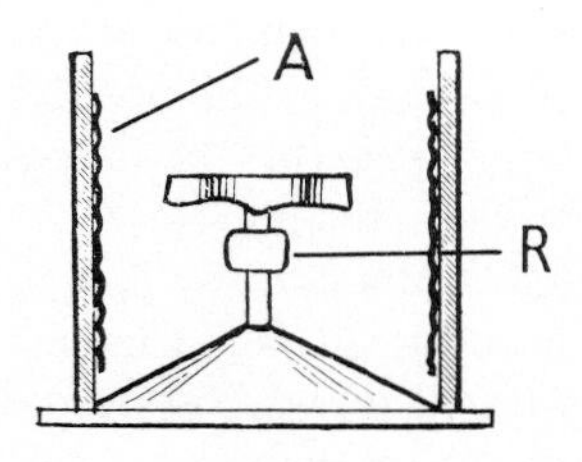

Fig. 9.15. The length of the sprue is adjusted to bring the pattern to the centre of the eventual investment mass; a reservoir, R, is added to the sprue. A: asbestos liner.

For an inlay of the size of the medium Class I cavity the use of a gold reservoir is necessary. Using a warmed plastic instrument, wax is run on to the sprue to build a spherical mass of wax, about one-third of the bulk of the pattern. At its nearest point it should approach the pattern to within 3 or 4 mm. Care must be taken not to touch or to distort the pattern.

Another way (Fig. 9.16) of building a reservoir before the sprue is set in the crucible former is to wrap a narrow strip of pink casting wax around the sprue near the free end. When the correct size is reached the wax may then be slid up the sprue to the correct position and sealed firmly there with a warm instrument applied to the side farthest from the pattern. This is perhaps an easier method and the reservoir can be placed close to the pattern without risk.

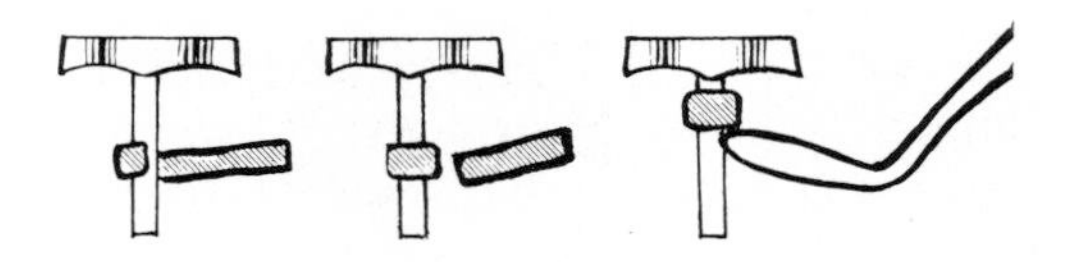

Fig. 9.16. A method of applying a reservoir with a strip of pink wax.

During all manipulations of the sprue and pattern it is safest to hold the sprue firmly between pointed pliers, and the greatest care must be taken to avoid damaging the delicate feather edge.

The sprue, reservoir, and pattern now being firmly mounted on the crucible former, the pattern is gently painted with a weak detergent solution, using a soft camel-hair brush. This wets the wax surface and should also remove any trace of oil if the cavity has been lubricated before insertion of the wax.

The inlay investment powder is now mixed in a rubber bowl, using 10 ml of room-temperature water and 28 g of investment powder, or such other proportions as are recommended for the investment chosen. The addition of the powder and the spatulation of the mix should be carried out in such a way as to reduce to a minimum the inclusion of air and the formation of bubbles in the mix. A slow sprinkling of the powder into the water and the use of a mechanical spatulator probably produces the most satisfactory mix.

With a clean camel-hair brush the creamy mix is applied very lightly to the pattern and the reservoir until all surfaces are covered with an even layer. With a dry brush investment powder is sprinkled on to this layer, the crucible former vibrated, and further powder added until the adherent layer is considerably more viscous than the main mix. This will prevent the displacement of this applied layer when the main mix is added. The actual increased proportion of powder resulting is insufficient to affect the properties of the mix as a whole.

The crucible former is placed on the bench and the casting ring, lined with wet strip asbestos, placed in position. The asbestos strip is narrower than the height of the casting ring, and it should be adjusted on the inner surface so that a small margin of the inner surface of the ring remains uncovered at both ends (Fig. 9.15 A). In this way the investment is held securely in the ring and, if pressure casting is used, no loss of pressure occurs. The ring is now filled with investment by tipping and tapping the bowl, again with due care to avoid air inclusions.

As an alternative to the investment core technique a vacuum technique may be used. The mix of investment is painted on the wax pattern with all due care, and the ring is filled. The complete assembly is transferred to a vacuum chamber and exhausted to 750 to 760 mm of mercury. This has the effect of enlarging air bubbles which may have been formed and allowing them to float to the surface. It affects particularly those bubbles which may have been left on the upper or fitting surface of the pattern.

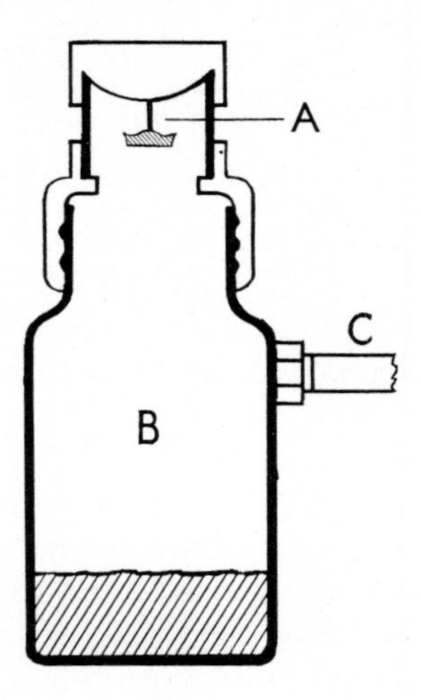

Fig. 9.17. Form of vacuum investing apparatus. A: sprued inlay mounted in casting ring. B: bottle containing mixed investment. C: to exhaust pump. When the bottle is tipped and vibrated, the investment fills the ring.

Many forms of vacuum-investing apparatus have been described. One of the simplest and most effective (Morrant 1956–7) allows the mixed investment to be poured into a bottle (Fig. 9.17), which is then closed and exhausted. By tipping the bottle the investment is poured into the casting ring, still under vacuum. This process is assisted by an electric vibrator and completely eliminates any bubbles formed in the mix. The vacuum is released as soon as the ring is filled. There are also various machines available which combine mixing and pouring under vacuum, in one process Fig. 9.18. It

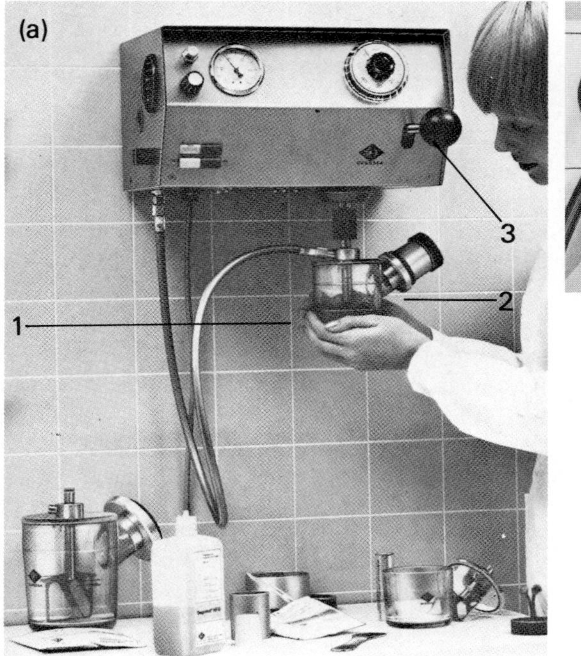

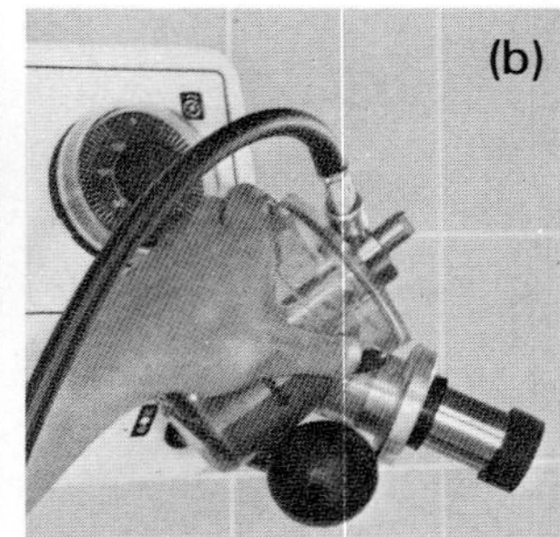

Fig. 9.18. Degussa vacuum investing apparatus.
(1) investment material mixed under vacuum;
(2) investment ring containing inlay pattern;
(3) vibrator used when filling the ring with investment, B.

should be understood that a good operator may get equally good results by investing without vacuum as with it. The latter method is more certain in its results.

When the investment has set for thirty minutes, the crucible former is removed and the free end of the sprue heated in a flame. Holding the ring with the *crucible downwards*, the sprue former should now be gripped with pliers and removed with a downward rotating motion, taking care to remove any loose flakes of investment round the orifice and the margin of the ring before it is brought into the upright position. The ring is now transferred to a drying furnace for ten minutes at 150 to 200 °C and then placed in the furnace at 750 °C for 25 minutes before being transferred to the casting machine.

For an occlusal inlay of this type, 2.5 g of medium hard 20-carat gold would be chosen. This allows 1 g of gold for the inlay and 1.5 g for reservoir and sprue. Whichever type of casting machine is used, preheating of the gold may be used but over-heating is to be avoided. The properly sited reservoir eliminates the risk of porosity in the inlay around the point where the sprue enters, such as might otherwise be caused by the contraction of the cast mass.

The casting may be allowed to cool, or it may be quenched in tap water according to the physical properties required of the completed inlay. With a fine stiff brush all traces of investment should be removed. It may now be heated to well below red heat and dropped into a pickling bath of 50 per cent hydrochloric acid. This procedure may have to be repeated before the gold is clean. It is then washed and inspected.

At this point it is not out of place to reinforce the principle that, whenever an inlay or other small object is washed in a sink without a tray, and this may sometimes occur in the surgery, always insert the waste plug *before* the inlay is dropped.

The cast inlay should now present all the details of the wax pattern and it may be inspected under a lens for any sign of imperfection. The gold should look clean and unblemished by carbon particles. The fitting surface should show no sign of small spherules, the result of bubbles in the investment. If these should occur on the occlusal aspect, they can be removed with ease at a later stage.

If small 'air-blows' are detected on the fitting surface, they may be removed with the blade of a sharp excavator. No further adjustment of this surface is allowable. Finally, the excess button is removed by cutting the sprue with a fine metal saw 3 mm above the inlay, so that a small tag of gold is left to assist in handling.

Fitting and cementation of the inlay

When removing the temporary dressing from the prepared cavity, it is necessary to ensure that all particles of the dressing are removed. **To do this the cavity should be cleaned with an atomizer spray, dried, and inspected from all angles in a good light, paying particular attention to the floor–wall line angle**. If the cavity is clear the inlay may now be inserted and should seat itself with only moderate pressure.

With a fine pointed probe the margins are now explored. The gold feather edge covering the enamel bevel should be intact and the probe encounters only the smallest catch when passing either way across the margin. The cast is tested for rocking, the result of distortion of the pattern, by stabilizing one end with a heavy pressure. The other end is then pressed in an attempt to elicit a rocking motion.

If the technique has been carefully carried through, virtually no finishing of the margins should be required. There is everything to be said, however, for bringing the marginal gold into the closest possible apposition to the enamel. If a medium soft gold, such as 20 carat, has been used, a small contra-angled point in an automatic mallet may be used to apply a series of light blows, in a contiguous series, along the gold margin at a distance of 1 mm from the edge of the gold. It is important that only moderate force should be used, and that the point should not be applied at the margin or on the adjacent enamel. This operation achieves a very good apposition of the marginal gold.

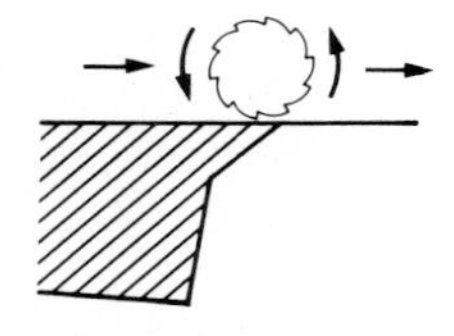

Fig. 9.19. All rotary instruments must rotate from gold to enamel surface.

A fine steel finishing bur is now used with a light touch along all the margins. This bur must rotate in such a direction that, at its point of contact, it passes from gold to enamel (Fig. 9.19). This is the process of 'spinning' and uses the ductility of the gold to maintain close adaptation of the margin. If the wax pattern was correctly finished, there are very few surface defects to be removed, and the margins may now be smoothed with a rubber-pumice disc.

It should be recognized that **this is the stage when adaptation of the margins must be brought to a high degree of perfection**. Whilst 'burnishing' of the margins may still be possible during cementation, the margins are obscured and the time available is too short to carry out this operation effectively. *To attempt to adapt margins after cementation is futile and, if it were effective, could only result in the fragmentation of the cement lute at its margin.* This is clearly most undesirable.

Using the base of the sprue for purchase, the inlay is now removed from the cavity and prepared for setting. Holding the inlay between the finger and thumb, or with the help of an inlay holder (Fig. 10.16) the remains of the sprue can now be removed by the use of small mounted carborundum wheels or the round finishing bur. This stage may be delayed until the inlay is set, but is perhaps better done before setting, though this choice may be left to the operator, particularly if the inlay is rather smaller than normal.

The instruments, mixing slab, and bottles of cement are made ready for mixing. The cavity is again washed with the atomizer spray, isolated with cotton rolls, and the saliva ejector placed in position. The crown of the tooth and the surrounding area is dried with pledgets of cotton wool and a jet of warm air, just sufficient to remove surface moisture. The cavity is lightly wiped with a pledget moistened with chloroform; this removes traces of oil or wax remaining from the earlier temporary filling.

The preparation of zinc phosphate cement of the correct consistency and setting time is a matter of importance. Too thick a mix and too rapid setting time may prevent the inlay from seating home. Too slow a set will require a long period of isolation from contact with moisture. A thin mix results in cement which is weak and more liable to solution in oral fluids.

The slab is best used at room temperature or only slightly chilled and the mixing proceeds at a slightly slower rate than normal. This means that smaller additions of powder are incorporated and the setting time is slightly prolonged. The mix is a little more fluid than normal; it should almost drop from the spatula.

If significant areas of fresh dentine have been opened in the process of preparing the cavity and still remain unprotected by lining, EBA or polycarboxylate may well be the cement of choice. The grain size, flow, strength, and solubility seem to lie within satisfactory range for practical purposes.

With a small flat plastic instrument cement is applied to cover the fitting surface of the inlay and the cavity surface with a thin even layer. The inlay is then carried in forceps and correctly positioned in the cavity. A contra-angled burnisher or other blunt instrument is used to seat the inlay firmly in position and this is checked by quickly passing a probe over its margins to confirm their close adaptation.

It is advisable to maintain pressure upon the inlay until the cement reaches an initial set. This may be done by continued pressure with an instrument, or, if the mouth can be closed, by placing a small piece of unvulcanized rubber, silicone rubber, or rubber dam on the occlusal surface and having the patient maintain firm biting pressure for about two minutes.

When the cement is hard set, the excess may be cleared away with a probe and finishing commenced. First the occlusion is checked with articulating paper and any slight excess at the site of the sprue removed. By the use of small finishing burs the definition of fissures may be increased and any small irregularities removed. Polishing proceeds by the use of small abrasive rubber wheels and points in grooves and fissures. The rounded contours of cusps and interproximal surfaces are best finished with medium, then fine, abrasive discs.

The final polish is applied with Tripoli in a soft rubber cup followed by rouge on a soft mop. As an alternative to the use of rouge, a final lustre can be obtained by hand-burnishing with a clean and highly polished burnisher.

Class V cavity

Cavity preparation

The Class V inlay bears the same relation to the Class V amalgam cavity as exists in the case of Class I cavities (Fig. 9.20 (a)).

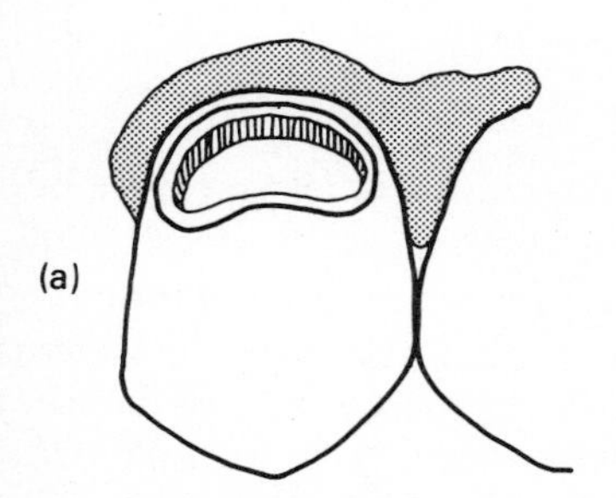

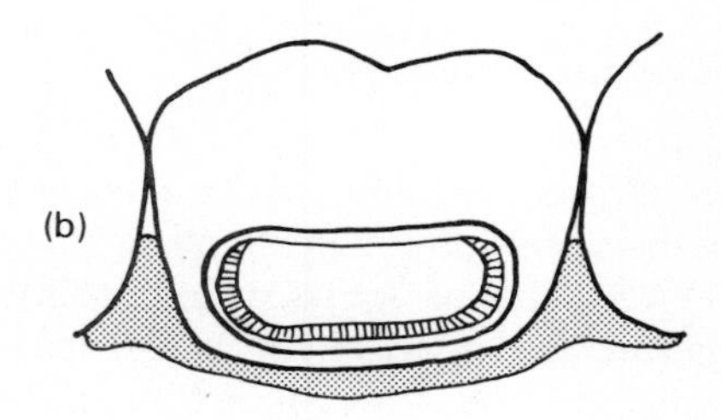

Fig. 9.20. (a) Typical Class V cavity in premolar prepared for inlay. (b) An extensive Class V preparation for an inlay. The mesial and distal walls are widely divergent; elsewhere they are nearly parallel. The cavo-surface angle is bevelled to 45 degrees.

In the cervical cavity the floor is evenly curved rather than flat; the floor–wall line angle is as close to 92–93 degrees as is possible but the augulation of the cavity walls presents this problem. The occlusal and gingival walls must carry a 5 degree taper, but as the walls approach the mesial and distal extremities of the buccal surface they increasingly diverge from one another and so cease to contribute in any way to retention. It is therefore of first importance that the occlusal and gingival walls be cut with full regard to depth and taper in order to provide greatest retention.

The factors controlling outline form and the management of the gingiva have been discussed in Chapter 8. In inlay technique the interval between the taking of impressions or pattern, and the insertion of the inlay, should be sufficient to allow complete healing of any slight injury inflicted upon the gum. This being so, the problem of maintaining dryness during cementation is greatly simplified.

It is difficult to distinguish between resistance and retention form. There is no direct occlusal force applied to this restoration. Probably the most effective dislodging force is that provided by tooth-brushing, particularly brushing in a transverse direction. To this type of lateral force the depth of the cavity is the most effective means of resistance and near parallelism of the walls assists effective retention.

The depth of the cavity is limited by anatomical considerations, chiefly the position of the pulp. The lining of these cavities is a matter for careful judgement. Not infrequently they are very sensitive to thermal change and sometimes to chemical irritation. The extent of likely areas of sclerotic dentine and the amount of fresh dentine necessarily exposed in preparation are material factors in the assessment of each case.

Wherever the depth of the cavity allows, a non-conducting cement lining should be placed in the floor of the cavity. This may best be done before the completion of the gingival margin, since the use of instruments on this margin must necessarily give rise to bleeding.

The marginal bevel is again achieved by the use of fine rotary instruments as previously described. The gingival margin, however, may be bevelled with the cervical margin trimmer or a hoe, using a light pulling stroke, if access and the level of the free gingival margin allow.

The wax pattern

Before the wax pattern is attempted, gingival bleeding and seepage must be controlled. In some cases it may well be justified to dismiss the patient with a temporary dressing, to allow the gum to heal for seven to ten days.

All traces of debris or temporary dressing are removed from the cavity before the insertion of inlay wax. The cavity may be moist, or lightly lubricated. A wax cone of size and shape suitable to the cavity is fashioned from a stick of wax. If the cavity is elongated, as for example in a molar (Fig. 9.20 (b)), the cone can be laterally compressed to adapt more closely to the shape of the cavity.

Fig. 9.21. The use of a spatula blade for the insertion of a wax cone into a buccal cavity in the lower molar. This is followed by finger pressure.

With cavities in posterior teeth it is often difficult, when the cone is held in the fingers, to find sufficient space to carry the cone into the cavity at the correct angle. In this case the base of the cone may be lightly melted on to the blade of a warm spatula, cooled, and the point of the cone rewarmed for the impression (Fig. 9.21).

As the cone is carried to position and pressed into the cavity, the spatula is removed and the wax rapidly forced into position under pressure of a fingertip. This pressure is maintained until the wax cools to near mouth temperature.

One of the chief difficulties in carving this type of wax pattern is the tendency for the pattern to become dislodged from the cavity. This is, of course, directly related to the general lack of retention of the design as already discussed. The removal of excess wax is best achieved with a warm flat plastic instrument. Thereafter a Ward's carver is used to reduce the surface to correct contour. The stability of the pattern may be achieved by leaving for the time being some excess of wax at an easily accessible margin, for example, the occlusal margin, and by supporting the pattern with a small moist pledget of cotton wool held in tweezers.

The surface of the wax must not be left too convex, particularly in single-rooted teeth. In fact it helps the cleaning of the area to have it slightly flatter than the normal convexity. Care is exercised to retain the marginal feather edge of wax covering the bevel. If the cervical margin encroaches upon the area of concavity leading to the bifurcation of molar roots, this feature must be faithfully reproduced. The final finishing of the pattern is carried out using eucalyptol or hot water on a cotton pledget as described on p. 171.

Spruing the wax pattern

The smaller cervical wax patterns common in anterior and premolar teeth should be sprued with a smaller gauge of wire than those in molar teeth. A large-size domestic pin, with the point cut off, is suitable for the former; it is about 0.5 mm thick. For larger inlays 1 mm wire can be used and its length is adjusted to allow for the limitation of space, due to the proximity of the cheek or the tongue.

In posterior teeth the sprue is held in forceps when being placed, and great care is necessary to ensure that the sprue is not moved during the cooling period, for this would dislodge and distort the pattern.

Immediately upon removal the pattern is inspected for detail, and to detect any adherent blood which may be on or near the cervical margin. If blood is present, it must be washed off with room-temperature water before it clots. If this is not done immediately, although the blood seems to disappear, careful inspection shows the persistence of a fine fibrin clot. This clot leads to blurring and distortion of the margin to which it adheres, and it is unnoticed until the pattern is cast.

Investing and casting

The general procedure is as described in connection with Class I patterns and should proceed without delay after removal from the cavity. Usually Class V patterns are of smaller bulk; many are considerably so. For this reason a reservoir is often unnecessary, but if it is dispensed with the sprue should be mounted on the crucible former so that it is short, not more than 5 mm. The excess gold in the crucible then acts as a reservoir (Fig. 9.22).

As the cervical inlay is not subjected to direct occlusal stress, a softer gold may be used—22-carat gold is suitable and allows easier marginal adaptation.

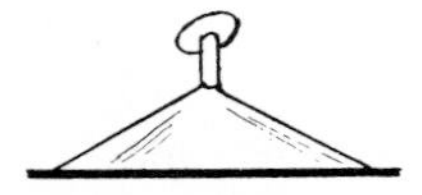

Fig. 9.22. Small cervical inlay mounted on short sprue, without additional reservoir.

Cementation and finishing

A short length of sprue is retained on the inlay to facilitate handling. Inlay and cavity must be scrupulously clean and on insertion the inlay should seat accurately in place.

The relatively shallow inlay will again give some difficulty in keeping its position during adaptation of its margins. In such a case the inlay must be steadied with a firm instrument, a blunt scaler or an old cohesive gold instrument serve the purpose, while the margins are adapted and smoothed (Fig. 9.23). As with all inlays, the more finely finished the pattern, the less finishing is required at this stage.

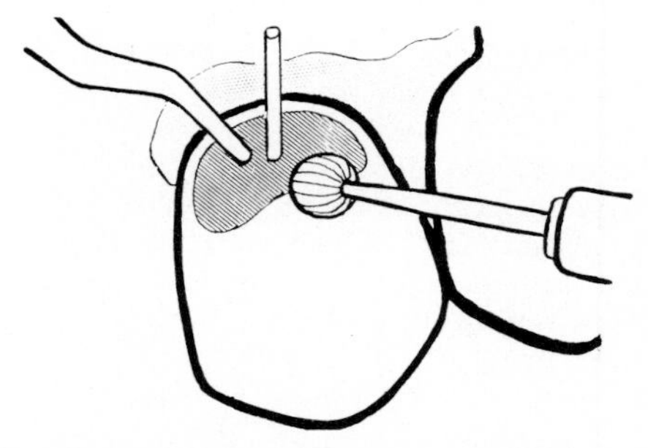

Fig. 9.23. The stabilization of a cervical inlay with a gold plugger whilst the margins are smoothed with a finishing bur.

Some of these inlays are surprisingly symmetrical and the exact location of the cast in the cavity should be determined so that replacement during cementation is not delayed. It is even worth while to make a visible but fine scratch mark near one margin to assist in rapid location.

Before cementation the sprue is cut short and smoothed down, except when this is particularly required for handling in a difficult position. A slightly slower-setting mix than normal is again used and the inlay maintained under pressure till the cement reaches its initial set. Non-irritant cements are particularly relevant to these cavities, which are frequently extremely sensitive.

When possible, finishing of this type of inlay should be delayed until another visit. If the sprue has been retained it must, of course, be removed, but this must be done with the lightest touch, having given the cement as long as possible to set.

Eventually finishing is done with fine stones or finishing burs, or with fine sandpaper discs if the restoration is accessible to these. The successive use of medium and fine abrasives followed by Tripoli, rouge or hand-burnishing imparts the final lustre to the inlay.

Finally, in closing this chapter it must be remarked that the Class V inlay is now a rarity in practice with the development of composite and glass ionomer materials, which are easier and quicker to use—and look better.

Summary

Gold inlay; cast gold restoration cemented into prepared cavity. Close fit essential. Cement seals crevice, secondary retention. Particular indications; extensive restoration; coronal protection.
Direct technique, wax pattern from tooth.
Indirect technique, wax pattern from model; indications, size and access.
Retention. Parallel, near-parallel and tapered walls. Length of walls, depth of cavity. Floor–wall line angle.
Longest single line of insertion.
Class I and Class V cavities. Essential features, 3–7 degree tapered walls; cavosurface 30–45 degree bevel; cuspal slope; choice of instruments. Class V retention from occlusal and gingival walls. Lining.
Wax pattern. Separating fluid. Conical wax form. Warming and insertion; direction

and pressure. Carving, sharp instruments, direction along margin; define feather edge. Wax addition. Smoothing pattern surface. Sprue wires, solid, hollow. Removal and inspection.
Investing and casting. Crucible former. Addition of reservoir. Casting ring. Vacuum mixing and investing. Drying, heating, casting, cleaning.
Finishing and cementing. Clean cavity. Check marginal fit. Closest marginal apposition; swaging, spinning, burnishing.
Preparation, cement, slow setting. Pressure maintained.
Final finishing.

10

Direct gold inlays in compound cavities

The general characteristics of gold inlays as referred to on p. 162 – apply to the compound restorations described in this chapter. The cavity design and techniques are modified only on account of the larger size of the restorations and by the increasing need to protect the remaining natural tissue.

Class II gold inlay

Cavity design and preparation

The design of the Class II cavity conforms to the general principles stated in connection with amalgam restorations as regards outline and resistance form. As regards retention, here again the principle of the longest single line of insertion is interpreted in terms of near parallel walls and the depth of the cavity (Fig. 10.1).

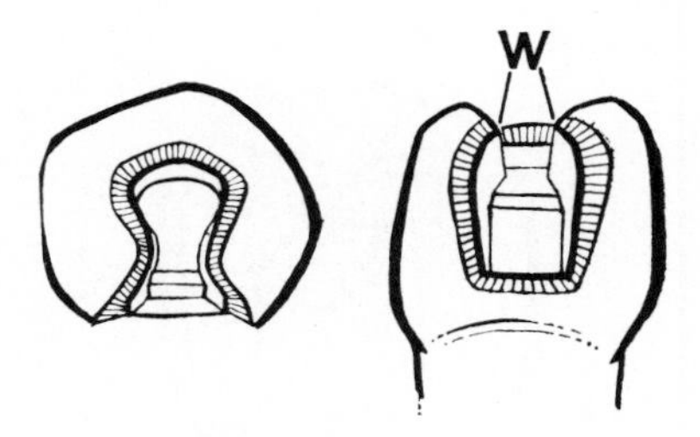

Fig. 10.1. Class II direct inlay cavity shown from the occlusal and mesial views. W: width of neck.

Resistance to lateral displacement towards the interdental space is effected by an occlusal lock in the form of a dovetail as in the design of the amalgam restorations. This part of the cavity can be treated in all respects as the Class I inlay cavity. It should have a flat floor, near parallel walls, and a bevelled cavo-surface angle.

The interstitial cavity preparation proceeds in its early stages as has been described for a cavity prepared for amalgam. It is an advantage to use tapered fissure burs, carbide or diamond instruments, since these more easily give the required taper to cavity walls. The angulation of the handpiece must be carefully controlled to ensure that undercuts are not inadvertently made nor the taper unduly increased.

In cutting the occlusal lock in a fissure which is, for practical purposes, non-carious, the balance between the dimensions of the inlay and the strength of the residual tissue must be maintained. The width of the neck of the lock in an inlay (Fig. 10.1 W) could be somewhat narrower than is required for amalgam because of the greater strength of the gold. On the other hand, the depth of the lock is an important factor in the retention of the inlay in an axial direction. Further, it must be remembered that a wax pattern of the inlay must be strong enough to withdraw without distortion at its weakest point. This puts a practical limit upon the reduction in size of the neck of the lock.

Having established the outline form, the retention, and the resistance form, which in this case consists of the occlusal and cervical floors, the operator's attention is now turned to residual caries.

If an inlay is used in the first instance for the restoration of a minimal cavity, there may be very little carious dentine remaining at this stage. In many cavities, however, a larger area of caries will be present and its eradication proceeds by means of excavators, first clearing the periphery, then cutting from the centre towards the periphery. Where necessary a sublining is used and the main lining of phosphate cement restores the internal form of the cavity and is used at the same time to fill in any rounded undercuts which the removal of caries may have created (Fig. 10.2). This is an example of the dual function of a lining, which insulates the pulp from thermal changes and also restores the structural form of the cavity.

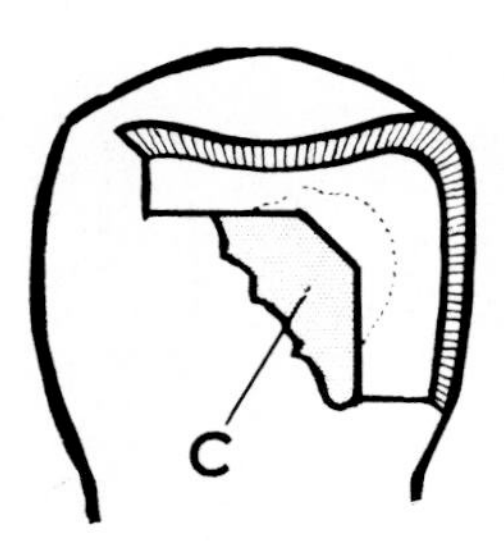

Fig. 10.2. The use of phosphate cement, C, to restore internal form and to obliterate undercuts.

Undercuts which are necessarily made in the removal of carious dentine may be treated in either of two ways:

1. Cutting back (Fig. 10.3 (a)). In this case the strength of the enamel and dentine above the undercut is not strong enough to support the pressure which it will meet. This wall must therefore be cut back till the undercut is eliminated. The cavity outline is thereby enlarged.

2. Blocking out (Fig. 10.3 (b)). Here the tissue will be more than strong enough to withstand the pressure applied to it. This undercut may be blocked out with cement.

Generally, undercuts near the occlusal surface are better dealt with by cutting back. Those deep in the cavity can well be filled with cement.

There are two qualifying factors to these general rules. First, overhanging dentine, D, in Fig. 10.3, is separated from the pulp. It will therefore be, or will become, more brittle than normal vital dentine. This must be remembered when assessing its strength. Second, shallow depressions do not hold cement well. The cement should always be applied as a thick creamy mix, but even so, it may be detached from a shallow undercut and cause inconvenience in later stages. The shallow undercut is the one which it is usually easiest and most expedient to obliterate by cutting back.

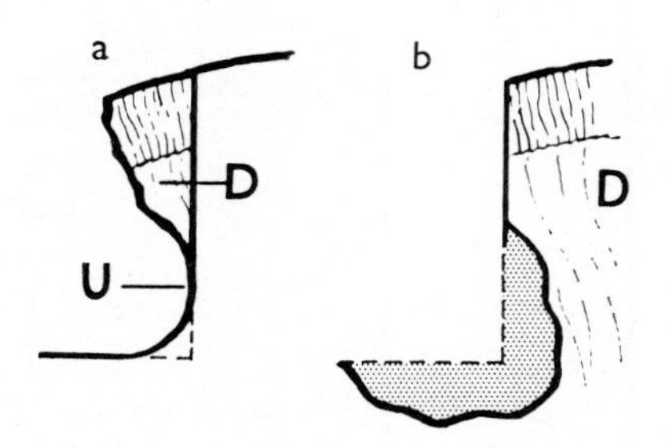

Fig. 10.3. (a) A superficial undercut, U, which is eliminated by cutting back the overhang, D. (b) A deeper undercut, which may be filled with cement.

There are many occasions when extensive undercuts are treated by both methods; cutting back provides a margin of adequate strength, filling with cement conserves tooth structure.

In the Class II cavity under consideration it sometimes happens that, because of the extensive destruction of dentine, cement comes to form a considerable part of the occlusal floor, the axio-pulpal wall, and some part of the lateral walls of the cavity (Fig. 10.2). The management of this cement lining is important and is worthy of detailed consideration.

Detail of lining technique

The cement used should be adhesive and hard so that it can be worked with finishing burs. Fortified zinc oxide and eugenol, EBA, and polycarboxylate cements can all be used but, though non-irritant, they are rather less satisfactory as structural linings because of their softness immediately after setting. Phosphate cement is certainly more satisfactory to work when it forms an extensive part of the deep cavity surfaces. In larger cavities which have been previously restored, *the basal dentine is sclerotic and underlaid by secondary dentine.* Thus the use of phosphate cement carries little risk; if doubt exists a thin sub-lining may be used.

The cement of choice should be mixed at normal speed to a thick creamy consistency.

Using a flat-ended probe, a small portion is teased into the deepest part of the cavity to cover as much surface as possible (Fig. 10.4 A).

Later portions are larger as the viscosity of the setting cement increases. They are applied first to the floor and the walls of the cavity and then pro-

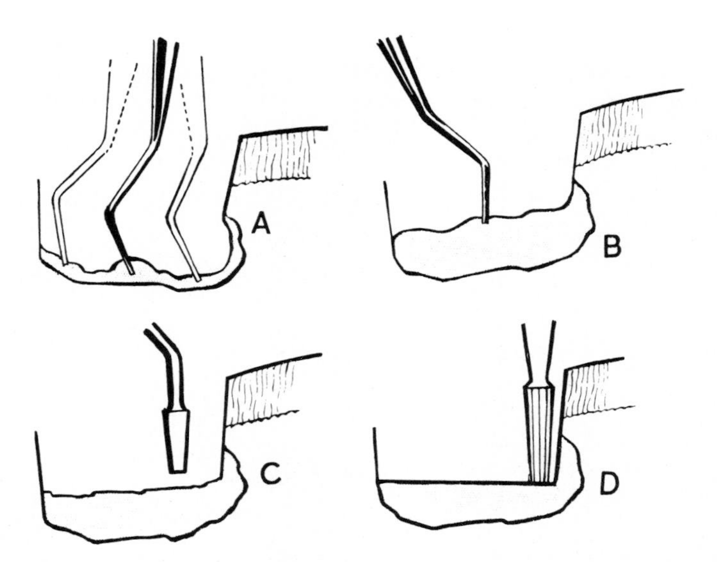

Fig. 10.4. (A) The application of soft cement on a blunt probe to floor and walls to cavity. (B) Filling of the bulk of the cavity to required extent. (C) Conforming the setting cement with a plastic instrument. (D) Finishing with a plain tapered fissure bur.

gressively fill the bulk of the area to be covered, to the required extent and no more (Fig. 10.4 B).

A flat-ended probe, made by cutting the terminal 2 mm or so from the point of any worn out probe of useful pattern, is the most effective instrument for placing small increments of cement accurately.

When the cement has set to the stage of firmness but is still conformable, a plastic instrument, moistened in alcohol to prevent sticking, is used to create in fair detail the form required of the cement (Fig. 10.4 C). Any slight excess should be removed with a sharp excavator. As soon as this is achieved, further manipulation should cease.

When the cement has set quite hard, the final form and detail may be given by a plain tapered fissure bur (Fig. 10.4 D). Usually this may be included in the finishing of the cavity as a whole.

The common errors in technique are these:

1. The **wrong thickness of cement** is used, leading to failure of the cement to adhere, or to formation of voids which may later be opened in the finishing stage.

2. The **use of too large additions** of cement and **failure to place the cement accurately** where it is needed. Inaccuracy allows the cement to stick to parts of the cavity where it is not needed and obscures the part of the cavity where it is required.

3. **Overfilling** of cavities also leads to obliteration of detail and loss of time in removing excess.

4. **Failure to conform the cement** at the right stage with a plastic instrument; this greatly increases the time and work required to produce a form and finish which is an integral part of the cavity design.

Review of cavity detail

The detail of the cavity may now be reviewed. The walls of the interproximal portion and the occlusal lock are carefully inspected from several angles to exclude the possibility of undercuts. For this purpose, and for inlay work generally, a plane mirror is preferable to a concave mirror. The latter produces distortion which makes it difficult to assess the angulation of various parts of the cavity.

Any surface which appears either undercut or, alternatively, too widely flared should be aligned by the use of the plain tapered fissure bur, which is equally effective on dentine and cement. It should be used at moderately high speed with a light touch, to flatten the occlusal and cervical floors and to true the cavity walls. If required, the floor–wall line angle may be sharpened by the use of a small chisel or hatchet (Fig. 10.5).

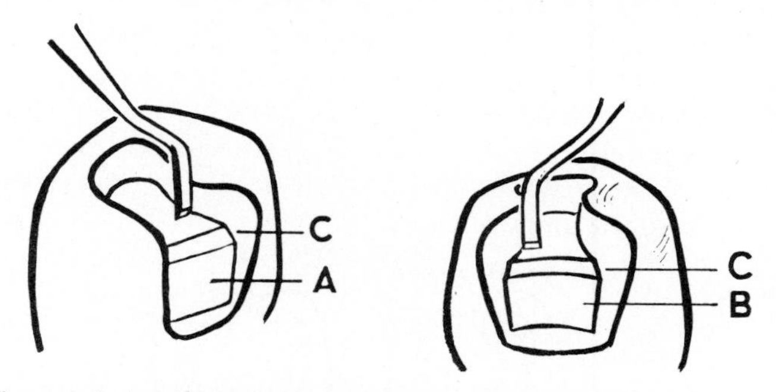

Fig. 10.5. The use of a hatchet to sharpen the floor–wall line angle. The axio-pulpal wall may be flat, A, or concave, B; its junction with the occlusal floor is bevelled, C. The cavo-surface bevel has not yet been made.

The axio-pulpal wall of the cavity may be formed partly of dentine, partly of cement, or preponderantly of cement. It is frequently represented to be a flat surface (Fig. 10.5 A), but when it is composed of an adequate thickness of cement there is no reason why it should not be concave (Fig. 10.5 B). This concave form has the advantage that it is easily produced by the tapered fissure bur referred to in the course of finishing the cement lining. The solid line angle between the axio-pulpal surface and the occlusal floor should be bevelled (Fig. 10.5 C), to encourage the flow of wax during impression taking and to reduce the risk of fine fracture of the investment mould during casting.

As to the cervical floor, this may be made flat, that is to say, at right angles to the axis of the tooth, or with a slight inward slope. It may not slope outwards, since this would produce a faulty resistance form (Fig. 10.6).

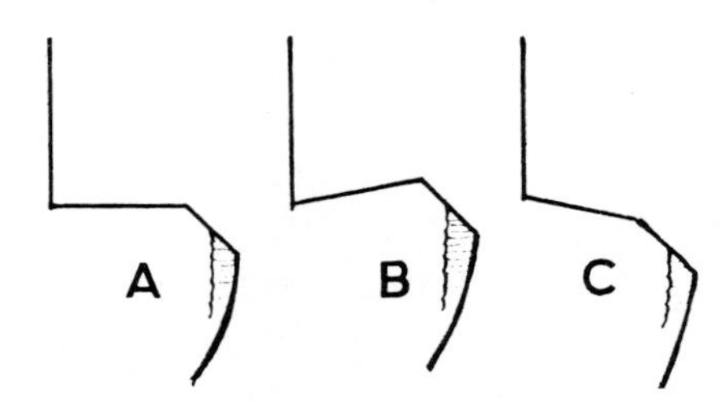

Fig. 10.6. The cervical floor may be flat, A, or with a slight inward slope, B. A floor which slopes outwards, C, has a faulty resistance form.

The preparation is to be completed by bevelling the cavo-surface angle at 45 degrees. The treatment of occlusal margins follows that of a Class I inlay cavity. Where the occlusal outline joins the axial margins of the approximal box, the bevel continues in the same manner, but in the case of these axial margins there is a new factor to be considered.

It will be seen that the mesial and distal aspects of molars and premolars are convex from above downwards; the greatest convexity is at the level of a survey line whose axis is the line of withdrawal of the pattern. Below this the enamel surface falls away to the cervical margin. If a bevel of normal width were continued along the whole length of the axial margin (Fig. 10.7 A), that part of the bevel below the level of the contact area would be undercut with regard to the normal line of withdrawal of the pattern. This would not, however, apply to the gingival bevel.

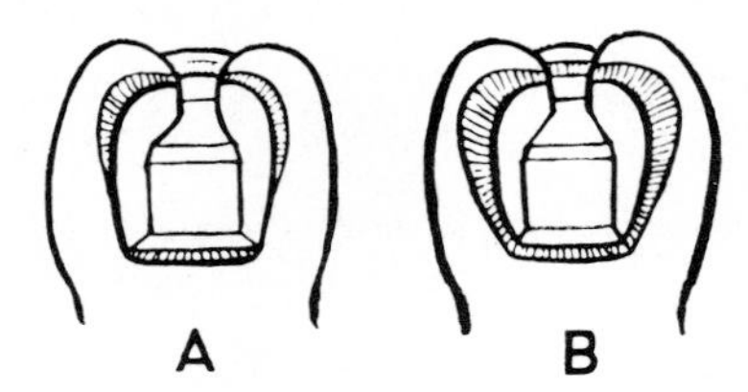

Fig. 10.7. (a) Shows axial marginal bevels applied only to those parts of the margin above the level of greatest convexity of the proximal surface. (b) By flaring the bevels well back, they may be extended to the level of the cervical margin.

The implication of this is that a bevel of normal dimension must diminish to nothing at the level of greatest convexity, leaving the lower part of the axial margin unbevelled so that the gold inlay would there meet enamel at a butt joint. Although this margin is not subject to direct occlusal stress, such a joint reduces the possibility of obtaining the closest adaptation of the gold and is therefore a point of potential marginal weakness.

This defect may be overcome by the use of an exaggerated, or flared, axial bevel (Fig. 10.7 B). In this case the bevel is much more marked and is laid well back towards the lingual and buccal aspects so that, in its cervical half, it is no longer undercut with regard to the line of withdrawal. The

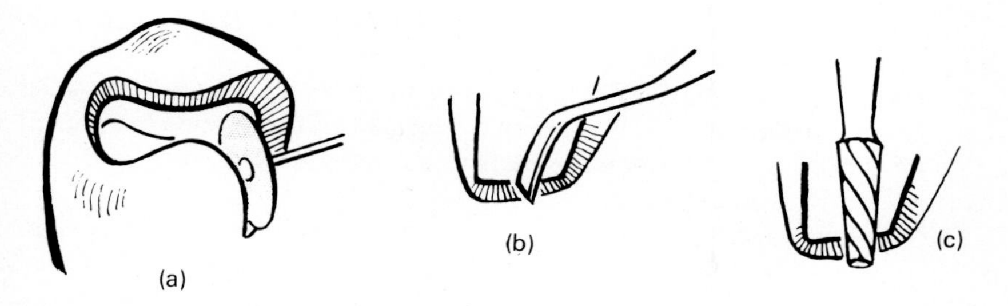

Fig. 10.8. Formation of marginal bevels, use of: (a) sandpaper disc; (b) tapered fissure bur; (c) reversed cervical trimmer.

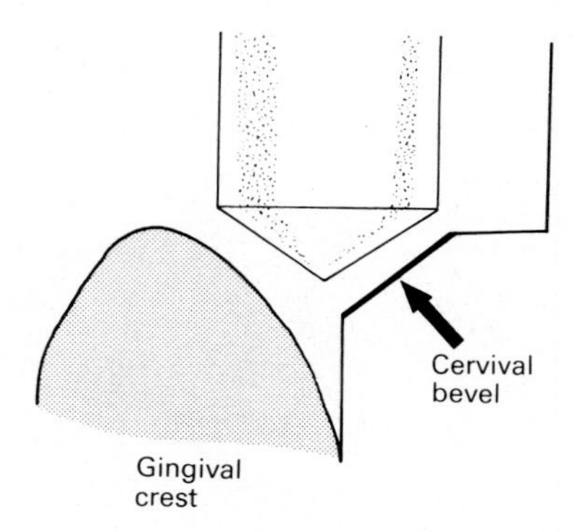

Fig. 10.9. The use of 135 degrees pointed fissure bur, shown in Fig. 9.8, for the formation of the cervical bevel.

Fig. 10.10. Low-power scanning electron micrograph of a cervical margin finished at an angle of 135 degrees by a conical end, non-bladed bur in an air-turbine handpiece. (Micrograph by courtesy of Mr I. E. Barnes. In Baker and Curson (1974) *Br. Dent. J.* **137**, 391.)

disadvantage of this alternative is that on the buccal aspect it brings the margin into a position where it is more easily seen.

Bevelling occlusal margins is considered in detail in Chapter 9. Axial margins down to the level of maximum convexity can be bevelled by tapered or flame-shaped burs. For a more widely flared margin a fine grit sandpaper disc may be used (Fig. 10.8).

The cervical margin presents difficulty because it is so often inaccessible. A sharp cervical trimmer, reversed angle, has long been advocated and used for this purpose; it is difficult to use with light touch and control, it can cause gingival damage and it does not produce a particularly smooth surface. Tapered or flame-shaped burs can occasionally be used at the correct angle. The 135 degree Baker–Curson bur (Fig. 10.9), is capable of producing a good surface, correctly angled and with little or no gingival laceration.

The cavity should be reviewed and, if satisfactory, thoroughly cleansed to remove all debris before proceeding to the wax pattern (Fig. 10.11).

Fig. 10.11. Completed Class II inlay preparation in premolar.

The wax pattern

The tooth is isolated and a matrix band is applied in the manner described for an amalgam filling. Two details require attention. The cervical margin of the band, having been contoured to the gum, should be closely and firmly fitted to the tooth, using a wedge if necessary. The height of the band may be adjusted to allow the teeth to meet in normal occlusion.

A thin film of lubricant or separating medium may now be applied to the cavity and the surrounding surface of the crown. A wax cone of suitable size is prepared in the way described, and chilled. The point of the cone is gently warmed by holding it in fingers over the flame. It is then inserted into the cavity, parallel to the axis of the tooth, but directed at its deepest part, that is to say, at the gingival floor. The wax is carried home with firmest pressure that can be brought to bear, and the patient is then asked to bite in centric occlusion until the wax cools. As this takes place he is instructed to make sideways and forwards 'grinding' movements.

Carving the inlay starts on the occlusal and the matrix band should be kept in position until the cavity margins and the general form of this surface are clearly defined. The matrix retainer is removed and the band taken off, preferably by drawing it laterally through the contact point.

It may well be found that the inlay is now easily removed along its intended line of withdrawal, and in some cases it is necessary to hold it in place, during further manipulation, by a soft pledget of cotton held in forceps, and applied to the occlusal.

If the cavity is well designed, there is little harm at this stage in removing the pattern with a probe (Fig. 10.14) in order to examine the detail of its impressed surface and margins. Care must be used in handling the pattern which should seat cleanly and accurately into position when replaced.

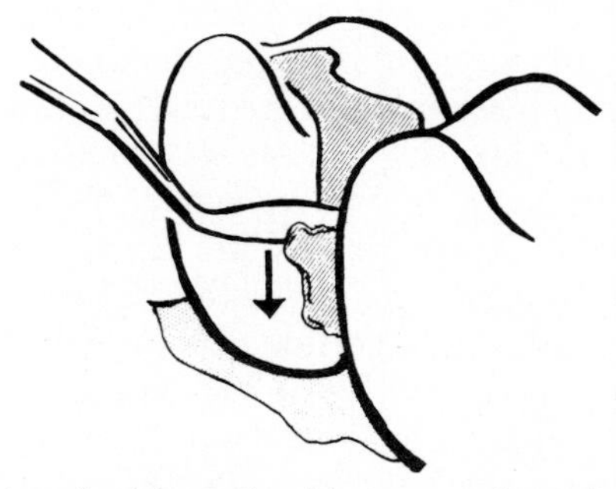

Fig. 10.12. Trimming of axial margin with a carver directed towards the gingival.

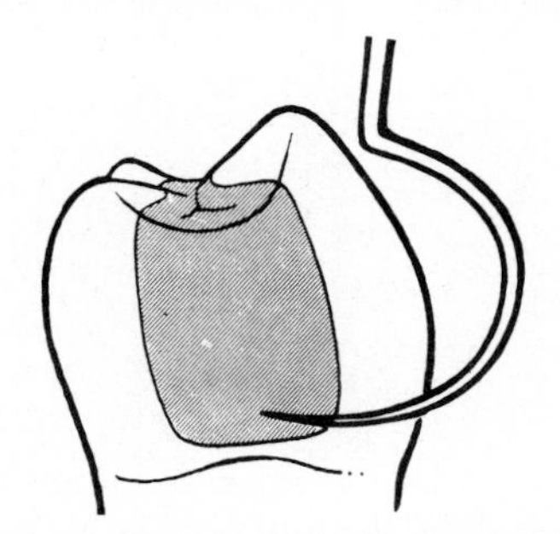

Fig. 10.13. The use of the sickle-probe for trimming cervical margin.

The axial margins are now trimmed with strokes towards the gingival (Fig. 10.12). If adaptation of the matrix to the gingival has been correct there should be no gingival excess, and if the pattern was earlier removed for inspection this can be confirmed, or the presence of an excess noted. In any case, the margin can be carefully explored with a sickle probe, Ash No. 54, and trimmed as necessary, with a stroke which must be guided by the adjacent tooth surface (Fig. 10.13). Instruments used for interdental carving must be fine and sharp. Ward's carvers 1 and 2 are suitable, but smaller-scaled instruments of the same pattern as these are even more suitable. Long-bladed trihedral scalers, (Fig. 7.31, p. 125), when fined down by repeated sharpening, are also effective and can be used with both push and pull action.

It will be noted at this stage that the wax may, or may not, be fairly accurately contoured as to the contact area. This is not a matter of importance, for the contour must later be overbuilt. The proximal surface of the pattern is now smoothed by passing a plain glazed linen strip through, or under, the contact area and gently passing if from side to side across the wax.

The final wax finish is now completed by a pledget of wool lightly moistened with eucalyptol or with hot water (see p. 171).

Spruing the pattern

Patterns of this type may be sprued in the cavity, or in some cases out of cavity. In the former case, a 1 mm diameter sprue wire is used and inserted in the marginal ridge obliquely at about 30 degrees to the axis of withdrawal (Fig. 10.14). To avoid undue melting of the wax and obliteration of occlusal carving of the sprue should not be overheated. When chilled the wax is withdrawn along the axis determined by the cavity preparation. The purpose of the obliquity of the sprue is to feed the molten gold, when cast, into both occlusal and interproximal portions of the pattern. If difficulty of withdrawal of the pattern is expected it can be anticipated by having the sprue wire more nearly parallel to the axis of withdrawal.

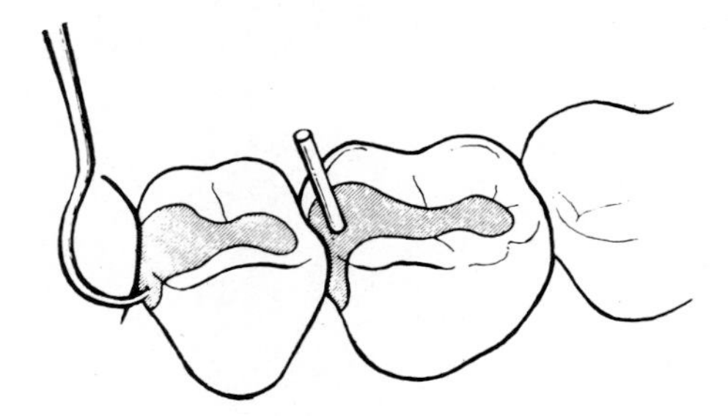

Fig. 10.14. The insertion of a sprue placed obliquely into the marginal ridge. The alternative removal with a curved probe is also shown.

Some operators prefer to remove the pattern from the cavity for spruing. The inlay may be chilled with room-temperature water and a sharp curved probe gently inserted into the interproximal portion of the pattern (Fig. 10.14). The pattern is removed on the probe point from which it is detached by gripping the probe close to its point with a pair of pliers, the points of which have been well heated. It is allowed to fall lightly on to a gauze square. The sprue wire may then be inserted at any desired angle into the contact area. The pattern must, of course, be handled with delicacy to avoid damage, but if this precaution is taken, the method has the advantage of not disfiguring the occlusal carving.

In passing it should be noted that every operator should have certain instruments reserved for heating in the open flame. For example, carrying forceps or tweezers, flat-bladed and ball-ended plastics and a straight probe are the common ones; they can be heated to redness if required. This prevents the heating of instruments used for the handling of materials, the loss of temper and general deterioration of shape and surface which results from casual and repeated heating of all instruments.

Investing and casting

This should proceed without delay, immediately upon removal of the pattern. A wax reservoir should be applied to the sprues of all but the smallest Class II patterns (p. 174). The investing and casting follow the principles described in the previous chapter. A well-balanced investment will control expansion adequately and 20-carat gold is hard enough for the great majority of these inlays. Inlays which are unusually extensive and subject to a much heavier occlusion than normal may be cast in 18-carat platinized gold; this is not often required, but it is important that it should be used in those cases where the indications exist.

Finishing and cementing

When the inlay is cleaned after casting, the sprue may be removed and the surface at its point of entry smoothed. If, in spite of care in investing, a small airblow exists on the fitting surface, this is carefully removed with a sharp excavator. No further modification of this surface is acceptable. The cavity is cleaned, dried, and inspected to ensure that it is free of temporary filling or debris, before trying the inlay in.

If the pattern has been sprued in the contact area, this area may be a little excessive and will prevent the inlay from seating home. It should be reduced with an abrasive disc, trying in as necessary, until the inlay fits snugly into place. A 16 mm (⅝-in) cup-shaped, outside-cutting, vulcarbo disc mounted on a straight mandrel is a very convenient instrument for this type of adjustment.

A piece of dental floss should be used to test the contact, now and later, and at this stage the contact should be left slightly tighter than normal. This means that the patient, with upper and lower teeth apart, should be aware of light lateral pressure on the teeth adjacent to the inlay. He will probably say that the inlay feels 'a little tight'.

In most cases where the pattern has been sprued on the marginal ridge, the contact will be found to be looser than normal and this must now be corrected. The area is built up in the following manner.

The interproximal surface of the inlay is lightly abraded with a disc to ensure that it is clean. The area to be built upon is outlined with the point of a soft lead pencil; this confines the flow of solder. A small piece of 18-carat gold solder of a bulk sufficient to produce the necessary overbuilding is dipped in flux and placed on the contact area. The inlay is then held in soldering tweezers in the upper part of a bunsen flame (Fig. 10.15) until the solder flows evenly over the area contained by the pencil mark. Care must

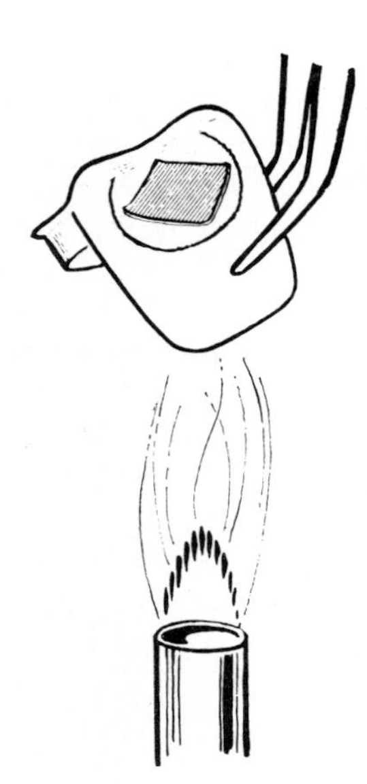

Fig. 10.15. The application of solder to the contact area. The flow of solder is confined by a pencil line.

be taken to avoid over-heating and 'sweating' fine feather edges. The inlay is cleaned in acid and washed.

When tried in the cavity again the contact, if properly built, should be a little too tight to allow the inlay to go home. This is now gradually reduced till the contact is slightly tighter than normal to the passage of dental floss. Whenever the inlay is removed from the cavity, this is done by passing a fine trihedral scaler or a curved probe under the contact area and exerting a firm pull in the line of withdrawal.

The adaptation of edges has already been described. The use of light mallet blows may be adopted for use on accessible margins, all of which, because of their accessibility, are relatively easy to finish.

The accuracy of the cervical margin must inevitably depend upon the care expended on this feature in the stages of preparation and the wax pattern. The margin is first tested with a curved probe for fit. Fine irregularities may be smoothed with a concave interdental finishing bur, Meisinger 28A used very lightly. It is also possible to trim the gingival margin and the approximal surface out of the cavity, with a fine sandpaper disc, trying-in at intervals to check the margin with a probe. Larger defects call for the remaking of the inlay since it is impossible to correct them and check the final fit of this margin.

The occlusion is tested. If the occlusal surface of the wax pattern was formed by the opposing teeth as described, there should be little adjustment to perform. It may be observed that a small degree of opening of the bite will render the inlay exceedingly noticeable to the patient (see also p. 202).

Using occlusion paper, the high spot should be progressively reduced with finishing burs and stones until the inlay cannot be detected by the patient in centric and in lateral excursions. At this stage the occlusal fissure form may be emphasized, using a blunt round bur, No. 3 or 4, to define fissure pattern if this is necessary. It is, however, preferable that all characterization should be carved into the wax pattern. The marginal pit usually requires emphasis, for this may have been partly obliterated by placing the sprue.

The polishing proceeds with graded abrasives, with the inlay in the cavity. Before cementing, the inlay is removed and, holding the inlay in the fingers or in an inlay holder (Fig. 10.16), a mirror finish is applied to the contact area, because this part will no longer be accessible to polishing when the inlay is cemented. The cervical margin bevel may be lightly burnished with the shank of a plastic instrument, in order to invert the feather edge to the slightest degree.

For cementing, a mix slightly slower than normal is used. Cement is applied to the fitting surface of the inlay and to the cleaned dried cavity surface. After insertion the inlay is pressed hard home and pressure maintained either by an instrument or by biting pressure on a small piece of rubber, until the preliminary set of the cement has occurred.

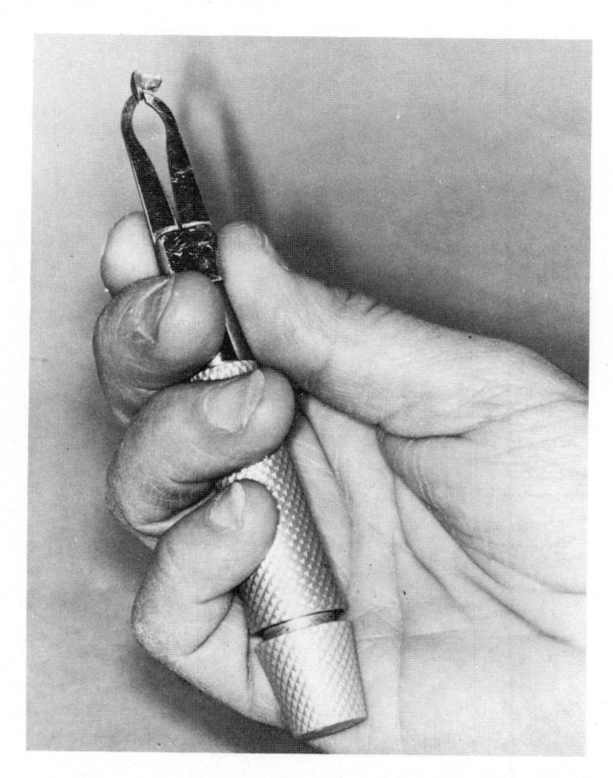

Fig. 10.16. Inlay holder. A small vice controlled by turning the bottom end of the handle.

When set, excess cement is removed with a probe, with particular attention to the clearance of the interdental space below the contact point. Any final smoothing with the finest cuttle disc is applied to accessible margins. The polish is applied by a mounted soft brush and rouge stick or by hand burnishing.

The MOD gold inlay

Cavity preparation

The MOD preparation is in most respects similar to that of mesiocclusal or distocclusal design. More care must be exercised in alignment of walls and, in particular, the line of withdrawal of the approximal portions must be parallel.

In this type of preparation it sometimes happens that the buccal or lingual cusp in premolars, or two cusps in the case of molars, is obviously weakened by previous removal of dentine. In mandibular molars this affects the lingual cusps more frequently; in maxillary molars the buccal cusps. When this is so, it may be judged expedient to protect these cusps by covering them with gold and the preparation may be modified accordingly (Fig. 10.17).

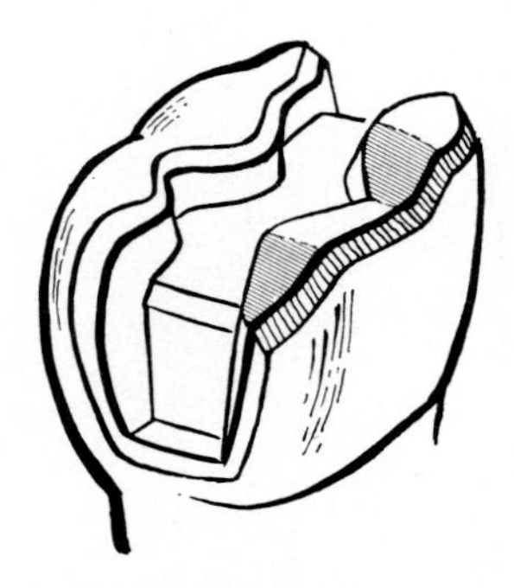

Fig. 10.17. The modification of MOD preparation to cover lingual cusps of a lower molar.

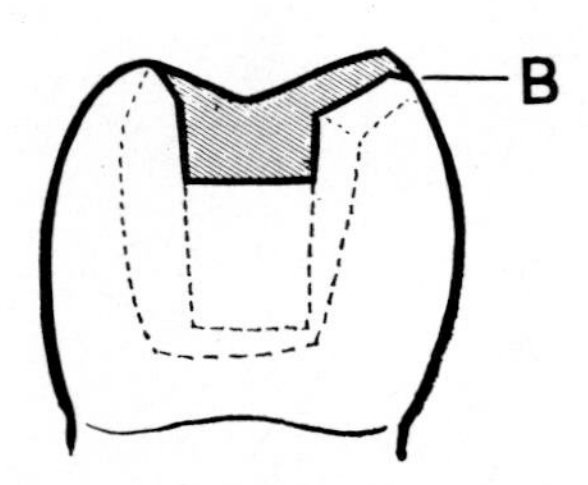

Fig. 10.18. Sectional view. Reverse bevel, B, can be seen running along the occluso-lingual aspect of the cusps.

The enamel covering the cusps is removed with a fine diamond instrument, or a plain tapered tungsten carbide bur if high speed is used. The general cuspal form is retained by flattening the existing cuspal planes. *Clearance of not less than 1 mm must be allowed* between the prepared surface and the cusp of the opposing tooth.

These cuspal planes finish just over the peaks of the cusps, at the 'reverse bevel' which runs along the lingual or buccal aspect allowing not less than 30 degrees at the edge (Fig. 10.18). In this manner the cusp, or cusps, are protected from direct occlusal stress and the gold margin is removed to a position where, whilst adequately cleaned, it is not very vulnerable.

The wax pattern

A matrix band is used for the MOD pattern since it helps to contain the wax and assists in its adaptation to the prepared surfaces. The cone of wax used for the impression is larger than that used for a simple Class II cavity and its pointed end is moulded into a bifid form. It is then chilled and gently reheated at the tips. Upon insertion into the cavity, the two projections are so placed that they enter the mesial and distal boxes of the preparation. The occlusal of the teeth again helps to carry the wax home under pressure, and to shape the occlusal surface correctly.

Trimming the wax follows the methods described, defining first the outline and bevels, the occlusal and then the approximal surfaces. Occlusal pattern and cuspal form are restored and the wax finished with eucalyptol or warm water.

The sprues of the MOD inlay are used, as before, to remove the pattern, but in this case there is a greater risk of distortion due to its size and complexity. There are variations in the manner of spruing and removing the pattern; the two described illustrate the essential principles.

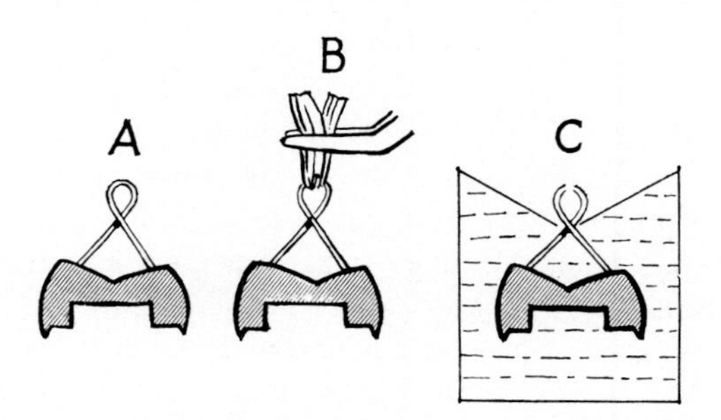

Fig. 10.19. (A) Use of an adapted paper clip as compound sprue. (B) Linen strip looped for removal of pattern. (C) Invested, with sprue wire cut for removal.

A stout paper clip is cut and bent to the form shown in Fig. 10.19 A. Its two limbs are adjusted so that their free ends will engage the marginal ridges of the pattern. This compound sprue is inserted with the usual precautions and the pattern may be gently cooled with room-temperature water. Excessive cooling is unnecessary, for a good inlay wax should be rigid enough for removal at mouth temperature. A narrow linen strip is looped through the top of the sprue wire, held in forceps, and the pattern withdrawn by axial traction upon this loop (Fig. 10.19 B). When invested, the point of crossing of the sprue is adjusted to meet the bottom of the crucible. Thereafter, the wire loop may be cut and the two portions of wire removed, leaving two sprue ways (Fig. 10.19 C).

In the second method, a similar paper-clip form, or a staple of any suitable wire, can be used, with a linen loop, to remove the pattern (Fig. 10.20 A). The staple is then used to hold the pattern for the insertion of a 1 or 1.5 mm tubular sprue into one end of the pattern (Fig. 10.20 B). Finally the staple is removed by holding it with hot pliers, and the pattern is allowed to drop on to a square of gauze (Fig. 10.20 C).

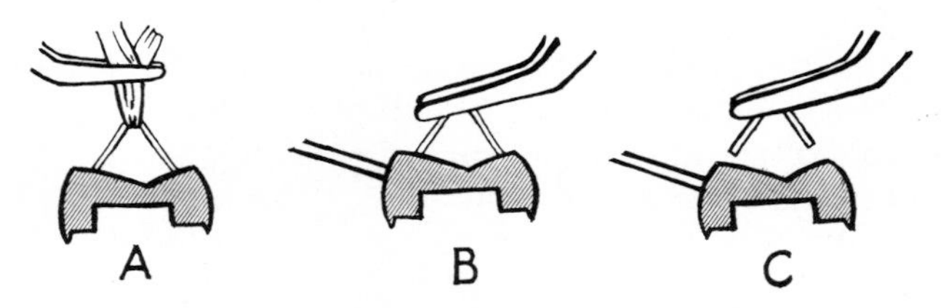

Fig. 10.20. (A) Wire staple and linen loop for withdrawal of pattern. (B) Staple held for insertion of tubular sprue. (C) Removal of staple by use of hot pliers.

The principles involved in both methods are that the inlay is held rigidly at the two points of insertion of the sprue and an axial pull is provided by the linen loop. Further, the sprue or sprues are adjusted in number, size, and position to allow unimpeded flow of gold into the investment matrix.

The interval occurring between the removal of a wax pattern from the cavity and the cementation of the gold inlay should always be as short as possible. In practice, twenty-four hours may elapse, but occasionally circumstances may make a longer interval necessary. During this period protection of the prepared tooth from heat and cold and from mechanical damage is required.

With small restorations a simple zinc oxide dressing suffices. In the more extensive forms, zinc oxide reinforced with cotton wool is more effective. When an extensive MOD preparation is to be protected and the comfort of the patient assured *for more than a few days a closely fitting copper band, trimmed to the gingival and turned in at the occlusal*, will be found to retain a reinforced zinc oxide dressing for a period long enough for most purposes.

The need for care may be explained to the patient and he should be told to avoid heavy biting in that area, but that normal cleaning should continue.

The design of cavity margins should be such that they will withstand normal stresses, but a chipped margin requires repreparation and a new inlay.

Before investing, a reservoir to each sprue is usually required and, according to the investment used, it is sometimes necessary in patterns of this size to use a method of controlled expansion. This is achieved by increasing the water–powder ratio, thus increasing setting expansion, or by the addition of a determined amount of water to the set investment, thus using its hygroscopic expansion. These methods must be used with understanding, since it is possible, by excessive expansion, to reduce the frictional retention of a MOD inlay to a serious degree.

The point made on p. 176 about the risk of loose fragments of investment falling down the sprue way is particularly true of more complicated sprues. These fragments will always gravitate to the lowest wedge-shaped part of the mould, namely the cervical bevel. This is an obvious explanation of cast marginal defects which were known not to have been present in the wax pattern. All sprue formers should have polished sides and smooth ends without a burr. Make sure that the ring is always inverted when removing the sprue, that all visible loose fragments are removed, that the ring is never turned crucible upwards unless, and until, casting makes it necessary.

The gold suitable for these inlays, even involving cuspal coverage, does not often need to be harder than a good 20-carat alloy. In the case of patients with a particularly heavy occlusion, in a case where cusps have been only thinly covered, or where margins have perforce been left in positions exposed to stress, a platinized alloy of 18 carat may be indicated.

Fitting and cementation

The importance of a clean cavity and a clean fitting surface, devoid of 'air blows', has already been stressed. The insertion points of sprues should be trimmed smooth and the cast tried in. It should seat directly into position with firm finger pressure. Failure to seat home may be due to a number of faults; some of them are obvious, others more obscure. The commonest is distortion of the pattern due to failure in cavity preparation, usually in alignment of cavity walls and axial surfaces.

Provided that the contact points are clear, which is normally so with the direct method, the use of an automatic mallet may be permitted to attain the *very last degree* of adaptation of the cast of the cavity. The more accurate the technique, the less likely is the use of the mallet to be required, and in no case may it be used unless the inlay is a visibly good fit. No amount of malleting will enable a distorted cast to be brought to a condition or fit. To attempt to use force in this way would cause pain to the patient and serious damage to the tooth.

The use of the mallet for closer adaptation of the margins, where these are accessible, is effective if 20-carat gold has been used. Even with harder gold it can produce tangible improvement.

Contact points may be rebuilt with solder as required. They are overbuilt and then reduced by trial. To begin with their excess will prevent the inlay from fitting. By successive reduction, and retrial with dental floss of *both* contacts on each occasion, the inlay is seated again. Further adjustment brings the pressure of the contacts to within normal range. This implies that the floss is subjected to the same degree of resistance to passage through the new contacts as is needed to pass through normal contacts elsewhere in the mouth. The patient is then unaware, or only just aware, of any sense of lateral pressure on the adjacent teeth.

Unless the inlay is correctly fitting to the cavity it is clearly impossible to check or adjust its occlusal relationship. At the impression stage the patient was asked to 'bite and grind' upon the soft wax. Thus the lateral and protrusive occlusal shape has been effectively defined. This has been carried a stage further in the process of carving.

Now, with the inlay correctly seated, the patient is asked to close his teeth. If the occlusal surface is correctly contoured, he will be unaware of the presence of the inlay. If it is high, the patient will be aware that he is biting much more heavily on that tooth than elsewhere. Even though the discrepancy may be small, say 0.1 mm, he may think that his teeth are widely apart, and he will say that he is biting on that tooth and nowhere else in the dental arch. Such a statement may be quickly checked, first by inspection of occlusal contacts in the premolar and canine areas, and then with a plastic matrix strip between the teeth on the opposite side of the mouth. This gives a useful measure of the degree of opening produced by the inlay.

If, due to failure to check adequately at the wax pattern stage, the articulation is inaccurate to this extent, adjustment is made thus. The occlusion is registered with thin articulating paper and the high spot or spots on the inlay reduced with bur or stone. After trial this process is repeated as necessary till the pressure on the inlay becomes less obtrusive, and that upon the other teeth increases.

Because of the gradual closing process the first stage of reduction of the inlay appears to the patient to produce little change, but later contact on teeth elsewhere in the mouth becomes apparent. The degree of this contact can be assessed by asking the patient to bite on a plastic strip placed between adjacent or opposite teeth and judging the force required to pull the strip from between the occluded teeth. The occlusion should be tested at several points.

As the load on the inlay is progressively corrected a fairly sudden reduction of pressure is observed by the patient as the end point is approached. Correction continues until pressure on the inlay is equal to that on all other teeth, but it is permissible to leave it with a *barely perceptible* increased load.

A good subjective method of testing the accuracy of the final adjustment is to ask the patient to close his eyes and concentrate. He should then try to bring his teeth into the lightest possible contact and to observe at which point contact is first made. If the occlusion is correctly adjusted, he will report contact 'on both sides' or 'all around'. Any sensation of premature contact on the inlay indicates that a small further reduction is required. **A good general criterion for any fixed restoration of moderate size is that the patient should have forgotten about it six hours after cementation.**

It is unlikely that occlusal correction of the degree described above will occur in direct inlay technique unless a serious error has been made in the occlusal registration at the wax pattern stage. Correction of an error of this degree can cause great damage to the occlusal form of the restoration and may call for repetition of the pattern.

After the margins are smoothed with stones and finishing burs and the occlusal contour adjusted, polishing proceeds. The contact areas must receive a final mirror finish whilst out of the mouth, for after cementation they are no longer accessible.

With an inlay which is complicated in shape it may be necessary to allow extra time for manipulation of cement and setting. The liquid may be slaked and the slab slightly cooled to allow adequate incorporation of powder and sufficient setting time.

On the rare occasion the operator may want to stop the process of cementing, once the cement has been applied either to the inlay or the cavity surface. For example, the cement may set too quickly, or perhaps the cement applied to the tooth has been allowed to get moist, or the inlay may be dropped. The application of an alkali will rapidly neutralize the cement reaction and make its removal much easier. A small bottle of 50 per cent aqueous solution of sodium carbonate should be kept for the rare occasions when this happens.

It is better to suspend the cementation of a difficult inlay than to persist when a delay or fault has developed. To proceed in these conditions brings to nought all the effort and skill so far expended in the final achievement of a defective restoration which will fail in its purpose.

Gold inlays in anterior teeth

Gold inlays in anterior teeth have been almost completely superseded for practical purposes by composite restorations. It is true that a well performed gold inlay would be more durable than any of the currently available plastic restorations, but the appearance of gold may not be tolerated by the patient, and this is not unreasonable. An extensive gold labial surface can be effectively faced with a tooth-coloured material but, apart from the added complication of the method, this introduces one of the very shortcomings the gold inlay is chosen to avoid; the facing will have to be replaced relatively frequently.

Pin retention and acid-etching have provided a simple method of using composites for Class IV restorations and the full crown has come into more common use. The Class IV inlay has always presented problems of design and appearance, whereas the porcelain or bonded crown can be one of the highest functional and aesthetic achievements of dentistry. Whether the inevitable destruction which the crown preparation demands is in the best long-term interest of a relatively young patient will always be open to question.

For those interested in inlay work the direct anterior inlay remains a challenge of skill. The brief notes which follow concern three designs of inlay which were fairly common a decade or two ago.

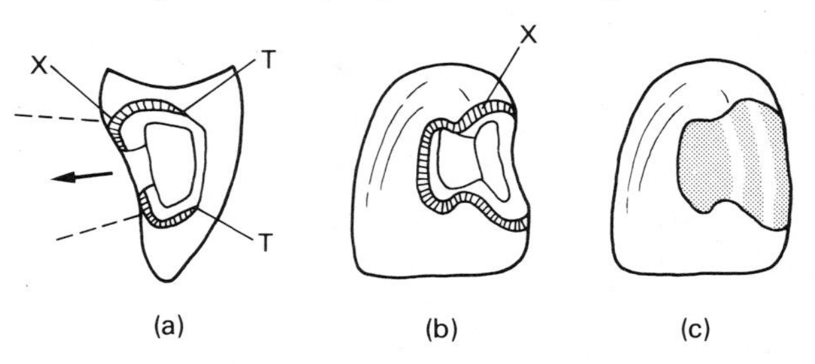

Fig. 10.21. Class III inlay. (a) Proximal view. Arrow shows line of withdrawal; gingival and incisal margins diverge from this. T: tapering bevels. X: Position of wide bevel. (b) Lingual aspect showing lock and bevelled margins. (c) Completed inlay, with lock in cingular region.

The Class III inlay (Fig. 10.21) derived from a medium-sized Class III cavity has the added retention of a palatal lock. The size and depth of the cavity and the lock will determine its retention. The taper of the walls and sharp internal line angles are all-important to a cavity which cannot in any case be very deep. The labial margin must be carefully placed to reduce its visibility from the front. It cannot be bevelled because the pattern is to be removed palatally; all other margins should carry the appropriate bevel.

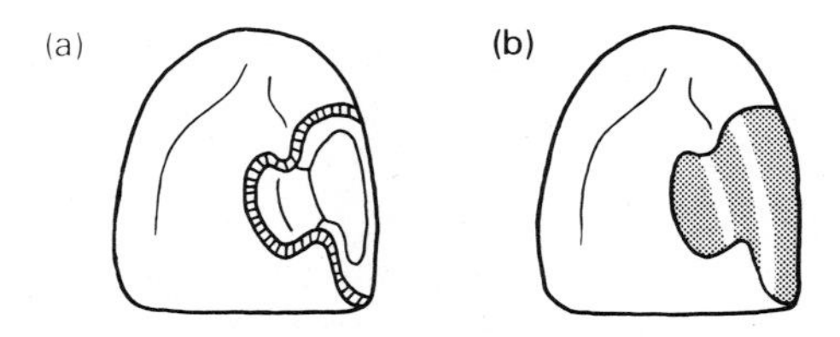

Fig. 10.22. Class IV lingual lock: (a) preparation; (b) completed inlay.

The Class IV inlay (Fig. 10.22) is derived by extension from the Class III design. Although commonly recommended and fairly commonly used, its design confers very poor resistance and retention. A moment's consideration of the occlusal forces acting upon the incisal and palatal surfaces will convince us that a dovetail lock in this position is very ill-adapted to the purpose it is expected to perform. The Class IV restoration (Fig. 10.23) which also protects the incisal edge is a design which opens obvious possibilities of providing surfaces which oppose some major forces, and pins, slots, and pits which oppose others and increase retention. The pin in this case must be easily withdrawn with the wax pattern, so its resistance to dislodgement is

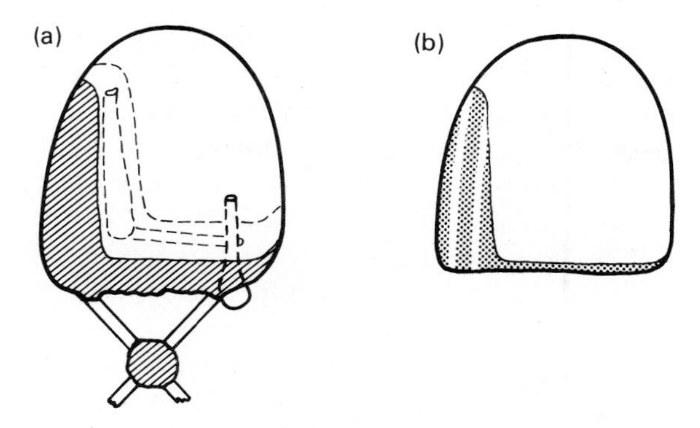

Fig. 10.23. Class IV inlay with retentive pin and incisal protection: (a) direct wax pattern; (b) completed inlay.

chiefly at right angles to its long axis. The size, rigidity, and visibility of this restoration are all matters for the judgement of the operator, but all the principles discussed in the preceeding chapters apply quite simply here. This restoration is not one of the most difficult direct inlays but when neatly done it can be not only one of the most satisfactory, but also one of surprising durability.

For further details of these and other anterior inlays the interested reader is referred to earlier editions of this Manual and to earlier standard texts upon inlay work.

Additional retention in direct inlays

In Chapter 9 reference was made to the significance of the long single path of insertion of an inlay and the derived principle of near parallelism or taper (p. 164). The additional features which can be used to increase retention and which conform to this principle are related to one another. They are the slot, the pin, and the pit.

The slot is a rounded groove, usually of elongated conical form, which is placed in an appropriate wall of the cavity in the line of withdrawal (Fig. 10.24 A). It serves three purposes: it limits the line of insertion of the inlay, it resists lateral displacement in certain directions, and it increases the

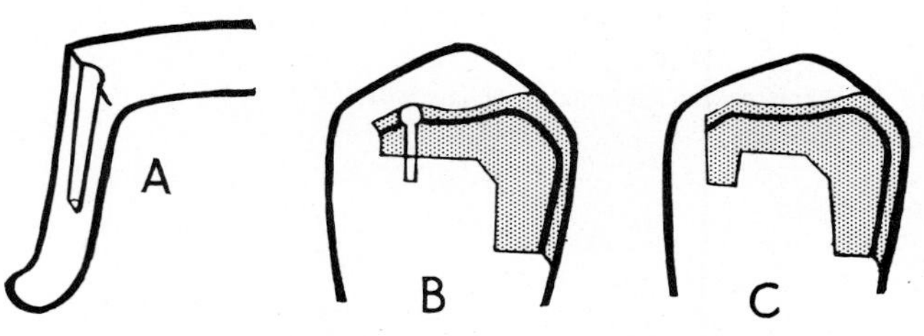

Fig. 10.24. (A) A slot in the axis of withdrawal; (B) pin; and (C) pit, used in the Class II type of inlay.

rigidity of the portion of the inlay of which it forms part. It will be clear that the two slots on opposing walls of a preparation will exert this effect to a much greater degree than will one alone (see p. 211).

The smooth pin is an effective method of increasing retention. It may be of wire which is incorporated in the cast or it may be cast as part of the inlay. In the former it is parallel-sided (Fig. 10.24 B) and in the latter a truncated cone. In both cases it provides parallel, or near parallel, surfaces which control the path of withdrawal, and insertion. It also has great resistance to *lateral* displacement.

The pit is clearly an extension of the cast-in pin referred to above. The diameter of the hole is larger and it can be impressed in one piece, with the wax of the pattern (Fig. 10.24 C). It may vary in depth according to its purpose, and its value is again related to its near-parallelism and its resistance to lateral displacement.

Experience with the larger types of inlay cavity, where much tissue has been lost, may show that the retention provided by near-parallel walls and occlusal locks, if such can be constructed, is insufficient. It is in these cases that the additional retention provided by slots, pins, and pits is particularly valuable.

Summary

Class II gold inlay
In general similar to amalgam cavity.
Tapered walls, bevelled margins, except axial below greatest convexity.
Undercuts; cutting back, blocking out, cement technique; four common faults.
Axio-pulpal bevel.
Cervical margin design; cervical bevel; choice of rotary instruments.
Wax pattern; matrix band. Occlusal surface.
Axio-cervical trimming. Spruing and removal.
Investing and casting.
Finishing; contact area; soldering. Marginal adaptation.
Occlusal adjustment. Polishing. Cementation.

MOD gold inlay
Combined mesial and distal designs. Cuspal coverage; clearance at least 1.0 mm.
Reverse bevel.
Wax pattern. Spruing to avoid distortion.
Investment. Risk of loose particles.
Fitting. Mesial and distal adjustment.
Occlusal check; progressive adjustment; criteria.
Cementation arrested.

Anterior gold inlays
Largely superseded by composites and full crowns. Restorations compared. Brief details of design Classes III and IV.
Role of palatal lock. Application general principles.
Incisal protection allows effective resistance and retention.
Additional retention; slots, pins, pits.

11

Indirect gold inlay technique

The indications for the indirect technique in the construction of gold inlays lie chiefly in the avoidance of prolonged and difficult procedures in the mouth and their replacement by those which can be carried out more effectively by a skilled technician in the laboratory. The very considerable saving in chairside operating time is of importance to the busy practitioner whose clinical work must be supported by a technician who appreciates fully the properties of his materials and the details of technique both in the laboratory and in the surgery.

The larger the size and the greater the complexity of the restoration, the stronger the indications for the indirect technique. This is all the more true when factors such as accessibility, visibility, and control of saliva are considered. The application of the indirect method to the simultaneous construction of several extensive restorations in adjacent teeth, places it as one of the most important methods of treatment in the restorative dentistry. The indirect techniques are, moreover, the basis of all the more extensive and complicated procedures of crown and bridge work. For this reason they are particularly important.

Class II, MOD, and large Class V restorations are the inlays in which indirect methods are used to best advantage. Classes I, III, and IV are generally better done directly.

As an example of the application of the technique, we may take a fairly extensive MOD restoration in a molar. The crown has probably been restored with amalgam on several previous occasions and now has fairly deep approximal cavities. It requires restoration of the occlusal surface, but probably not to the extent of replacement of the cusps. When the cavity preparation is completed it will be necessary to take an accurate impression of the cavity itself and of as much of the surrounding tooth surface as is required to ensure that all the inlay margins can be correctly constructed and finished.

Next, a record will be required of the crown of the prepared tooth in its exact relationship to the adjacent teeth. The very least that is necessary is the contiguous interproximal areas of the adjacent teeth and their occlusal surfaces. Finally, a record of the opposing teeth in centric occlusion is needed in order to build the correct functional form.

There are very many minor variations in technique and materials, but in general there are two patterns of cavity design—the slice preparation and the box preparation. There are also two types of impression techniques. The copper-band method, by which a rigid composition impression is taken of the isolated prepared crown; and the elastomer technique, which uses an elastic material to record the prepared and adjacent teeth in the same impression. Although several combinations of these methods are used in practice, probably the most successful is the slice preparation with the elastomer impression technique. For the single restoration, however, there are still indications for the copper band and composition impression and as this is still widely used it will be described.

The slice preparation

The essential characteristic, from which the design takes its name, is the treatment of the mesial and distal surfaces of the crown by slicing or abrading away the enamel and dentine to give two plane surfaces which are nearly parallel to the long axis of the tooth and transverse to the mesio-distal axis (Fig. 11.1). The long axis of the tooth is generally the line of insertion of the inlay. It is therefore clear that the mesial and distal planes must converge occlusally. The nearer they approach to parallel, the greater the retention, and it is to this application of the general rule concerning the longest single line of insertion (p. 165) that the slice preparation owes its efficacy. As with the internal taper of the cavity, a convergence occlusally of about 5 degrees is acceptable. With angles greater than this, retention of the inlay declines very rapidly.

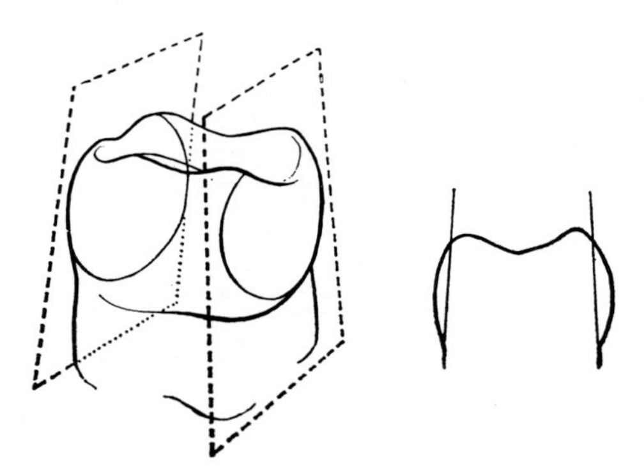

Fig. 11.1. The planes of mesial and distal slices are near-parallel, converging occlusally, and transverse to the mesio-distal axis of the crown.

Near-parallel planes can only serve retention as long as the inlay is rigid. If it is deformed under functional stress, its retention is immediately destroyed. For this reason, the design of the cavity must include features which impart rigidity; these are the mesial and distal slots and the occlusal channel which joins them together with the choice of a sufficiently hard gold, they are just as important as the correct siting and paralleling of the slices.

The proximal slice may also be regarded as a *modified bevel of the axial and cervical margins*. The problems involved in establishing these bevels, the cervical bevel in particular, have been discussed. A correctly sited slice bevels all interproximal margins and allows a knife-edge gold margin to be placed with much greater accuracy than can commonly be achieved by other means.

To sum up. The indications for the slice preparation are: **the need to use the indirect method to which it is particularly suited: the large cavity with deeply sited cervical margins: the need for a high degree of retention** and occasionally, economy of the tissue of the natural crown. This last factor is often given undue prominence, particularly in the case of 'the minimal cavity'. In fact, gold inlays are practically never chosen for minimal cavities.

The contra-indications to the slice preparation are few: the chief is seen in the distal aspect of a tooth with a pronounced distal curve of the root running sharply away from the cervix. This sometimes occurs on tilted second and third molars (Fig. 11.2). It is easy to see that an attempt to form near-parallel slices would lead to difficulty with the distal cervical margin.

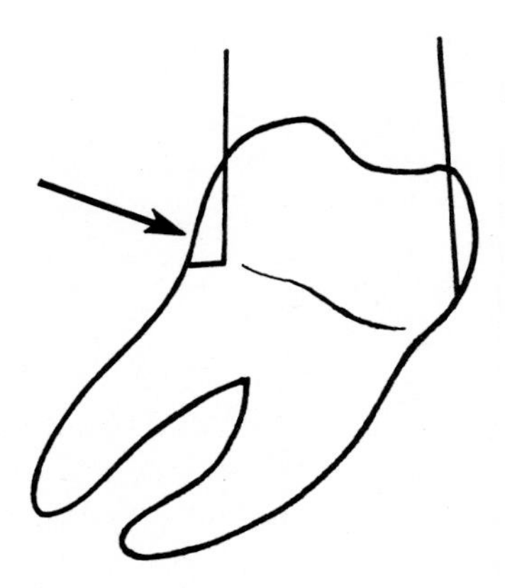

Fig. 11.2. The distal aspect of a tilted tooth is generally a position where the slice is contra-indicated. Note also the shortness of the mesial slice and the ledge, indicated by the arrow, formed by the distal slice.

The concave cervical margin, as is found in the canine fossa of the maxillary first premolar, may also give trouble. This type of case is overcome by a combination of the slice and the more normal 45-degree bevel of the cervical margin. The visibility of the buccal aspects of mesial slices on premolars is a disadvantage which can be mitigated by slight angulation of the slice so that the buccal embrasure suffers less encroachment (Fig. 11.3).

Steps in preparation

The cavity is first rendered caries free and any residual amalgam from previous restorations removed. The contact areas, if not already accessible,

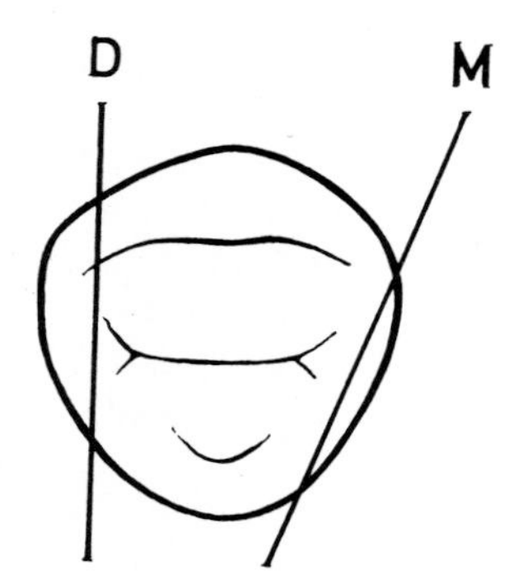

Fig. 11.3. To reduce visibility of a mesial slice it may be angulated.

are freed by the use of metal abrasive strips so that a disc may be passed down the mesial and distal aspects. The abrasive discs used for the initial slices are either lightning discs, if the contact areas are still close, vulcarbo, or diamond discs. These must *always* be used with a disc-guard to avoid the very real danger of cutting the tongue or the cheek.

The initial slices are made by passing the disc, revolving at moderate speed, through the contact area and lightly abrading the interproximal surface, keeping correct angulation and avoiding any risk of producing a step at the cervical margin. At this stage the slice should extend to the free margin of the interdental papilla but, in order to avoid causing haemorrhage it should not go beyond this point (Fig. 11.4).

Attention is now turned to the remainder of the cavity. Light undercuts are removed by cutting back, deeper ones remain to be blocked out; the occlusal outline is established. The lining, which serves both structural and protective functions, can now be placed. With the cavity completely dry and the cement mixed to thick creamy consistency, the depth of the cavity is covered, undercuts are filled and also the mesial and distal boxes, which are usually large and irregular. Now, whilst the cement is still conformable,

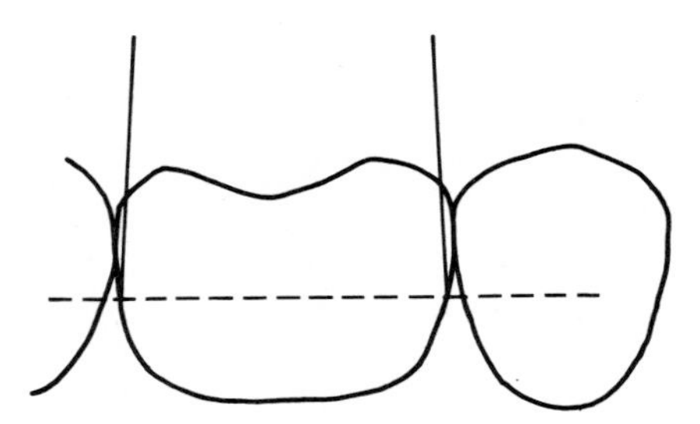

Fig. 11.4. As a first step in preparation, slices are taken to, but not below, the gingival margin.

using instruments moistened with alcohol, the cement is brought rapidly to shape in three particulars:

1. The occlusal channel is established with a flat floor at about 2.5 mm depth.

2. The initial slices are flattened so that the cement forms part of the plane surfaces.

3. The outline of the approximal slot is established in the centre of the slice area. These slots are a feature of great importance, for, with the occlusal channel, they give the inlay rigidity which is essential to retention.

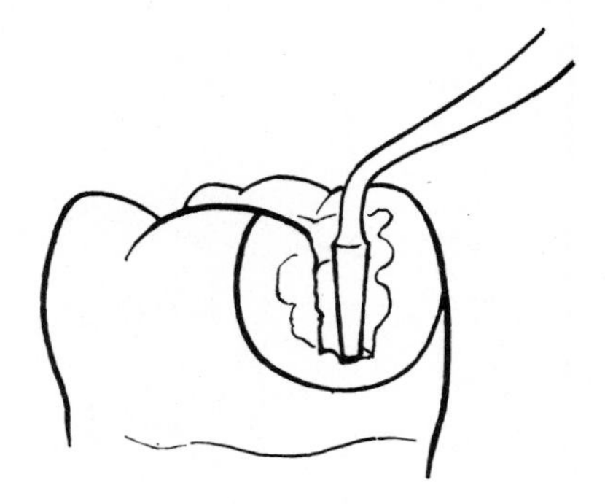

Fig. 11.5. The soft cement is shaped to the plane of the slice and the outline of the slot.

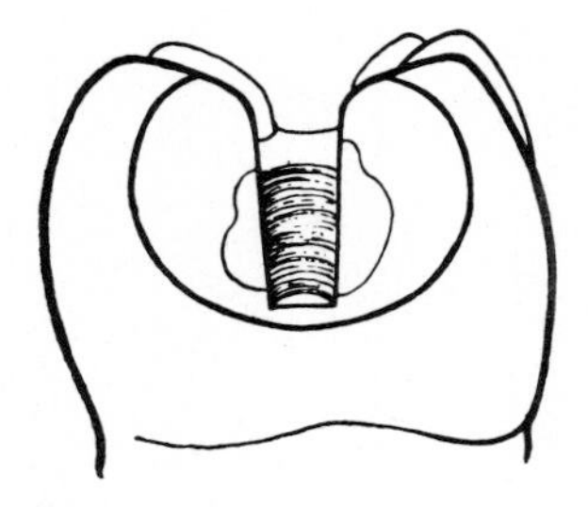

Fig. 11.6. To demonstrate the final relation of the slot to the slice. Its base lies in dentine and about 1 mm inside the cervical margin.

Each slot is shaped as an inverted half-section of a truncated cone. It is in fact the shape which would be made by a tapered fissure bur. It is first established in soft cement by the use of a tapering plastic such as an Ash 49, or an instrument specially made for this purpose, Porro Special A (Fig. 11.5). The shape and position of the slots must ensure that, when completed, *the cervical margin is in dentine, but it must lie at least 1 mm within the outline of the slice* (Fig. 11.6). The size of the slot depends upon the size of the inlay and the rigidity needed, and it is outlined as accurately as possible in the cement. The slot should be made deeper in the pulpal direction than is eventually intended, for some of this will be reduced by the final slice.

The next step is to finish the internal design of the cavity, now outlined in cement as described. This is done with rotary instruments at low or moderate speeds. Plain tapered fissure and end-cutting burs are the most useful. The occlusal channel is first redefined, establishing the outline, the correct flare of the walls, a smooth floor and a clearly defined line angle. In finishing the slot, a plain cut, tapered fissue bur will be used at medium or low speed. In the case of an extensive box where great rigidity and retention are required, the slot can be enlarged in a transverse direction so that it becomes semi-elliptical in cross-section (Fig. 11.7). The walls of the slot should not be more tapered than is necessary to ensure withdrawal and, of course, in the MOD preparation the mesial and distal slots *must be parallel to one another and to the line of withdrawal.* It is at this stage that the siting of the floor of the slot in dentine is checked.

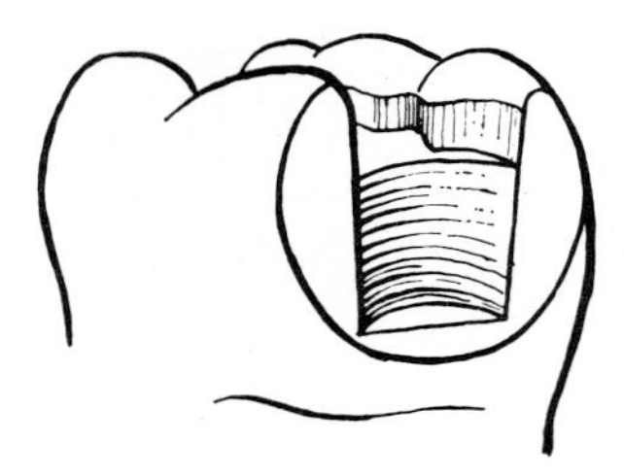

Fig. 11.7. In a wide cavity the slot may occupy the greater part of the slice.

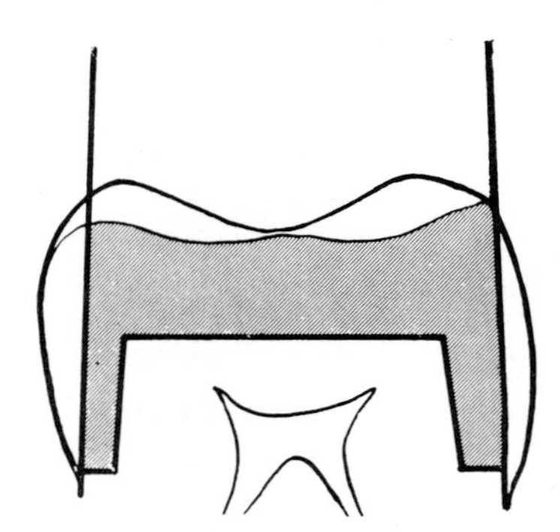

Fig. 11.8. Rigidity is essential to retention and is gained by reinforcement of the gold by the channel and interproximal slots.

The definitive slice may now be made. A fine-grit disc is carefully aligned with the previous slice plane and, with the disc turning at about 1500 r.p.m. and with a slow side-to-side movement, the slices are re-established in their final position, particularly with regard to near-parallelism and the position of the cervical margin (Fig. 11.8). The slice margin must lie below the floor of the slot by at least 1 mm for it is to act as a cervical bevel. In achieving this final slice, especially in the deep cavity, the disc usually cuts deeply into the gingiva. Indeed, in the very deep cavity, when the root contour is suitable, it may be necessary to carry the disc to the level of the interdental alveolar crest. The laceration normally caused produces considerable brief haemorrhage and this is the reason for delaying the definitive slice till after lining. Healing occurs very rapidly and the transient periodontal damage is a small price to pay for a good cervical margin to the restoration. Gingival damage *and* a poor margin are an unacceptable combination!

At this point the relationship of the floors of the slots and the margins of the mesial and distal slices is reviewed. It is sometimes difficult to define these with a probe and it may be impossible to determine them accurately without a trial impression, which for this purpose can quickly be done with a copper band and composition, even though the final impression may be intended in elastomer.

The last stage of preparation is the formation of the occlusal bevel. This follows the same principles as those governing the direct preparation. The slices should be continuous with the occlusal cavo-surface bevels at their mesial and distal extremities.

As with all inlay preparations, the prepared surfaces should be as smooth as possible: roughened and scored surfaces do not contribute to retention. Finishing burs, plain tungsten carbide burs, Baker–Curson burs, fine carborundum points, and fine discs all have a part to play. This reduces distortion of non-elastic impressions and of wax patterns and so contributes to the accuracy of fit of the cast inlay. The essential characters of the completed preparation are shown in Fig. 11.9.

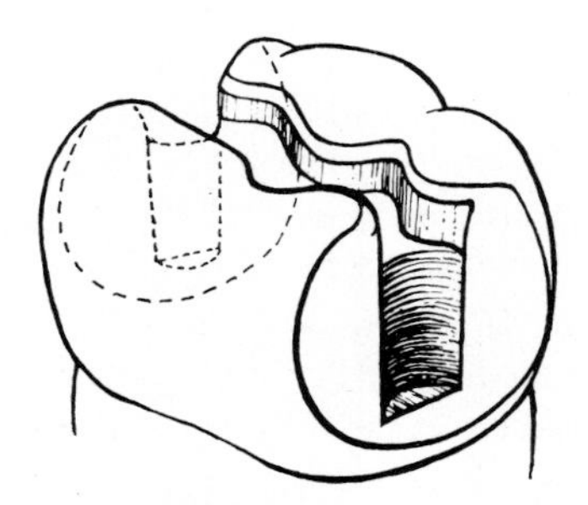

Fig. 11.9. Completed MOD slice preparation on a molar.

Elastomer impression technique

The characteristic elasticity of the silicone and thiokol elastomers allows an accurate impression of undercut areas. Some of these areas of the crown lie below the gingival margin and form one wall of the gingival crevice. To take an impression of these areas the elastomer must fill the crevice and to ensure that this can be done it is usually necessary to open the crevice by inserting a pack carrying an astringent or vasoconstrictor drug. The pack is formed of

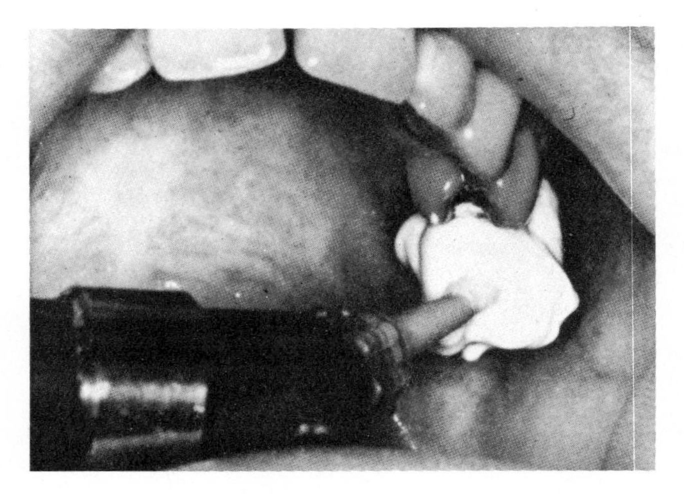

Fig. 11.10. The injection of silicone elastomer into the detail of a preparation.

a cotton wick of a thickness suitable to the size of the crevice and long enough to encircle the cervix if required. It may be placed in position whilst preparations are made for the impression. The astringent most commonly used is 8 per cent aqueous solution of zinc chloride. As an alternative to this, 0.1 per cent solution of adrenaline hydrochloride can be employed: it is of course chiefly active where absorption takes place through areas denuded of epithelium, and for this reason a combination of both drugs has found favour.

With the crown and gingiva isolated with cotton-wool rolls and dried, the pack is inserted into the crevice with two square-ended flat plastic instruments about five minutes before the impression is to be taken. The blade of the first instrument stabilizes the pack already in the crevice whilst the second, working away from the first, packs the next section. The question which inevitably arises is whether this process, which sometimes must be fairly forcible, produces any permanent damage to the periodontal tissues. Using the technique and drugs described and no more force than is needed to achieve the objective, it is now agreed that recovery is rapid and permanent changes are virtually unnoticeable.

Both thiokol and silicone materials require the use of an impression tray to cover the prepared crown and the adjacent teeth: there are numerous methods of providing this. There are various proprietary disposable trays made of polythene and other plastics and designed for the purpose. For most thiokol techniques a rubber adhesive is needed to stick the thiokol to the tray or composition. With silicones, a perforated tray is adequate.

When a model is available a tray may be constructed from any of several materials. Polystyrene sheet can be softened by heating and then moulded to shape, trimmed and perforated with a hot pointed instrument. Quick-

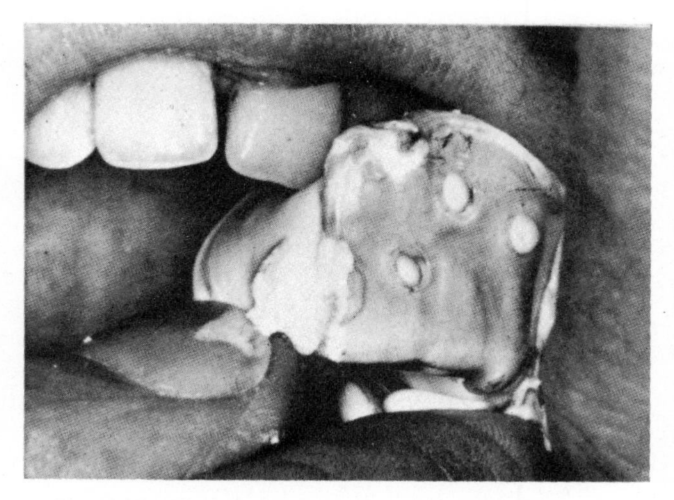

Fig. 11.11. The main body of material in a polystyrene tray.

setting acrylic resin can also be used to produce a similar tray. Expanded aluminium sheet is easily cut and shaped. These trays may be dammed at each end with composition, in order to locate them more accurately on the teeth.

As to the impression materials, suffice it to say that each proprietary make, and sometimes each batch, seems to have its characteristics which are modified by many variable factors. The amount of catalyst used, the speed of mixing, the ambient temperature, the age and shelf-life of the product—all these have an effect upon setting speeds. Some techniques involve the use of two mixes. The first, so-called 'light-bodied', more fluid and slower-setting, is used to fill the gingival crevice and the detail of the cavity by using a plastic instrument or a special syringe. The second—'heavy-bodied', less fluid, and quicker-setting — is placed in the tray and when positioned becomes fused with the first and helps to carry it into all the details of the impressed area (Figs. 11.10 and 11.11).

When tray and impression material are ready, the gingival pack is carefully removed, the area dried and kept dry. The elastomer is injected and the tray carried to place and held in position till well set, after which it is removed and closely inspected. The points for particular attention are these:

1. All cavity surfaces and line angles must show fine detail with no air-blows.

2. The cavo-surface bevel must be clearly shown throughout its length and must be seen to be continuous with the slices.

3. The margins of the slices must be traceable throughout their full extent and must embrace the whole of the interproximal parts of the cavity preparation.

At the cervical, the margins must be beyond the floors of the slots by at least 1 mm. The reason for the insistence upon this relationship between the slot and the slice can now be appreciated, for if this relationship exists and the cervical floor of the slot is in dentine, then in the great majority of cases the slice will be correct in other respects.

As well as the prepared crown, the impression of adjacent teeth must be checked. Here, the interproximal surfaces next to the preparation and all occlusal surfaces must be clear in detail.

With practice, the habit of reading the impressed surface is acquired very effectively, and it is often easier to detect faults of design from the impression than from examination of the tooth, especially where a deep cervical margin is concerned. If a serious fault in design or failure in impression is found at this stage, repetition of the impression is unavoidable. This emphasizes the value of a trial impression when this can conveniently be included. It will be described later.

When the operator is satisfied that he has an accurate impression, the next step is to record the occlusion. There are many complicated methods of doing this, but for a restoration limited in extent, the wax squash bite is probably as good as any. The disadvantage of all such records is that they register centric occlusion and not lateral positions, which must be checked in the mouth. The record of the occlusion must be given as much care and attention as any other part of inlay technique. The fact that attention is so often focused upon a difficult preparation and impression, that the bite recording comes late in the appointment when patient and operator are tiring, and the latter may be pressed for time, all these are circumstances which make the temptation to accept something that 'will do' almost irresistible. It is a cause of great waste of time when the occlusal relation is found to be wrong, and a good inlay spoiled by excessive grinding and reshaping is a sad sight for any operator.

For the squash bite a piece of sheet pink wax about 3.5 × 4.5 cm is required. At one end of this is placed a strip of thinnest gauze or cellophane, 3.5 × 1.5 cm and the warmed wax is folded in three so that the result is a rectangular bite wax having a layer of wax, a layer of gauze or cellophane, and two layers of wax. The centric occlusion of the patient is checked, noting particularly the cuspal relationship on the *opposite* side of the mouth, for with the wax is position the prepared side cannot be seen.

The wax is evenly warmed and placed so that the prepared tooth is centrally located, with the one layer of wax towards it and two layers towards the occluding teeth. With finger and thumb the wax is gently moulded to the lingual and buccal aspects of the teeth *opposing* the preparation, and the patient told to bite centrally and the occlusion is checked on the opposite side of the mouth. When chilled and removed this squash bite should show a detailed record of occlusal and lateral surfaces of the opposing teeth just beyond the survey line of greatest circumference and an impression of the occlusion of the prepared quadrant into which the working model can be fitted. From the elastomer impression two models are made, one of which is

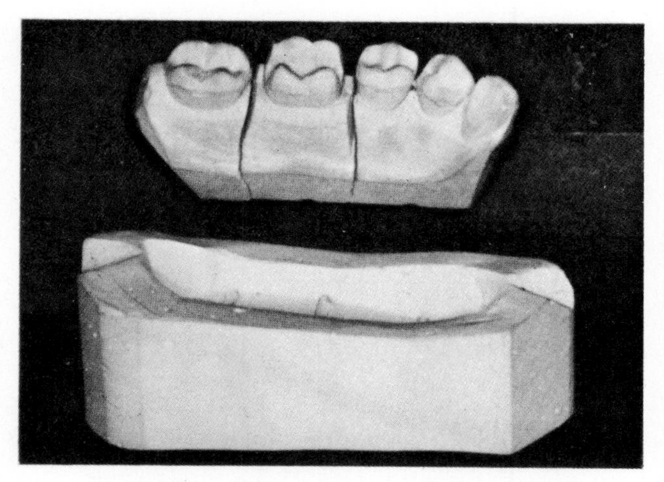

Fig. 11.12. Split working model obtained from an elastomer impression, with its locating bite block.

used with the wax bite to give an articulated model: this working model is later split and trimmed to be used as a working model on which the inlay is made (Fig. 11.12). The second model is used as a master for the final fitting of the inlay. The purpose of the gauze or cellophane in the wax bite is to keep the wax together, but both cotton and cellophane have a thickness which must be compensated by the use of a thin tin, gauge 4, carefully burnished to the articulated model, *but only on the tooth which will occlude with the wax pattern* of the inlay.

The more important points in the laboratory preparation of the inlay are followed on p. 202.

Copper band and composition impression technique

In the case of a single-tooth preparation, there are times when the depth of the cervical margin makes this technique useful, and in multiple-crown work the method is still of value. For these reasons ability to use the technique quickly and accurately is still a useful acquisition in restorative practice.

The selection of the correct size of copper band is of primary importance. **More time is wasted in useless attempts to secure a good impression with a band which is too big, than from any other single cause**. It is therefore, necessary always to choose a band which, when contoured and seated, will fit snugly around its periphery. It follows that if the exact size is not available, it is better to select a size too small and enlarge it to the correct size. The temper of the copper should be moderately soft. If too soft the band is too easily bent and distorted: if too hard there is a tendency to spring which may distort the impression without giving any sign that this has happened.

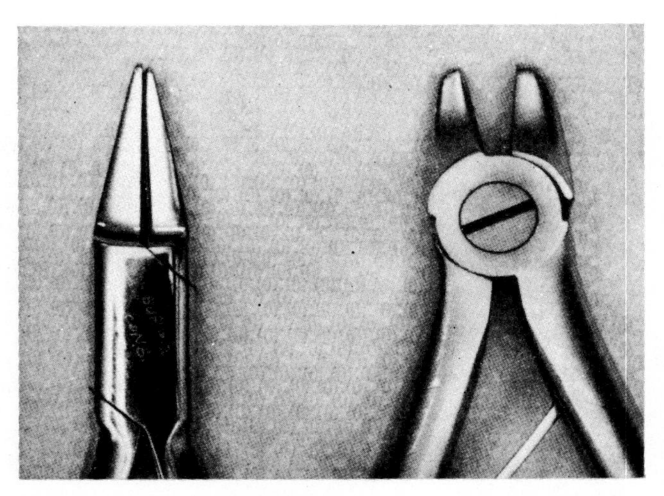

Fig. 11.13. *Left*: collar pliers (Ash 118). *Right*: band expanding pliers (Ash 130), suitable for manipulation of copper bands.

The manipulation of copper bands is much facilitated by the use of correct pliers. Two patterns are suggested (Fig. 11.13). The collar pliers, Ash 118, are generally useful in conforming the band to shape, turning edges and contouring without roughening. The band expanding pliers, Ash 130, have short powerful jaws. They are used to compress the periphery of the band and so to increase its circumference.

Having selected the correct band it should be trimmed at its cervical end so that it covers both slices and extends approximately 1 mm beyond the slice margins. On the lateral aspects of the crown the band extends to the bulge of the surface, so that when seated it is steady and firmly located, but nowhere covers an undercut area. When completely trimmed and shaped it has the appearance shown in Fig. 11.14. If the band is incorrectly trimmed and allowed to cover undercut regions of the crown, when the impression is taken it will be impossible to remove it from the tooth unless it is distorted or 'dragged', or broken. If dragging or breakage occurs the fault is often apparent, but distortion is usually imperceptible and is therefore a much more serious cause of failure because the fault is not discovered till the completed inlay which looks correct and *fits the model*, is found not to fit the tooth in the mouth of the patient.

It is useful to regard the first impression as a trial impression to check upon cavity detail and fit of the band. If all preceding preparations have been satisfactory, the first impression will be correct, but it is easy to repeat if required.

The impression is taken by filling the copper band with composition. Two grades of this material are available, one of brown colour which is harder and has a higher softening range of temperature, and one green which

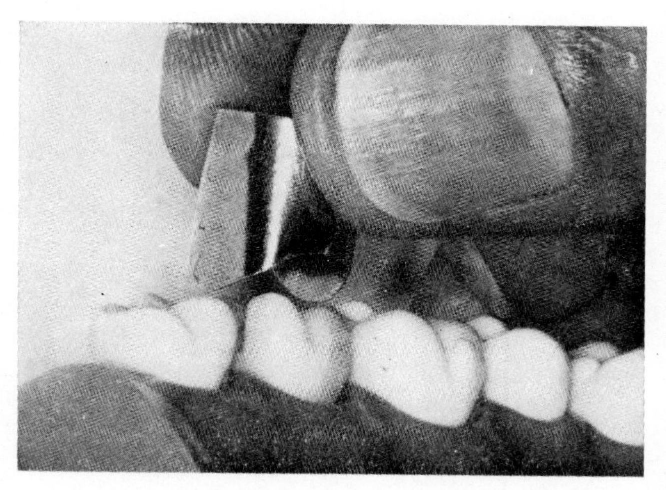

Fig. 11.14. The general form of the copper band trimmed for composition impression of a MOD cavity.

softens at lower temperature and is not so hard. On the whole, the former is preferable because it favours breakage rather than distortion. The prepared cavity may be lightly lubricated with liquid paraffin. When the band is filled, the cervical end should present a smooth, unwrinkled surface of composition and the whole contents should be uniformly soft and conformable, but not runny. The band is correctly located on the tooth, and with firm finger pressure carried home to its previously checked position. Further pressure would risk carrying it into undercut areas. The composition is chilled with water and removed. If the impression has been properly carried out the band should be removed with no more than a moderately firm pull. If great force or the use of instruments is required to remove the impression, this is generally an indication that the band is incorrectly trimmed and has overlapped an undercut area. The impression will be broken or distorted.

Inspection of the impression follows the points already mentioned, cavity detail, slices, slots, and cavo-surface bevel. The commonest causes of failure at this point is the inadequately trimmed copper band. Now the relationship of cavity to the band may be closely studied and necessary modifications in its contour made with greater precision.

The greatest difficulty in this, as in all other techniques, is the cervical margin. An advantage of the copper band method is that when properly used it is more effective in obtaining a clear impression of the deeply placed margin. As a result of study of the trial impression it may be necessary to adjust the preparation by deepening the slice or the slot, or in other details. The band may also need adjustment to allow it to reach the cervical margin. As a rule, one slice is deeper than the other and this calls for adjustment of contour. In order to deepen one margin of the band, all the other margins are reduced evenly by the same amount. Any areas of band overlying undercuts must be carefully eliminated before the second and final impression is taken.

In certain cases the depth of the cervical margin makes it difficult to place the band fully loaded with composition. In such cases the empty band may be carried to a position which is checked by inspection and the use of a probe. The impression is then taken by using a bullet-shaped piece of composition about two-thirds of the diameter of the ring and softened at its rounded end. This method is often successful when the other fails.

The wax bite record is similar to that previously described, yet different in some important details for it is necessary to obtain more detail of the prepared and adjacent teeth into which the die of the preparation may be placed.

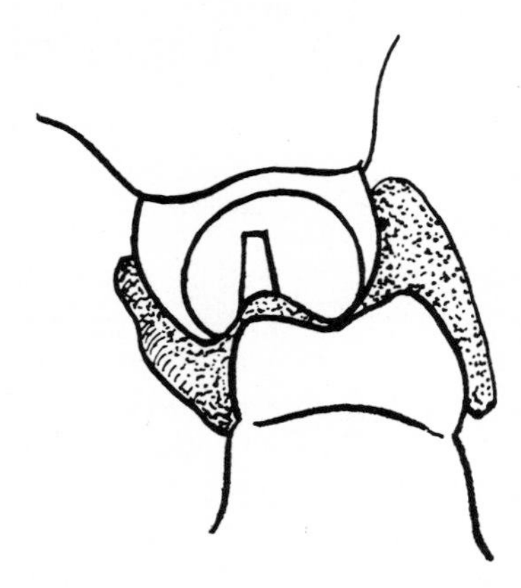

Fig. 11.15. Sectional view of wax bite in position and teeth in centric occlusion.

When the wax, evenly warmed, is placed in the mouth the two layers of wax should be placed towards the *prepared tooth* and it is adapted first to the preparation and to the lateral surfaces of the adjacent teeth; this is the reverse of the procedure used with an elastomer technique. Then, as the patient closes, the wax is conformed to the occlusal surfaces of the opposing teeth (Fig. 11.15). The cuspal relationship is checked, the wax chilled and removed and dried. Inlay wax is now dripped into the wax impression of the prepared tooth only. As soon as it has begun to solidify the wax bite is replaced in the mouth and the patient instructed to close gently and 'find his way into position'. In this manner an accurate record is obtained for localization of the model registration and of the occlusion.

Laboratory procedures

A brief account of the laboratory procedure will now be given, paying attention to the most important points of technique in the successful production of an MOD inlay.

When an elastomer impression has been used, the working model will be cast in a hydrocal plaster (Fig. 11.12) compensated for expansion. In the case of the copper band and composition impression this is most likely to be copper-plated and formed to provide a root portion which localizes the preparation (Fig. 11.16). The occlusion may be registered either in a plaster block articulator as shown or in one of several designs of hinged articulator suitable for inlay construction.

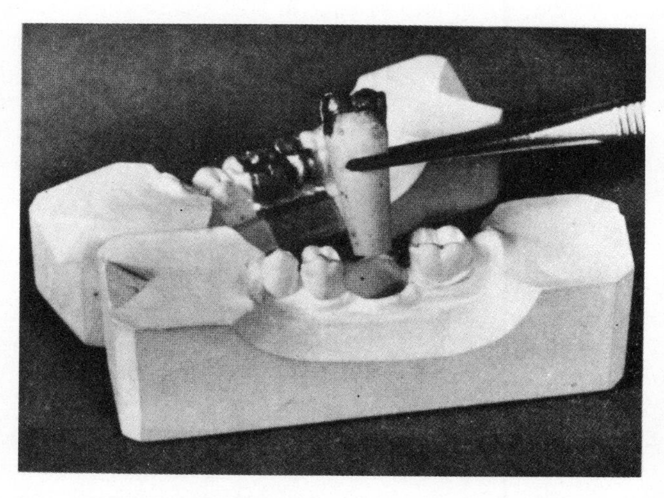

Fig. 11.16. Copper-plated die being placed in articulator block. Note the tin foil burnished to the occluding teeth, shown in the background.

There is one very significant generalization which should be recognized at this stage. It is that, in recording occlusion and constructing the inlay *each of the small technical errors which are commonly made can only contribute to raising the occlusal level of the inlay.* As a result, the commonest fault when the restoration is fitted is a high bite and the occlusal surface must be reduced to allow the teeth into normal occlusion. The careful registration of the occlusion and attention to the following points during articulation of models and pattern production will help to avoid this defect:

1. The *opposing* side of the wax bite should be cast first.

2. The hydrocal die of the prepared side or copper-plated die of the preparation, must be localized in the wax record with just as great care as has been shown in taking the impression. If necessary the wax must be cut back if it is over-extended.

3. The die must be very firmly pressed into the wax bite.

4. The tooth opposing the inlay should be relieved with thin tin foil, if the registration was performed in the manner described.

There is something to be said for using thin tin foil in all cases, for it eliminates the possibility of wear and loss of the opposing cusps during the working of the wax pattern and the inlay. Each of these points influences the occlusal level of the inlay. If all are disregarded an inlay which is grossly high on the bite will result.

In building the wax pattern a lubricant should be used on a warm die and the initial layers of wax must be placed firmly in contact with the cavity surface. In contouring, attention is given particularly to contact areas, bevels and sliced margins. The approximal areas of *adjacent* teeth on the model may be lightly scraped to ensure that in the mouth the contacts will be slightly tight. The position and extent of the contact areas must be carefully reproduced in order to restore correct anatomical form. Bevelled margins on the occlusal surface may be trimmed to their final state, but slice margins should be overlapped by wax to the extent of 0.5 to 1 mm, no more. No attempt should be made to reproduce the actual knife edge of this margin at this stage, and the wax margin should be thicker than normal in order to avoid possible fracture.

An experienced technician will be able to work to a sliced margin with a high degree of accuracy given a good model. For many, however, a small latitude is desirable; a gross overlap is very undesirable and will, if as a wax pattern it was capable of being withdrawn without breaking, result in a casting which cannot be replaced on the die without damage to the die margins.

When the angle between the slice and the adjacent surface of the crown is more than 5 degrees it is helpful to finish the slice margins of the pattern in a slightly softer wax in order to achieve a 1 mm overlap which can be withdrawn from the model without fracture. Pink casting wax has the correct properties for this purpose. The pink wax periphery undergoes a small distortion on withdrawal, without breaking. It is desirable, if not essential, that the fitting surface of the wax and casting should present the appearance shown in Fig. 11.17, so that the true margin of the slice can be clearly seen.

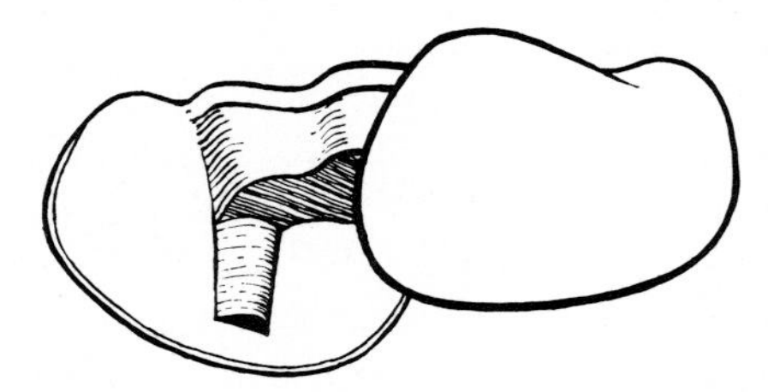

Fig. 11.17. View of fitting surface of cast inlay before trimming, showing a small overlap to the slice margin, later to be removed in fitting.

The cast inlay is fitted to the copper or stone die for finishing. Care is taken to preserve the cavity margins of the die in order not to trim the inlay margins short. In the case of the stone die there is a master model to be used as a final check. When fit, margins, contacts, and occlusion are correct, the inlay is brought to a high polish and is ready for fitting to the tooth.

Fitting the inlay

The procedure in the surgery is much the same as that described on p. 201 but with special attention to the slices. First the contacts are adjusted, for until this is done the inlay cannot be fitted into the cavity, then occlusion and margins are adjusted. Here the slice margins are brought to their final state by successive trials; since this is the culmination of the technique the process must be closely followed.

1. When the inlay is fitted to the cavity an examination of the occlusal bevel should show very little excess, if any, but a slice margin may present a perceptible catch to the probe. The periphery of both slices is carefully explored with a curved probe to find out where the overlap exists.

2. With inlay off the model and the fitting surface under direct vision, the margin is reduced, using a cup-shaped vulcarbo disc in a straight handpiece. The disc surface should be at right angles to the slice plane (Fig. 11.18).

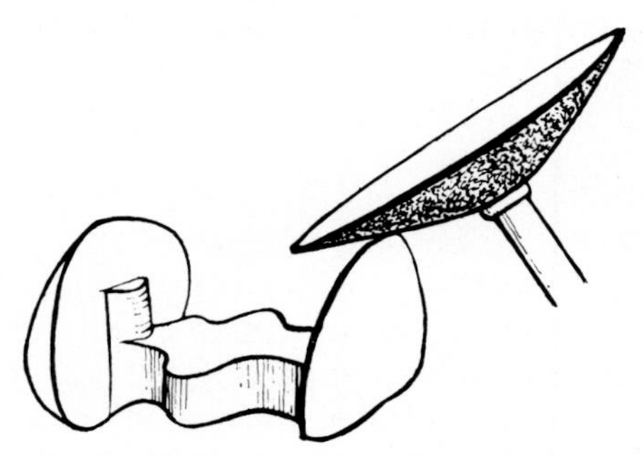

Fig. 11.18. Trimming the margin of the slice back to final contour, using a cup-shaped abrasive disc.

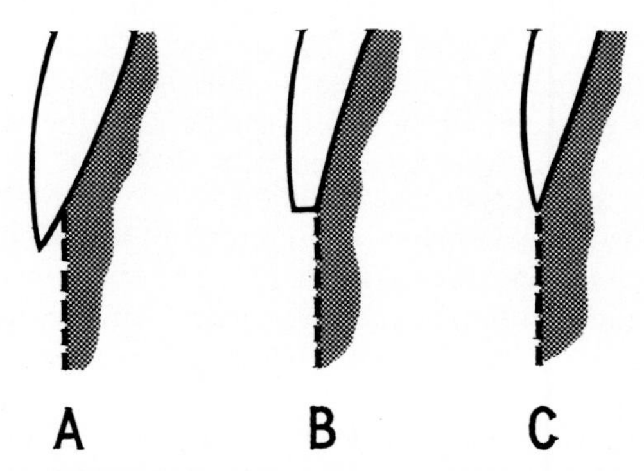

Fig. 11.19. Final adjustment of slice margin (A) inlay margin overlaps slice; (B) inlay trimmed off model as in Fig. 11.10 and replaced; (C) margin 'knife-edged' on model.

Watching the fitting surface, the excess is removed almost to the outline of the slice and the burr which forms is removed with a trihedral scaler before the inlay is placed back on this model. The disc is used to thin the gold margin down to a knife-edge (Fig. 11.19).

3. The inlay is now tried in and exploration of the margin with a fine probe should show only the smallest catch. The areas where an excess can still be detected are noted. The inlay removed, a small reduction is made with the inlay off the model, then 'knife-edged' on the model and retried in the mouth. This procedure continues until the slice margin is everywhere correct, presenting no catch to the exploring probe.

4. Finally, when the fitting of the inlay is complete and polishing done, the casting removed from the cavity for cementing, the smooth shank of a burnisher held at an acute angle may be used to produce the lightest inturning of the margin, so allowing the closest possible apposition of gold to tooth surface. Care should be taken for excessive burnishing may well prevent the inlay from being replaced on the tooth — an awkward situation when cement is setting!

When the set cement is removed and the inlay examined with a fine probe, the slice margins should be scarcely perceptible throughout, and a bitewing X-ray should show complete continuity of the interproximal contour.

The box preparation

The difference between the box and slice designs concerns the management of the axial and cervical margins. In discussing the box type of preparation (p. 190) it was pointed out that an axial margin bevel was impractical below the greatest circumference of the crown (Fig. 9.11) unless the margins were laid well back. This involves placing the axial margins farther out into the embrasures and enlarging the axial bevel to the extent that the pattern can be withdrawn without distortion of its thin edges. The axial margins may be prepared with a straight or bin-angled chisel and can be finished with a fine disc as shown in Fig. 9.13, (p. 191). The object of preparation is to form a 45 degree cavo-surface bevel, continuous, with the occlusal bevel at one end and with the cervical bevel at the other. The bevel of the cervical margin is dealt with in the manner also described on p. 192.

Both elastomer and copper-ring techniques may be used for impression taking. Of the two, the elastomer is preferable, for with the copper ring, which must extend beyond any cavity margin, the operator is more likely to meet distortion due to adjacent undercut areas. An elastomer will give an accurate impression of these. In all other details of inlay production, the two techniques vary little.

To sum up: it may be said that the box preparation is indicated in direct techniques but in the opinion of many it is superseded in indirect methods by the slice preparation. The slice preparation, however, is unsuitable for direct methods. Both box and slice designs can be made fully retentive and, with experience, combinations and modifications of both designs can be used to meet the numerous and varied forms of coronal breakdown found in practice.

Occlusal coverage

In describing the direct MOD gold inlay, it was mentioned that, if cusps were weakened, they could be protected by covering or 'overlaying' them with gold (Fig. 10.17, p. 198). This design is very useful and can be achieved quite easily by indirect techniques. Using high speed, the occlusal surface is evenly reduced, whilst retaining the flattened cuspal slopes, until a clearance of at least 1 mm is obtained between the preparation surface and its opponent. If in doubt, slightly more than 1 mm should be allowed, rather than less. The occlusal bevels now become reversed bevels along the buccal and lingual margins (Fig. 11.20). In forming these bevels the natural slope of the enamel surface is again immediately relevant. The angle of the gold margin should not be less than about 30 degrees and in many instances, for example the buccal margin shown in Fig. 11.20, no additional bevelling is needed.

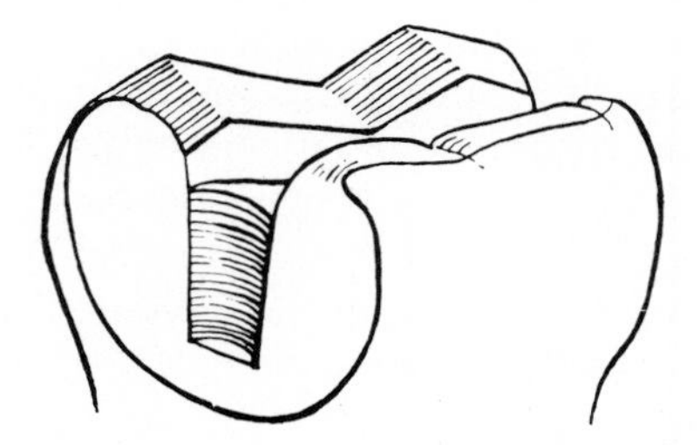

Fig. 11.20. MOD preparation for full occlusal coverage. The occlusal margins are converted to reverse bevels on buccal and lingual surfaces.

Besides protection of weakened cusps, there are other advantages to full occlusal coverage. It has just been observed that what would have been occlusal margins are now sited on lateral surfaces. Here, as reversed bevels, they are generally less exposed to failure due to occlusal wear. Complete coverage, by providing additional connection between mesial and distal, increases the rigidity of the restoration and thereby gives additional retentive strength. By suitable design and choice of hard gold, the full coverage MOD can be effectively used as a bridge retainer.

Finally, coverage of the occlusal allows the form of the surface to be rebuilt for functional purposes. If a tooth is not completely in occlusion with its opponent, the surface can be reshaped to improve its function. This is the principle involved in the so-called 'onlay', in which the occlusal surface alone is covered and remodelled. The onlay, the full-coverage MOD and the three-quarter crown—next to be described—are used in extensive restorative treatment aimed at reconstructing a disorganized dentition.

There is one very important factor to be aware of in all preparations in which the whole occlusal surface is reduced and the prepared tooth thereby taken out of occlusion. The prepared tooth, and presumably its opponent in some cases, erupts in an attempt to re-establish occlusion. The speed with which this occurs is not uniform; it is rapid in young subjects and slower in older patients, but it can never be disregarded. Even when an aluminium crown-form is fitted as temporary cover, it is unwise to delay insertion of the restortion for more than two weeks. *Whenever possible, full coverage restorations should be placed within five or seven days.* Failure to do so may lead to the restoration being high on the bite and adjustment could lead to thinning of the gold and perforation at the time of fitting or subsequently. In either case, modifications of the preparation and remaking may be required.

There is a quick and effective method of making a temporary replacement for these extensive occlusal restorations of all types, including crowns; **it depends upon taking an impression of the region before starting preparation.** If there is a significant coronal defect, quite often there is not, this is first filled with a quickly inserted material such as gutta percha or inlay wax, roughly to restore the coronal surface. An alginate impression is then taken of the tooth in question and the adjacent teeth. This impression is put aside until the preparation of the tooth is finally complete. Then a soft mix of plastic resin (*Scutan* or a quick setting acrylic could be used), is run into the coronal portion of the prepared tooth in the alginate impression, which is carefully repositioned in the mouth till the plastic is set. When removing the resulting temporary plastic restoration can be lightly trimmed and cemented into or on the tooth with thin temporary cement. It normally needs no further adjustment.

The three-quarter crown

This restoration can be used on canine, premolar, and molar teeth. The name was first applied to the canine restoration, in which case it is accurate enough. For convenience, the name is retained in the case of the premolars and molars although in these teeth it is used to describe extra-coronal restorations covering four out of five surfaces of the clinical crowns.

The derivation of the design and the principles involved are best seen in the development from the MOD inlay with full coverage in a premolar or molar tooth. In order to convert this to a three-quarter crown, the lingual enamel is reduced with a high-speed diamond instrument (Fig. 11.21) so that the axial curvature of the surface is abolished and a gentle inclination towards the occlusal is produced. In this manner the slices become continuous with the prepared lingual surface which must not, however, be reduced more than is necessary to place its cervical edge below the free

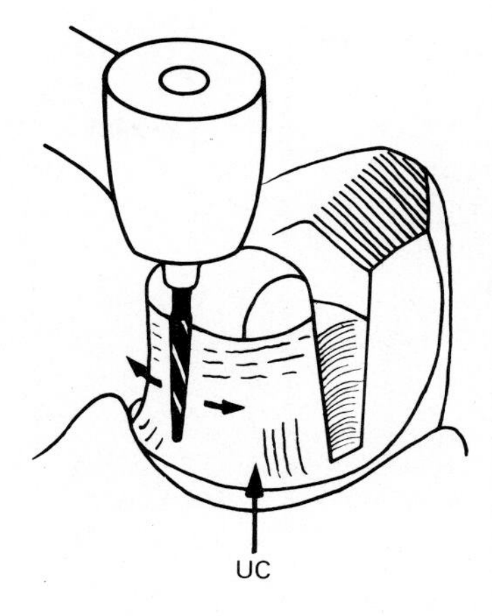

Fig. 11.21. The reduction of the lingual enamel in preparation of a three-quarter crown. Pay attention to the elimination of undercuts in area marked UC.

gingival margin and to eliminate areas which would be undercut with regard to axial withdrawal. In particular, there is a small triangular area near the cervical (Fig. 11.21 UC) lying between the lingual margin of the slice, and the lingual surface proper, which may be overlooked. It is important to ensure that this slightly undercut area is removed and the cervical margin of the preparation runs straight from mesial to distal aspects.

When completed the preparation should present mesial and distal slices with slots, occlusal channel, complete lingual and occlusal cover with a reversed bevel along the occluso-buccal margin. The cervical margin must be knife-edged and, in most cases will be subgingival throughout. In elderly subjects where the gingiva has receded there is little difficulty in providing a supra-gingival cervical margin, but in younger subjects pre-existing cavities, the shortness of the clinical crown, and the need for retention usually make a sub-gingival margin unavoidable.

The design of this restoration confers great strength and rigidity; with near parallel slices it is highly retentive. To facilitate removal of the crown during the process of fitting it to the tooth, the provision of a small cleat on the lingual to allow an axial pull is a step which saves much time and avoids risk of damage. When the wax pattern is complete, a small bead of lower fusing pink casting wax is placed in the centre of the lingual surface (Fig. 11.22). When cast in gold this allows the use of a hooked instrument to exert a strong pull. The cleat is removed just before final polishing.

Three-quarter crowns are not only strong and retentive, they also protect what remains of the natural crown and leave the buccal surface intact. They are often used to restore teeth in which destruction, though extensive, has

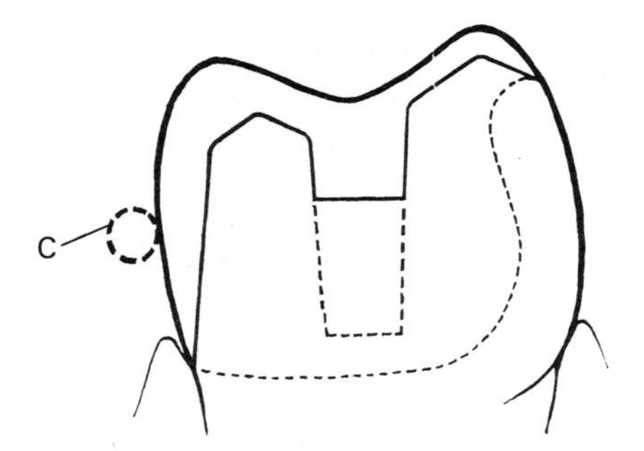

Fig. 11.22. Bucco-lingual section through three-quarter crown on a premolar showing occlusal and lingual cover and position of interproximal slots and occlusal channel. The position of the cleat for removal is indicated on the lingual surface.

spared the buccal cusp, and also in suitable circumstances as retainers in bridgework.

The three-quarter crown on the canine has, in principle, most of the features described above, but their application is modified by the different shape of the crown and by the need to reduce visible gold to a minimum in the interest of good appearance. There are a number of major differences and a few minor ones in the design of this preparation when compared with the premolar or molar restoration. The design now to be described is probably the simplest which meets the basic requirements.

The first stage is the establishment of mesial and distal slices. In this case, though near-parallel in an axial plane, they converge lingually (Fig. 11.23). The palatal enamel is then reduced to give at least 1 mm clearance of the occlusion, using a diamond wheel 1 cm in diameter. This instrument gives

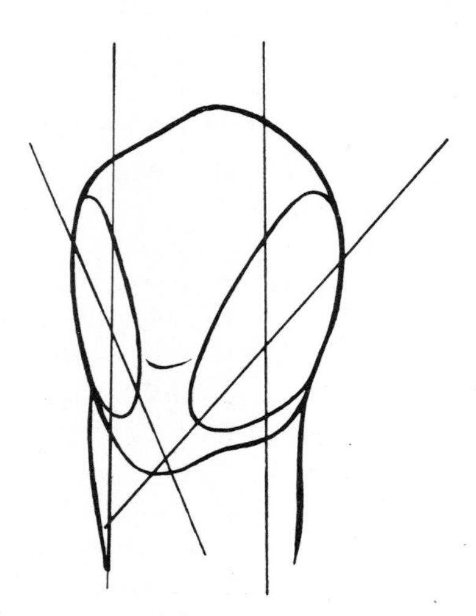

Fig. 11.23. Initial slices in canine three-quarter crown preparation, near-parallel axially, but convergent lingually.

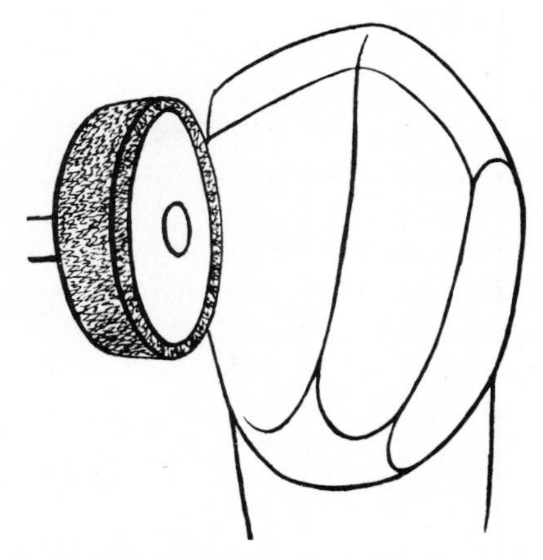

Fig. 11.24. Reduction of lingual surface of canine with diamond wheel, and the establishment of the incisal bevel.

the correct curvature from incisal to cingulum, and the lingual surface is often shaped to form two gently curving surfaces meeting at a very obtuse angle along a line running axially down the centre of the crown (Fig. 11.24). This is not, however, a feature of great importance. With the same wheel a bevel is imparted to the lingual aspect of the incisal margins. The lingual surface is completed by removing cingular enamel along the cervical, joining together the interproximal slices.

The remainder of the preparation is concerned with establishment of retention and strengthening the crown.

Mesial and distal slots are now formed, using 700 or 701 fissure bur. The slots, which must be parallel to one another, run from high upon the lingual and terminate just short of the cervical margin of the slice. They must be clean cut and not less than two-thirds of the bur diameter in depth.

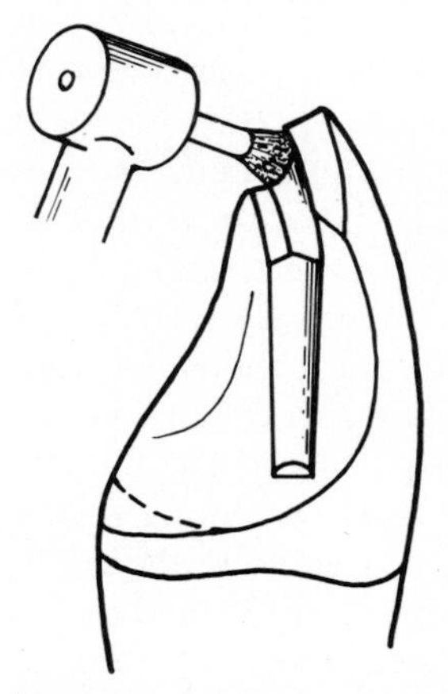

Fig. 11.25. The formation of the incisal groove with an inverted cone instrument.

The strength of the crown in a transverse direction is now increased by forming a groove, V-shaped in cross-section and also in outline, across the lingual surface near to the incisal bevel (Fig. 11.25). This groove joins the free ends of the slots and it must not be placed too close to the incisal, for to do so would invite fragility of the incisal edge and also cause discoloration due to loss of translucence.

In many cases this completes the preparation of the crown, but when additional strength is required still further rigidity can be gained by forming a small cervical shoulder on the lingual. It is at its widest in the midline and tapers to nothing as it passes on to the mesial and distal slices. The shoulder should lie just above the knife-edged cervical margin which must be below the free gingival margin (Fig. 11.26).

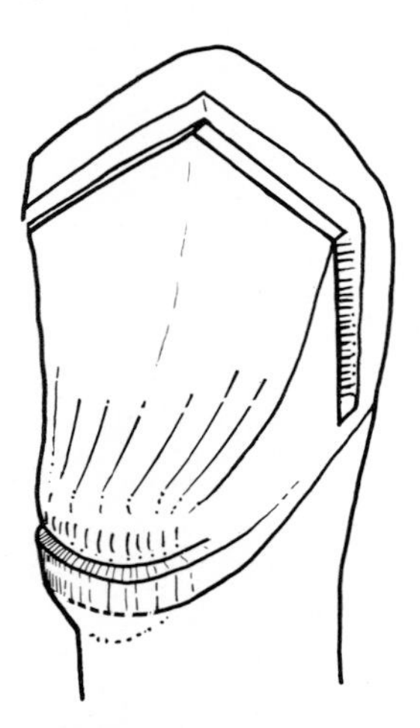

Fig. 11.26. The completed three-quarter crown preparation. Note the abbreviated lingual shoulder to impart rigidity and therefore retention.

The impressions for this preparation and its construction may be taken either with elastomers or by the copper ring method. In both cases the general principles described earlier in this chapter apply equally to this case.

The numerous small modifications in design which have been advocated in the past have been aimed at decreasing visible gold and at the same time increasing the retention of the canine three-quarter crown. These are to be found in works which treat the subject in greater detail.

The techniques described in this chapter are applicable to the Class II inlay, MOD inlay, with or without occlusal cover, and to the three-quarter crown. The loads which these restorations may carry vary widely. It is important to remember that their retention depends chiefly upon near-parallelism and depth of the slices, and ridigity. Rigidity may be increased by the addition of slots, grooves, and shoulders, all of which may be varied in depth as required. Further, rigidity may be increased by thickening the restoration or by using a hard gold alloy.

Note on cervical margins

It will have been noticed in this and all preceding chapters dealing with coronal restorations that much emphasis has been laid upon cervical margins. It is an inescapable fact that the cervical margin of any form of restoration is always the most technically difficult to achieve, and the point at which ultimate failure is most likely to occur.

On p. 67 it was inferred that, as even the best cervical margin has an injurious effect however slight upon the gingival crevice—and a bad one can be disastrous — from point of view of periodontal health it is best to site margins at or just above the free gingival margin. Three circumstances, which arise quite frequently, militate against this good intention, particularly in the case of extensive restorations such as MOD inlays, three-quarter and full crowns.

1. All too often caries or a pre-existing restoration is present at or below gingival level, in a Class II or V position.

2. A short clinical crown, or previous destruction of a crown of normal height, makes it impossible to approach 'the longest single line of insertion' essential to proper retention, without extending sub-gingivally.

3. In the front of the mouth a visible supragingival margin is usually unacceptable both to patient and operator. Passive eruption is to be anticipated and the problem of retention may coexist.

There are three forms of gold cervical margins commonly used, and one but rarely (Fig. 11.27).

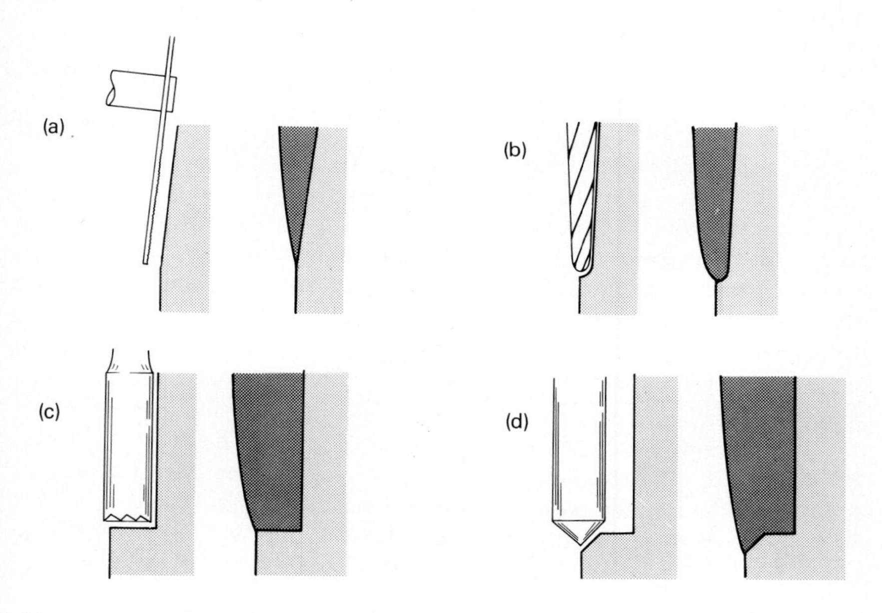

Fig. 11.27. Formation of four types of cervical margin: (a) knife-edge; (b) chamfer; (c) butt joint; (d) bevelled.

The knife-edge, which we have already met in the form of a slice. This can also be formed by the action of a long tapered bur or diamond instrument. It is the easiest to prepare and the most difficult for the technician, because it may not everywhere provide a clear finishing line. If the technician has difficulty so, in all probability, will the clinician.

The chamfer is similar but the instrument used has a rounded end, and rather more tissue is removed. This gives a curved shape to the cross-section of the margin. The result is a clearer finishing line, easier to define and work to on the model. It also thickens the margin, giving a less fragile wax pattern and a more rigid casting.

The butt joint obviously gives the clearest finishing line but does not allow spinning or swaging and, perhaps more important, exposes the widest cement line. It is not normally used in gold work when it can be avoided.

The bevelled margin, developed from the butt joint, we have met in the cervical of the Class II inlay. Like the chamfer it gives a good finishing line and exposes a narrower cement line than the butt joint. The advantages of the bevel have been described.

Those are the theoretical considerations. In practice, the technician *must* have a 'finishing line' to which to work. If it is not apparent on the die, then the clinician should draw it with the finest lead pencil.

No matter how good the preparation, a poor impression gives a poor die which produces a bad restoration.

The die can never be better than the impression; it usually looks worse.

The process of defining margins of the correct shape on the tooth, above, level with, or below the free gingiva calls for skill only to be acquired by practice. This must be accompanied by facility in the use of impression materials, a high degree of technical skill in the laboratory, and the ability to use the properties of gold to best advantage. The management of the soft tissues of the gingiva has been described on p. 214.

The full crown. This is a cast restoration covering the whole of the clinical crown, except where supragingival margins can be allowed. Of the many designs formerly used, the cast restoration is now universal. The crown to be considered here aims simply at the protection of the remaining tissue of a vital tooth and the renewal of the form and function of that tooth. However, a similar type of crown, with only small modifications of design and materials is the most important type of retainer for fixed and removable bridgework.

In the first instance, when clinical indications are in favour of a full crown, any extensive defect of the crown should be restored with amalgam, composite resin or glass ionomer cement, using pin retention where necessary. Pins must be so placed that they are not later uncovered. The crown preparation will be described as though the contours of the crown are for practical purposes normal.

The object of preparation is to remove the surface of the natural crown to a sufficient depth to produce a taper of retentive degree and, what is much

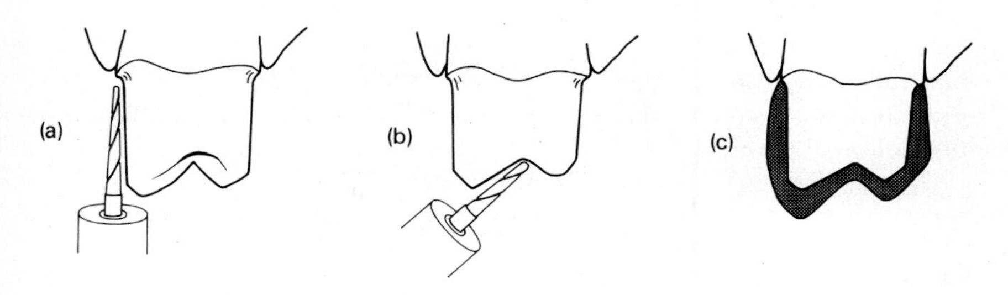

Fig. 11.28. (a) Reduction of lateral surfaces of maxillary premolar for full crown preparation. (b) Reduction of occlusal surface. (c) Sectional outline of finished crown.

easier, a casting of sufficient thickness and rigidity. In effect this usually means the removal of the full thickness of enamel from the greater part of the clinical crown, though this is not essential. Some enamel will probably remain near the cervix, depending upon the relationship of the cervical margin of the crown to the amelocemental junction and the gingiva.

Take as a simple example the preparation of a second upper premolar. From Fig. 11.28 it can be seen that without excessive removal of enamel and dentine from the lateral aspects of the crown, a uniform taper of about 5 degrees can be attained.

If a temporary plastic crown is to be made for this tooth, an alginate impression is taken at this stage and put aside, see p. 226.

The preparation starts with the opening of contact areas, by metal separating strips, and discs properly guarded. Safe-sided diamond or vulcarbo discs are used to remove the convexity of the mesial and distal surfaces down to the level of the eventual cervical margin, as in the first stage of the slice technique. Using a tapered tungsten carbide bur with a rounded tip, or a similar diamond instrument, at high speed, the lingual and buccal surfaces are progressively reduced, maintaining a 2 to 3 degree taper and forming a chamfered cervical margin (Fig. 11.28 (a)). This is extended into the embrasures, joining the lateral and approximal surfaces to produce a uniform taper right round the crown and a cervical margin placed 1.0 mm above the gingival margin.

If on account of recent or longstanding Class V restoration the cervical margin has to be subgingival on one aspect, it may still be possible to locate it at or above the gingival elsewhere. Unfortunately it is not uncommon in practice to have to establish a subgingival margin throughout, with the attendant difficulties this entails.

The third step is the removal of the occlusal surface to a depth of at least 1.0 mm and preferably 1.5 mm. The general form of cuspal slopes and fissures, where they exist, are retained (Fig. 11.28(b)). This is done on the principle of conserving coronal dentine where possible, not thickening the gold of the crown excessively, but particularly to preserve coronal height and therefore retention. Quite often, however, occlusal morphology has been lost by attrition or by early and inadequate restoration. If this is so the occlusal surface can be made flat, but with a minimal loss of height. The general form is completed by reducing any enamel and sharp margins between lateral surfaces and occlusal.

The preparation may now be reviewed with particular reference to the following points:

1. The occlusion must have the appropriate clearance in centric and excursive positions.
2. Lateral walls must be near-parallel and free of undercut.
3. The cervical chamfer must be correctly sited and show a clear finishing line.

The preparation is finally smoothed and polished with finishing burs and sandpaper discs. No roughness or scoring should be visible on prepared surfaces. The finished preparation is often very nearly symmetrical and it may be helpful to cut a shallow vertical slot on the mesial surface. This is reproduced on the fitting surface and makes localization quick and certain.

The alginate impression taken earlier is now used to produce a temporary plastic crown.

The impression may be either by elastomer or copper ring method, the former being more common. Occlusal recording follows the lines described earlier and must be carried through carefully and accurately. The temporary crown having been trimmed and fitted, is placed with a thin mix of temporary cement and should need no adjustment. As an alternative, an aluminium temporary crown, prefabricated, can be adapted to cervical margin and occlusal level and set with temporary cement.

Laboratory procedures and the final insertion of the crown are to those described for large MOD inlays and three-quarter crowns. It is very helpful to have a small spherical cleat cast on the lingual surface. This done, the crown can be easily removed from the prepared tooth without damage.

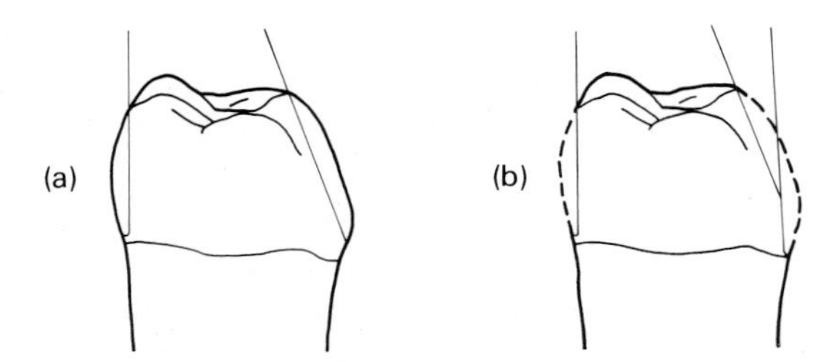

Fig. 11.29. (a) Shows the obliquity of the buccal surface of a lower molar; (b) the method of achieving a satisfactory degree of near-parallelism.

The preparation of a molar tooth presents an interesting problem. Inspection of the anatomy of a lower first molar crown (Fig. 11.29), makes it clear that in the bucco-lingual plane the slope of the buccal surface makes it impractical to achieve with the lingual two *single* surfaces with an included angle of less than 20 degrees, which contributes nothing to retention.

The solution to this is to prepare *two* surfaces on the buccal aspect, the gingival one-third tapered at 5 degrees to the lingual and the remainder more obtuse. This modification immediately increases retention quite markedly. This can also be used on the palatal surfaces of maxillary molars, and it illustrates a principle which can be widely applied to other non-retentive forms.

Common modifications of this basic full crown preparation include the provision of mesial and distal slots and a connecting occlusal channel. These impart additional rigidity and firmer localization of the casting on the prepared tooth; they are used chiefly when the crown is serving as a bridge retainer.

The function of all the restorations described in the preceding chapters depends, increasingly as they get larger, upon accurate reproduction of occlusal surfaces in functional conditions. For intracoronal inlays straight line articulation suffices, but when the whole occlusal surface is involved an articulator with some degree of lateral movement is desirable. The stage at which full mouth impressions and functional occlusal records are required will depend on the demands of the operator and the skill of the technician, as well as upon the extent of the restoration.

The operator who follows gold inlay, crown and bridge work will find a multitude of refinements of design and clinical and laboratory techniques. Mastery of indirect inlay work opens the door to the whole of advanced restorative dentistry.

Summary

Indirect technique. Indications, extensive, multiple restorations, difficult access, good technical support. Impression of prepared and adjacent teeth, opposing teeth; occlusal record.

Slice preparation. Near-parallel mesial, distal slices; modified bevel, good retention mesial, distal slots, occlusal channel, rigidity; choice of suitable gold.
Indications, indirect method; large cavity, deep cervical margins, good retention.
Contra-indications, tilting, marked distal curve, concave cervical.
Preparation. Initial slices to free gingival margin, remainder cavity, enamel margins, line, complete tapered contour. Full depth slices, 1.0 mm below slots.
Occlusal bevel continuous with slices.

Impression. Gingival retraction, packing.
Elastomer; ready made trays, constructed trays.
Two stage impression; light body injected, heavy body in tray. Requirements of good impression. Occlusal record.
Copper band method. Smaller band better than larger. Cervical trimming, loading, impression, inspection. Occlusal record.

Laboratory procedures. Model production. Errors leading to high bite. Wax pattern, management of slices.

Fitting inlay. Clean cavity; contact areas, occlusion.
Slice adaptation; bevels; final adaptation margins.

Box preparation. Axial, cervical bevels; otherwise similar cavity preparation. Elastomer preferred. Combination of box and slice.

Occlusal coverage. Protection. Minimal clearance 1.0 mm. Occlusal margins, reverse bevel, bevel omitted.
Importance of temporary cover; over-eruption; 5 to 7 day rule. Immediate plastic temporary restoration.

Three-quarter crown. Originally canine, now pre-molar, molar.
Indication, sound buccal, occlusal protection, approximal cavities. Preparation; slices, occlusal reduction, slots, occlusal channel, palatal reduction; knife-edge margins.
Canine; modifications, slices coverage lingually, V-shaped channel; possible lingual shoulder.

Cervical margins. Location supragingival, gingival, subgingival; conditions controlling. Types. Knife-edge, definition difficult. Chamfer, better definition, rigidity. Butt, unworkable, wide cement margin. Bevel, workable, close apposition.

Full crown. Cast restoration, clinical crown; protection, function; bridge retainer. Impression for temporary cover.
Preparation. Restore normal form. Approximal slices, lateral reduction, establish taper, connect slices. Occlusal reduction, cuspal or flat. Rounded margins, smooth surfaces.
Check cervical chamfer, taper, occlusal clearance.
Slots, occlusal channel if required.
Impression, occlusal record. Temporary crown.
Modifications, buccal, lingual of molars.

12

Inflammation of the pulp and the principles of endodontic therapy

Treatment of vital pulp

The soft tissue components of the dentine–pulp complex react to injury and infection in the same general ways as do soft tissues elsewhere in the body. In the case of the dental pulp, however, the outcome of this reaction is governed very largely by the vulnerability of the apical blood supply and by the enclosure of the pulp in a hard cavity.

Relatively mild degrees of injury which would resolve without incident in a soft tissue elsewhere in the body, rapidly proceed in the dental pulp to a non-vital tooth and acute or chronic osteitis of the apical alveolus.

Up to the point at which resolution is still possible, injuries to the pulp can be treated and resolution encouraged. Beyond this stage treatment is aimed at retention of the tooth by emptying the pulp chamber and root canal, eliminating infection, and re-establishing normal conditions in the periapical region.

Inflammation of the pulp

The pulp may be inflamed due to a number of causes of which the commonest are caries, thermal injury arising from operative procedures, particularly the injudicious use of rotary instruments, and chemical irritation caused by irritant restorative material. Excessive preparation of fresh dentine, failure adequately to line a metallic restoration, and fracture of the crown are other relatively common causes.

The early stages of pulpal inflammation, marked by mild hyperaemia, exudate, and an increase in defensive cells, produce characteristic clinical signs. The tooth is hypersensitive to heat and cold and the pain produced by these stimuli may be severe; it ceases, however, shortly after the removal of the stimulus. The pain does not occur without provocation in this way. The effect of sweet, and to a lesser extent sour, substances upon exposed dentine is similarly exaggerated.

Treatment. Treatment of the condition resolves into three lines:

1. Removal of the irritant where possible. If, for example, the inflammation has arisen as the result of injudicious operative procedures leading to the insertion of a restoration, this restoration should be removed. If it is necessary to use local anaesthesia for this purpose, great care must be taken to avoid further injury by the application of heat in the use of rotary instruments, and further extension into fresh dentine must be avoided. For this reason it is perhaps advisable to avoid the use of high-speed instruments

in these conditions. The deeper layers of the restoration are better removed with hand instruments.

2. Insertion of a non-irritant dressing. Zinc oxide and eugenol is the safest. Preparations containing chlorobutanol are useful in this condition.

3. Relief of the occlusion. The occlusal contact should be lightly relieved so that the patient can no longer feel discomfort when his teeth meet in centric or lateral occlusion.

Following this treatment the relief of pain is often rapid and dramatic. The patient should at least be aware of definite improvement within twenty-four hours. The persistance of symptoms, particularly without improvement, is of serious prognosis. In such cases the patient will complain of a more or less continuous toothache of low intensity with severe attacks of lancinating pain lasting from five minutes to half an hour. *These attacks are particularly significant when they occur without provocation of heat or cold, or other similar stimuli.* The occurrence of such symptoms indicates that the inflammatory changes in the pulp have passed the point at which resolution is possible. The hyperaemia and exudate are more severe and there in increasing infiltration of inflammatory cells, probably with focal abscess formation.

The treatment of this more advanced condition is either extirpation of the pulp, followed by root canal filling or extraction of the tooth. The principal factors governing this decision will be described later, but it is important that the decision should be taken without delay as soon as the clinical condition becomes apparent.

Direct pulp capping

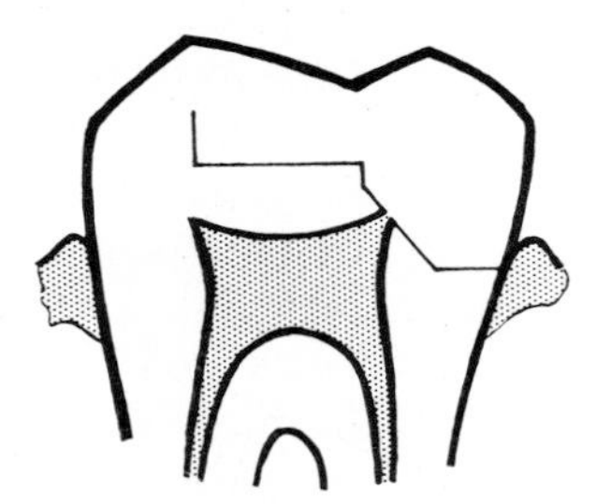

Fig. 12.1. Traumatic exposure of pulp during cavity preparation.

This procedure is one which may be used on some occasions for the treatment of the 'traumatic exposure' (Fig. 12.1). This term implies the exposure of an otherwise normal pulp in a cavity which is, for practical purposes, free of caries. It is the type of exposure which happens accidentally as the result of instrumentation in the later stages of cavity preparation.

It must be emphasized that there is no question here of existing carious infection of the pulp. The small breach of the pulpal wall frequently the tip of a cornu, is surrounded by normal dentine and the pulp is superficially infected only as it is laid open to the cavity.

Direct capping is only indicated in these circumstances when reasons for avoiding pulp extirpation exist, for example in a young patient, or in the case of a multi-rooted tooth. In the best cases the prognosis is fair, though most of the successful ones probably undergo pulpal degeneration or hyperaemia in time; but the treatment may nevertheless be justified.

In cases of doubt as to the original condition of the pulp, where infection is suspected and root treatment uncomplicated, immediate extirpation should be undertaken. In these cases prognosis is good.

Treatment. This involves the avoidance of unnecessary contamination and the application of a dressing. It is carried out thus:

1. Once the exposure is recognized, no further application of instruments to the exposed area is allowed. Further interference, even though the instrument be sterile, can only result in a further spread of the light infection now on the exposed surface of the pulp.

If the cavity preparation is incomplete and the crown is not under rubber dam, the tooth should be isolated with cotton-wool rolls, whilst any caries remaining elsewhere in the cavity is rapidly excavated. The cavity is then irrigated to eliminate debris, blood clot, and such superficial infection as may be removed. For this purpose it is preferable to use sterile saline, but in its absence a weak antiseptic solution such as 1 per cent phenol or 10 per cent Roccal (0.1 per cent benzalkonium chloride) could be used. Thereafter the crown must be isolated by rubber dam.

2. The dam applied, the cavity is dried with a sterile pledget of cotton wool and a sterile dressing is placed. The dressing may be calcium hydroxide in sterile water, and this is covered with a metal cap, when space exists to accommodate one. The use of calcium hydroxide cement is less satisfactory because, although it may contain no pathogenic organisms, its sterility cannot be assured.

The preparation should be mixed on a sterile slab with a sterile spatula. A thin slab may be boiled, or rubbed with spirit and flamed. If no sterile water

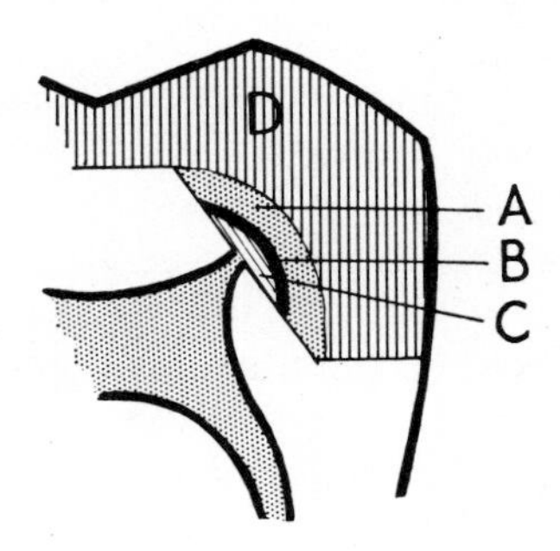

Fig. 12.2. Capping of pulp: A: phosphate cement; B: pulp cap; C: sterile dressing; D: temporary or permanent restoration.

is available, water may be boiled in a teaspoon over a small flame. If the sterility of the powder is suspect this may be heated on a spatula over the flame.

The powder and liquid are mixed to a consistency which will allow the mixture to be teased lightly over the exposure with a sterile blunt probe. If a tin pulp cap is to be used it is held in tweezers, lightly flamed, and its concave side filled with the mixture. It is then inverted and placed gently over the exposure (Fig. 12.2).

3. The pulp cap may now be sealed in place by the application of zinc phosphate or zinc oxide cement. If the cavity is small, and no pulp cap is used, the cement may be placed directly upon any of these preparations, but the presence of a metal cap precludes the transmission of pressure to the exposed pulp in the later stages of filling.

4. An amalgam restoration may be placed in the normal way using spherical alloy to reduce condensation pressure or, if it is preferred, a temporary filling of zinc oxide and eugenol may be inserted, to be replaced later with the permanent restoration. The tooth should be put on probation for a time and tested for vitality after three and six months. The patient should always be told what has been done and of the possibility of trouble at a later date.

Indirect pulp capping

This term refers to the treatment of carious dentine on the floor of a cavity closely approaching the pulp, in the absence of clinical signs of pulpitis. The condition and its treatment are described on p. 87.

Partial pulpectomy

This procedure implies the amputation of the heavily infected portion of a vital pulp in conditions which allow the remainder to survive and heal. The condition which favours resolution is one in which the root apex is not yet completely formed. At this stage the blood supply to the radicular pulp is plentiful and the risk of thrombosis of the apical vessels is reduced. A good example which meets these requirements is provided by a coronal fracture of an incisor, exposing the pulp, at an age when the apex is still unformed.

The object of the operation in this case is to allow the vital remainder of the pulp to complete the root formation. For the ultimate coronal restoration it may be necessary, at a later date, to extirpate the remaining pulp and fill the canal, a procedure which is made easier if the completion of the root has occurred.

In the case referred to, the coronal fracture of the incisor, the prognosis is good if the apex is open, if the fracture is recent, say within twenty-four hours, and when an aseptic technique can be followed. The operation is carried out in this way:

A general sedative such as quinalbarbitone sodium (Seconal Sodium) 50 mg, or diazepam (Valium) 5 mg is given by mouth half an hour before the operation and the tooth is anaesthetized by local infiltration.

Rubber dam is placed to isolate the fractured tooth. If the crown has been fractured in such a manner that the retention of the dam is doubtful or impossible, a copper band, cut down, contoured to the gingival and burnished, is cemented in position. This greatly facilitates the application of the rubber dam and the stability of the retaining clamp. The surface of the crown and surrounding rubber dam are now sterilized by painting with 2 per cent alcoholic solution of iodine, or 2 per cent benzalkonium chloride (Roccal) or 10 per cent alcoholic solution of chlorhexidine (Hibitane). With a sterile fissure bur the pulp chamber is widely opened, extending from the site of the exposure until the whole width of the coronal pulp is displayed. A sharp sterile excavator is now inserted along one side of the pulp chamber and used to cut across the pulp as high up as can be reached with the instrument. The detached fragment of tissue is removed and haemorrhage controlled with a small sterile pledget soaked in 1:1000 adrenaline. Pressure on the pulp must be avoided.

When the bleeding has ceased the fissure bur may again be used slightly to enlarge the entry in such a manner as to form a small step in the wall of the pulp chamber, about 3 mm from its orifice. The purpose of this step is to prevent pressure of the sealing materials upon the pulp (Fig. 12.3).

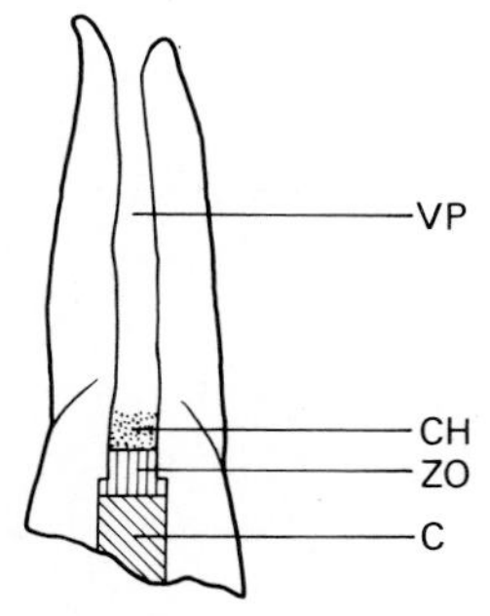

Fig. 12.3. Dressing used following partial pulpectomy: VP: vital pulp; CH: sterile calcium hydroxide; ZO: zinc oxide cement; C: composite resin.

A small quantity of calcium hydroxide mixed with sterile water is now teazed on to the surface of the remaining pulp tissue. The surface being covered, any excess is removed with an excavator and a layer of quick-settng zinc oxide cement applied. This is now covered with phosphate cement mixed to a creamy consistency. If the pulp is well covered with calcium hydroxide and space is limited, cement may be placed directly upon it. When this is set, the remainder of the cavity may be lightly undercut with an inverted cone bur and filled with composite resin.

A radiograph of the tooth taken at three and six months after operation should show the development of a calcific barrier a short distance above the level of original amputation. Further examinations at six-monthly intervals should reveal the progressive completion of apical contour and narrowing of the pulp canal to normal dimensions.

Absence of this radiographic evidence is an indication that the remaining pulp tissue has succumbed to infection, and in this case further treatment will be required.

Anti-inflammatory treatment

It will be apparent from what has been said that the chief problem is the control of infection of the pulp in order to reduce inflammation and promote healing. Attempts have been made to use for this purpose the anti-inflammatory properties of corticosteroids combined with the bacteriostatic effects of antibiotics. The most widely used preparation contains hydrocortisone and chloromycetin.

From clinical and experimental results available, there seems little doubt that the inflammation normally arising from infection such as results from carious exposure, is often controlled by these drugs. There is evidence that pain can be relieved and the vitality of the pulp preserved by an anti-inflammatory dressing for about one day, followed by capping with calcium hydroxide cement. In this way the vitality of the pulp can be preserved where otherwise it would be lost. Prolonged use of a corticosteroid preparation seems to lead eventually to pulpal death.

Principles of endodontic therapy

Before proceeding to a description of the commoner operations of root canal therapy the principles upon which this practice is based should be considered.

The immediate object of root canal, or endodontic, therapy is the evacuation of the contents of the canal, the elimination of infection, and the filling of the canal with an impermeable, non-irritant material. If this is achieved, the periapical bone, periodontium, and cementum can return to relatively normal conditions.

The ultimate object of endodontic treatment is the retention or restoration of a normal crown, for reasons of function or appearance, or both. *It follows that the first decision to be made when considering the possibility of root canal therapy is whether the crown of the tooth in question can be satisfactorily restored or replaced.* The method of coronal restoration should be planned in detail, particularly if a full crown is to be placed, for this may have a direct bearing upon the later stages of root canal treatment.

Clinical condition of the pulp

When treatment of a root canal is called for, the clinical condition of the pulp may vary between wide limits. It may, at one end of the scale, be vital and only very lightly infected, the condition which pertains immediately after a traumatic exposure. At the other end of the scale the pulp may have been entirely destroyed by necrosis, with spread of infection through the apex to form an acute or chronic alveolar abscess. Because the clinical condition of the pulp has important implications for the technique of treatment, the commoner conditions must be briefly described.

The infected vital pulp

These conditions may conveniently be grouped under two headings: **closed**, in which there is no direct connection with the oral cavity; and **open**, in which the pulp tissue is open to direct infection from the oral cavity.

Closed infection. These cases arise as a later development of the conditions which were described above, under the treatment of the vital pulp:

1. Injury due to injudicious use of rotary instruments short of traumatic exposure.
2. Irritation of the pulp by restorative materials such as silicate cement, acrylic resin, and unlined amalgam or gold.
3. The early stages of carious invasion of the pulp (Fig. 12.4).

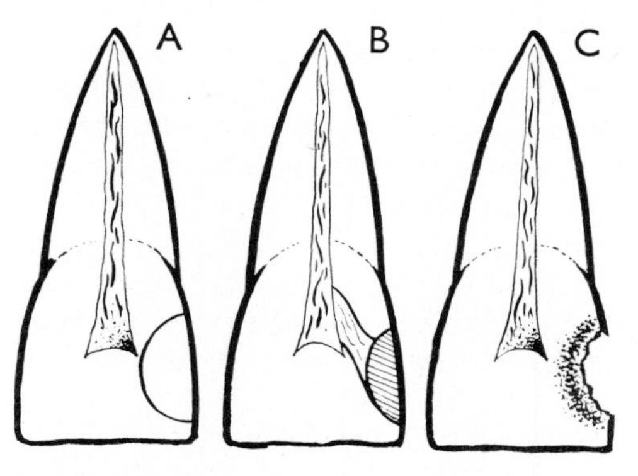

Fig. 12.4. Common causes of closed infection of the vital pulp. (A) Injudicious cavity preparation; (B) irritant restorative materials; (C) early carious invasion.

Pain and severe hypersensitivity to thermal change are generally prominent symptoms of this condition, but occasionally the pulp passes so rapidly through this stage that the patient is unaware of symptoms and total necrosis of the pulp supervenes without any sensations comparable to the severity of the injury (see below).

The pulp shows all the signs of acute inflammation with focal necrosis, for example, and the cornual abscess. Although the inflammation may in some cases be sterile in its initial stages, with necrosis, infection of haematogenous origin or through dentine must be assumed. Spread of the necrotic process will eventually involve the whole pulp, sometimes very rapidly, sometimes only after a considerable period of days or weeks.

Open infection. The common causes of these conditions are:

1. Gross carious involvement of the pulp, in which the pulp cavity has been widely opened.

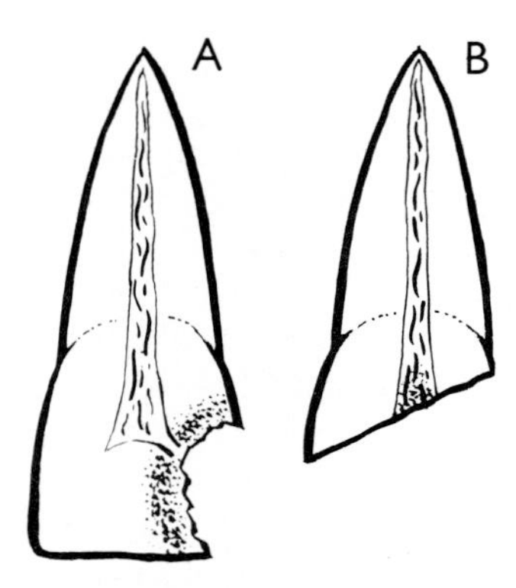

Fig. 12.5. Common causes of open infection of the vital pulp: (A) gross carious involvement; (B) coronal fracture with exposure of pulp.

2. Fracture of the tooth, exposing the pulp but without causing rupture or thrombosis of the apical vessels (Fig. 12.5).

Because the pulp is open to the mouth and the exudate resulting from its inflammation can drain away, severe and continuing pain is not characteristic of these conditions, in contrast to the closed infections. Direct stimulation by pressure and by heat and cold, sweet substances, however, give rise to severe pain as long as they are effective.

Histologically the pulp presents an ulcer at its exposed portion, below which in the main body of the pulp there is evidence of some degree of acute or subacute inflammation. The pulp tissue in the apical region is often normal in appearance. Eventually, as in the closed infection, necrosis of the whole pulp occurs, but in the open cases this is often long-delayed because of the absence of the pressure factor which is present when the pulp remains enclosed.

By definition the traumatic exposure falls into the class of open infections. This is, however, a special case in which the pulp tissue is normal and, at the time, only lightly infected at the site of the exposure. Extirpation of the pulp at this stage gives the best possible opportunity for successful root canal treatment. If the pulp is retained, treated by capping and subsequently succumbs to infection it is then in all respects a closed infection.

The infected non-vital pulp

It has been observed above that all infections of the vital pulp may, by neglect and the passage of time, lead to necrosis of the whole of the pulp. When this occurs pain, which has usually been present in varying degrees, ceases. There follows a period during which infection spreads through the apical canals and involves periapical tissue (Fig. 12.6).

According to the severity and nature of the infection, and the resistance of the patient, the periapical infection forms either a chronic alveolar abscess, which may be virtually symptomless, or an acute alveolar abscess

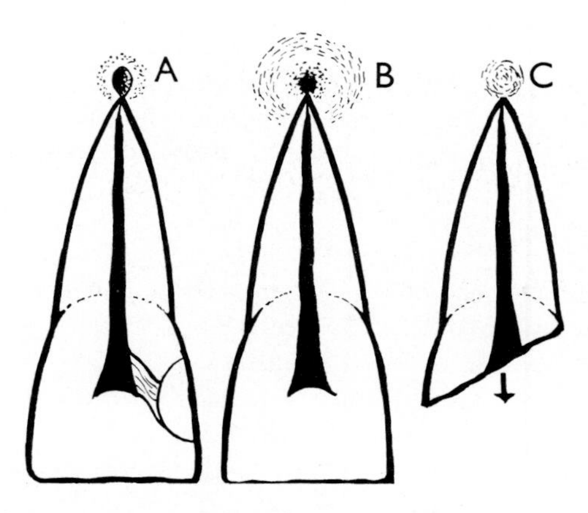

Fig. 12.6. Common causes of infection of the non-vital pulp: (A) closed necrotic pulp with chronic apical abscess resulting from irritation of an unlined restoration; (B) closed necrotic pulp with acute apical abscess following trauma of a direct blow; (C) open necrotic pulp following coronal fracture, with chronic apical abscess draining through root canal.

with the train of clinical signs and symptoms of that condition. The chronic abscess may at any time pass into an acute phase.

The distinguishing characteristic of these conditions is periodontitis of the apical region. Percussion of the tooth elicits pain, usually very mild in the chronic condition, increasing in more active infection, and very severe in the acute abscess. Swelling, sinus formation, lymphadenitis, and more general clinical signs, together with radiography, help to elucidate the stage of development of the infection.

In general, open infections may remain quiescent for long periods because the inflamed area is drained more or less effectively through the root canal. Closed acute infection, giving rise to severe symptoms, demands drainage by opening the root canal, incision of an abscess through the mucosa, or by extraction of the tooth.

Prognosis

Having regard to the stages of treatment which are described as evacuation of the canal, sterilization, mechanical preparation, and filling, it is clear that some conditions offer better prospects of satisfactory treatment than others.

Disregarding for the moment many other factors such as anatomical complexity of canals and difficulty of access to posterior teeth, it may be said that the vital, lightly infected pulp offers the best results. Moderate infection of the still vital pulp offers a good prognosis if care is taken to avoid spreading the infection to the apical canals and beyond. In the necrotic pulp the apical canals are heavily infected and this fact alone complicates what would otherwise be a simpler treatment. The prognosis is least favourable in the necrotic pulp with extensive periapical spread, either acute or chronic.

Here a wider area is to be controlled and induced to heal. The number of cases which are unsuitable for treatment by modern endodontic, therapeutic, and surgical methods decreases as the operator acquires skill and experience in their application.

Clinical assessment of cases

The decision to undertake endodontic treatment can only be made with detailed knowledge of the tooth to be treated, against the background of the full evaluation of dental and general health.

It follows, therefore, that if it has not already been done, the full history and clinical examination must be carried out, including radiography. In paticular, the radiograph of the tooth under consideration must show clearly the shape, size, and number of roots, the periapical and periodontal conditions, and the adjacent alveolus.

The possible systemic effects of chronic apical infection must be considered in relation to the general health of the patient. Whilst it is true to say that radiographic evidence of mild degrees of periapical sclerosis and hypercementosis may be tolerated in the healthy patient, it is sometimes impossible to place a non-vital tooth entirely above suspicion where focal sepsis is involved.

To summarize the approach to treatment:

1. History and full examination; general health.
2. Clinical condition of tooth; appearance and function of crown; periodontal attachment.
3. Radiograph of tooth—
   Root shape, number, and complexity.
   Periapical infection.
   Periodontal attachment.
4. Clinical condition of pulp; vital or non-vital; extent of infection.
5. Other methods of treatment; extraction; fixed or removable prosthesis.

Principles of technique

From a technical point of view, the straight, single-rooted canal is the simplest to treat. Multiple canals and canals of complicated anatomical form are proportionately more difficult. It follows that the treatment of anterior teeth is simpler than that of posterior teeth, where the problem of access still further increases the difficulty of operating.

The evacuation of a root canal, its cleansing and the mechanical preparation of the canal to render it straight, smooth, and ready for filling demands the use of special instruments, the more important of which will be described in Chapter 13, and an adequate supply of these instruments of various patterns and in good condition is necessary if preparation of the canal is to be carried out with certainty and efficiency.

The most important principle of endodontic treatment is the control of infection. This implies that all root canal treatment must be carried out under rubber dam. This is essential to avoid re-infection of the root canal from the patient's mouth and also to eliminate the ever-present risk of the patient swallowing or inhaling the small instruments which are frequently used. If the placing of a rubber dam should prove impossible, some other line of treatment should be chosen, and this usually implies extraction and replacement of the tooth under consideration.

It must be possible for the operator to rely upon the sterility of all instruments used in root canal treatment. Treatment systems using uniform sets of instruments in trays, packed, sealed, and sterilized, usually in dry heat, are the most common in use. Provision should be made for the re-sterilization of instruments which have necessarily become contaminated in the course of operating. In addition, a sterile technique of operating must be achieved in order to preclude the possibility of the introduction of extraneous infection. Bacteriological control of infection present in root canals is the most effective method of control and correlation with clinical signs; it should be used when available, rather than placing reliance entirely upon clinical signs.

There is a very wide range of bactericidal and bacteriostatic drugs available for use in root canals. Each operator acquires his own preferences and there can be very few antiseptics and antibiotics which have not, at one time or another, been advocated for use in this type of treatment. To be effective the drugs used must be highly antiseptic but at the same time they must cause a minimum of irritation to the soft tissue of the periapical region. For this reason antibiotics have proved powerful additions to the range of organic and inorganic antiseptics.

A permanent root canal filling must be capable of forming an impermeable seal, particularly in the apical third of the root canal, and of occluding as much more of the canal as may be desirable or necessary in the circumstances. It must be non-irritant and is usually mildly antiseptic. It should not discolour the dentine of the root.

Summary

Treatment of vital pulp

Pulpitis: commonest cause caries; also thermal, chemical irritation; fractured crown. Results, hyperaemia, increased response all stimuli. Treat; remove cause, non-irritant dressing, relieve occlusion.
Chronic progressive hyperaemia. Unprovoked attacks severe pain, increasing severity, frequency. Raised thermal response. Treat, extirpation or extraction.
Direct capping. Traumatic exposure; clean cavity, rubber dam. Calcium hydroxide paste or cement, accelerated zinc oxide cement, amalgam or composite restoration.
Partial pulpectomy. Recent coronal fracture, sedative, anaesthesia. Dam; excision coronal pulp, haemostasis. Dress calcium hydroxide, accelerated zinc oxide cement, composite resin.
Anti-inflammatory preparations. Limited role of steroid preparations.

Principles of endodontic therapy

Objective, restoration function and appearance of normal crown; prerequisites are normal periapex and complete canal filling.

Closed pulpitis. Caries, chemical, thermal irritation.

Severe continuous pain, thermal sensitivity. May be symptomless.

Open pulpitis. Gross caries, coronal fracture, allows drainage. Pain on direct pressure, heat and cold.

Proceeds to necrosis. May be symptomless.

Infected non-vital pulp. Necrosis, open or closed.

Symptoms apical periodontitis acute or chronic.

Prognosis depends on apical anatomy, periapical spread. Summary of approach to treatment p. 246.

Principles of technique. Evacuation of canal, cleaning, mechanical preparation. Control infection, rubber dam; antiseptics, antibiotics. Permanent seal apical canal.

13

Instruments and procedures in endodontic therapy

Endodontic instruments

A number of patterns of common instruments such as conveying forceps and burs have been slightly modified for the special purposes of root canal work. In addition there are a number of instruments which are designed specifically for use in root canals (Fig. 13.1). In general, they are tapering steel points of various shapes, ranging from very fine broaches to thick reamers.

Broaches (Fig. 13.1 (1)). These are fine, tapering, resilient steel points used for the initial penetration of the canal. The plain broach is smooth and may be either round or square in cross-section. The former, being very fine, is

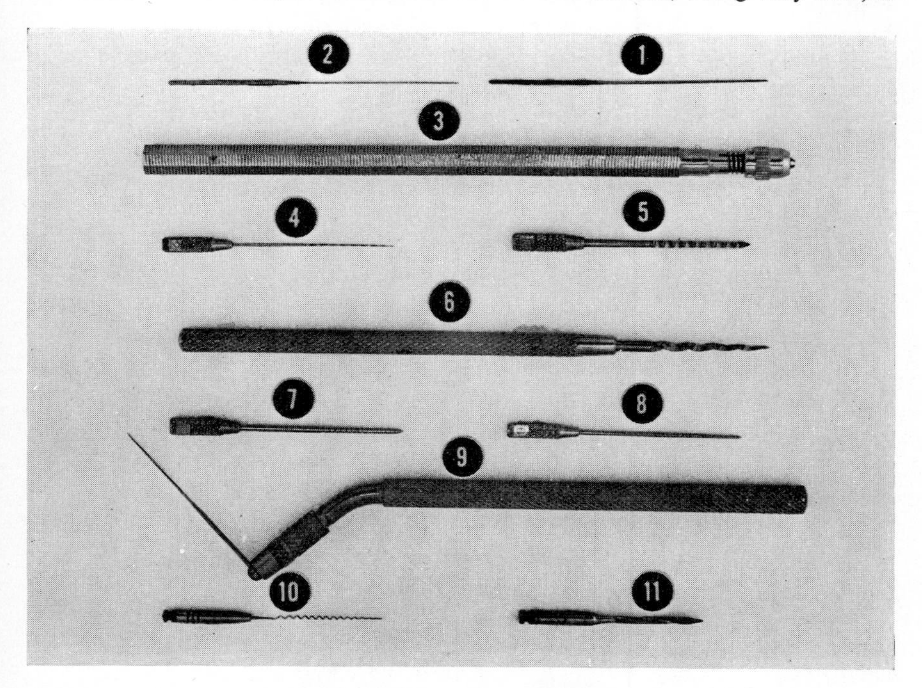

Fig. 13.1. Endodontic instruments: (1) smooth rectangular broach; (2) barbed broach; (3) broach holder; (4) hand root canal reamer No. 20; (5) hand root canal reamer No. 90; (6) hand reamer permanently fixed on straight handle; (7) root canal file, rat-tail; (9) root canal file on detachable contra-angled handle; (10) root canal filler No. 30; (11) engine reamer, Peeso.

used mainly as a seeker to explore very narrow passages. The latter, having rectangular edges along its length, can be used by rotation to enlarge, to a limited extent, a constricted opening. The fine-tempered square broaches are very resilient, but are liable to break if over-strained. The soft-tempered broaches bend and twist, but they do not break.

The barbed broach (Fig. 13.1 (2)) is round in section and smooth-sided except in the last 10 mm of its tip. Here about 25 barbs project. When the broach is used in the presence of intact pulp, the barbs engage the tissue which may then be withdrawn.

The shank of a broach is inserted into the small chuck of a broach holder (Fig. 13.1 (3)), which serves as a handle. If necessary, a broach may be bent at the junction of tine and shank, at an angle of 20 to 30 degrees, to enable easier access to the canal.

Reamers. These are side cutting instruments. They are triangular in section and are twisted to form three helical blades meeting at the tip, which may be fine, pointed, or bluntly bevelled. They run in sizes from very fine, No. 10, to the largest, No. 140 (Fig. 13.1 (4) (5)). These figures represent the diameter of each reamer at its tip, measured in hundredths of a millimetre. For example, No. 35 has a diameter of 0.35 mm and No. 140 a diameter of 1.40 mm. The sizes are also colour-coded, so that a certain size always has a handle of the same distinctive colour. Reamers with short handles are used for posterior teeth, but the long-handle pattern (Fig. 13.1 (6)), the use of which is confined to anterior teeth, gives much greater control.

The function of a reamer is to enlarge and smooth an existing canal by cutting shavings from the walls. It is not designed as a drill, and when properly used will not cut a hole where none previously exists. When wrongly used, a reamer may cut a notch in the wall of an apical canal. Into this notch all subsequent instruments find their way, greatly to the detriment of further preparation of the canal.

The precise method of use is of importance. The exact length of the canal to be reamed to the apex must be known, and from the anatomy of the tooth, the operator should have a fairly exact knowledge of the probable calibre of the canal at various points between the pulp chamber and the apex. The size of the first reamer to be used should be selected to engage the walls of the root canal at about 5 mm from the apex. For example, in the case of a maxillary central incisor, this could be size No. 25. To use too small a reamer at this stage would be useless, for it would not engage the walls of the canal and it also invites the possibility of passing the reamer through the apical canal into the periapical region. The reamer of the correct size, say No. 25, is inserted into the canal and advanced until it begins to engage the sides of the canal. When this happens, and *before it jams firmly*, it is rotated two or three turns in a clockwise direction and then withdrawn a few millimetres. Again, it is advanced until it engages and again rotated a few times and withdrawn. A few anti-clockwise turns during withdrawal help to clear the blades of debris and so make cutting more effective.

During clockwise rotation only light forward pressure should be used, otherwise there is a risk of jamming and fracturing the fine blade or penetrating the apex. Fracture of the reamer is a serious accident because the retained fragment is firmly wedged in the canal and is usually inaccessible. Its removal is always difficult and may prove impossible without root resection.

When the first reamer has been advanced to within 1 mm of the apical foramen, as predetermined by measurement, it is withdrawn and No. 30 substituted. This is used in the same way until, in its turn, it reaches the same point. In so doing it further enlarges the canal and smooths out irregularities of the walls. No. 35 follows, and so on in regular series until the whole length of the canal is enlarged to the required extent.

It is essential that none of the series be omitted, for if this is done by going, for example from No. 35 to No. 45, it will be found that the latter binds firmly and is much more difficult to turn, with an increased risk of breakage. Further, its point is liable to engage an irregularity and form a pit into which all subsequent instruments find their way.

If at any stage a reamer tends to bind excessively the operator should return to the size smaller and gently free the constriction. The large size, used with a light touch, will then be found to pass without difficulty.

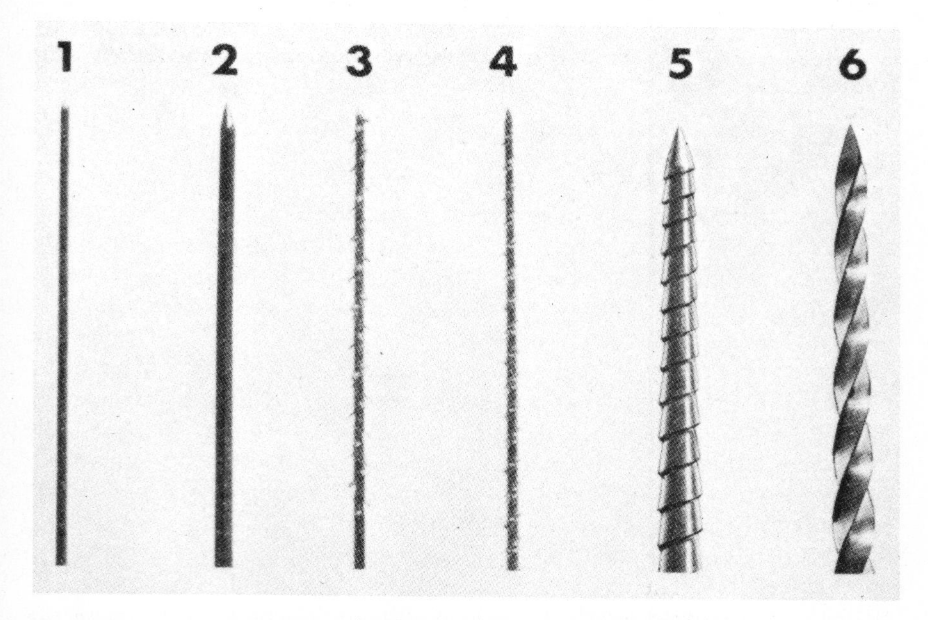

Fig. 13.2. An enlarged view of the working points of: (1) fine smooth broach; (2) soft broach; (3) barbed broach; (4) rat-tail file; (5) Hedstroem file No. 80; (6) reamer No. 80.

To summarize the rules for the successful use of these instruments

1. All reamers must be sharp; since they cannot be sharpened, blunt reamers must be replaced as soon as their edges are lost.
2. They must be free of rust, for this blunts them and increases the risk of fracture.
3. They must not be forced; if a reamer seizes, return to the smaller size.
4. They must be used in regular series up to the required size.

Files. The working surface of a root canal file is superficially similar to a reamer. There are two common patterns. The Hedstroem files (Fig. 13.1 (7)) have close helical blades; the rat-tailed file (Fig. 13.1 (8)) has numerous short barbs. They may have short handles or long handles. Long handles are available in right-angle and contra-angle patterns (Fig. 13.1 (9)).

They are used to supplement the action of reamers in further removing irregularities of the canal walls. It will be understood that a reamer must necessarily ream a hole, or canal, which is *round* in cross-section (Fig. 13.3). Most canals are irregularly oval in section and few can be regularized throughout their length by the use of reamers alone. In an oval canal two areas (Fig. 13.3 B) will remain untouched by reamers; it is these areas which are reached by files.

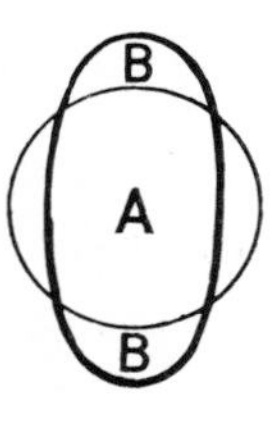

Fig. 13.3. A reamer reams a circular bore, A, in an oval canal, areas marked B are unaffected by reaming.

When the canal has been reamed to a suitable size, a file of size slightly smaller than the largest reamer used is selected. It is inserted as far as it will pass without jamming, then withdrawn with a firm stroke against the wall. With these files the withdrawal is the cutting stroke and this movement is repeated again and again.

A knowledge of the probable shape of the canal, derived from the morphology of the tooth being treated, helps the operator in directing instruments. Since the detailed shape of the canal may not be known, the file should be used on all the walls of the canal until the operator is satisfied by the 'feel' of the canal that all irregularities have been reduced. Only very few trials are required to demonstrate the superiority, especially in this instrument, of long-handled patterns over those with short handles.

Mechanical files. There is a type of mechanical file which is of value, particularly is first penetration of difficult, sinuous canals in posterior teeth, difficult of access. The *Giromatic* is a latch-type contra-angle handpiece

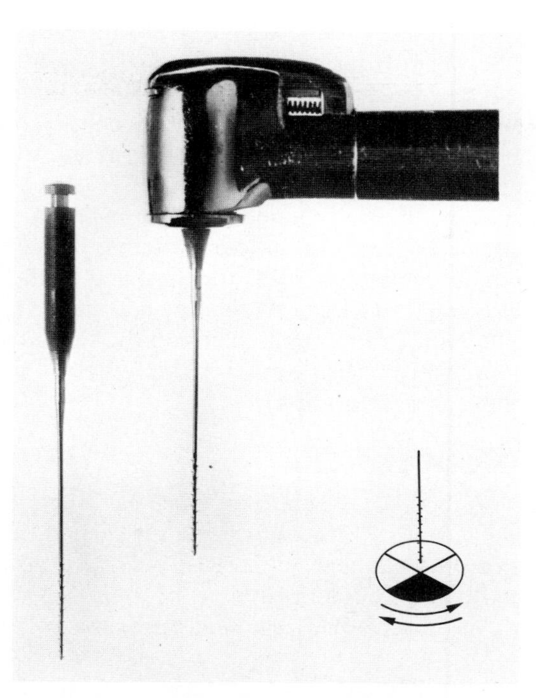

Fig. 13.4. The *Giromatic* root canal instrument has a 90 degree reciprocal rotation.

(Fig. 13.4) which takes a finely tempered file, similar to a rat-tail file, attached to a bur shank. The rotation of this instrument reciprocates over an arc of more than 90 degrees and achieves very active penetration; as it is flexible, quite considerable curves can be negotiated. When the canal has been penetrated to the correct extent preparation normally continues with hand reamers and files.

In considering the technique of using reamers and files remember that **the predominant factor in rendering a canal fit for filling, and in controlling infection, is the thorough use of these instruments and the removal of the debris which arises from their use.**

Rotary canal fillers. These instruments (Fig. 13.1 (10)), available in three sizes, are used for introducing plastic sealing cements into the canal. They are finely tempered, tapering wires wound in helical form, attached to shanks suitable for straight or contra-angled handpieces of conventional forms. They act upon the principle of an Archimedean screw.

The size suitable for any canal should be considerably smaller than the largest size of reamer used, to avoid the possibility of binding. On the other hand, a very small filler will not act efficiently in a large canal. The filler is loaded with soft cement, inserted into the canal and, with the engine running at its slowest speed, or using a speed-reducing handpiece, advanced gently up the canal.

If only high rotary speeds are available, these instruments can be used quite effectively between finger and thumb. Care is taken not to over-fill, for the cement can be forced through the apex. This is undesirable unless a resorbable material is used, when it may be a necessary part of the technique. When the canal is judged to be adequately filled, the filler is slowly withdrawn, still rotating in a *forward* direction.

Engine reamers. These are not commonly used in primary procedures of root canal treatment. They are of various patterns, for contra-angle (Fig. 13.1 (8)) and straight handpieces. Their function and method of use are, in general principles, similar to those applying to hand reamers. They are more difficult to control but are more rapid in effect; they are used chiefly in the preparation of a previously filled canal for the reception of a post for the retention of a full-crown restoration.

Access to root canals

The mode of access to single or multiple canals has a very significant bearing upon the ease with which instrumentation is carried out and upon the ultimate success of the treatment.

There are specific and well-defined areas of approach which give maximum accessibility in the axis of each canal; this is the only approach to be used.

The instrumentation of canals through existing Class II, III, IV, and V

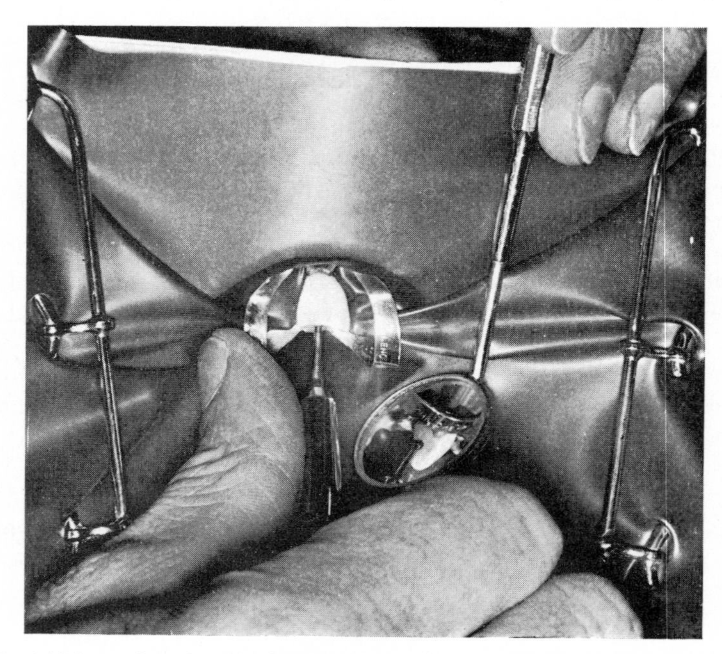

Fig. 13.5. Initial entry into the pulp chamber of a maxillary central incisor. Note the alignment of the bur and handpiece with the long axis of the tooth.

cavities is ruled out. *When these unfilled cavities are present they must be completely excavated, prepared, and filled with a restoration which is an hermetic seal. No carious dentine may be left, no possibility of leakage may remain.* If the crown contains a restoration the margins of which are suspect, the restoration must be replaced. If these rules are observed, the possibility of unsuspected re-infection of the canal from this source is avoided.

The mode of approach to incisors and canines will be considered in detail. That to posterior teeth will be indicated briefly and in more general terms.

Incisors and canines

The point of initial entry in these teeth lies midway between the most prominent part of the cingulum and the centre of the incisal edge (Fig. 13.6).

If the entry is to be made through intact enamel, as is usually the case, a medium-sized sterile round diamond or carbide bur, in a well-centred handpiece, may be used at medium high speed. The enamel once perforated, lower speed has the advantage of conveying a better sense of touch (Fig. 13.5).

The handpiece is carefully aligned in the axis of the canal in a mesio-distal plane and inclined about 15 degrees lingual to the pulpal axis in the labio-lingual plane (Fig. 13.7). It is thus directed at the incisal extremity of the pulp chamber. The bur is carefully advanced, checking alignment all the time, until it is felt to enter the pulp chamber. It is immediately withdrawn and the entry checked by insertion of a smooth broach.

If the tooth is severely periodontitic, as happens when an acute apical abscess has formed, the passage through enamel and dentine may be extremely painful. The pulp is necrotic and therefore insensitive, but vibration transmitted to the periodontium causes acute discomfort, particularly if the bur is not correctly centred. Local anaesthesia is contraindicated for this purpose in the presence of this infection. In these cases high speed is indicated, for here the lightest touch can be used. The disadvantage lies in the loss to the operator of tactile sensation. This makes it difficult to know with certainty when the pulp has been entered. In the circumstances this difficulty must be accepted (see also p. 256).

The next stage is that of enlargement of the opening; this may be done with a tapered fissure bur 702 or 703. This is inserted into the entry already made and with a gentle motion the walls are removed. The hole is enlarged more on the cingular aspect than on the labial side (Fig. 13.8), so that a funnel-shaped opening is formed, inclined slightly towards the lingual.

The risk of using the tapered, flat-ended bur is that of forming ridges on the side of the entry. This may be reduced by blunting the corners of the bur tip by running them for a few seconds against a carborundum wheel held in

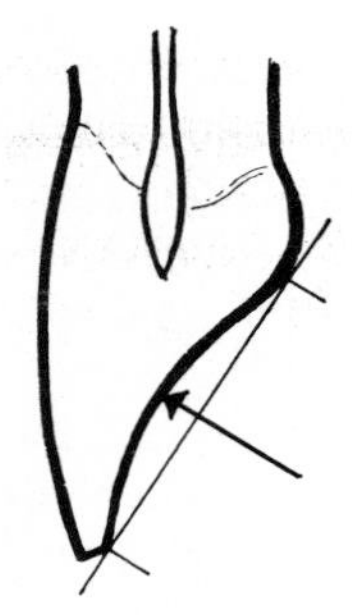

Fig. 13.6. Point of initial entry to pulp chamber in maxillary incisor.

Fig. 13.7. Direction of entry of bur towards pulp chamber.

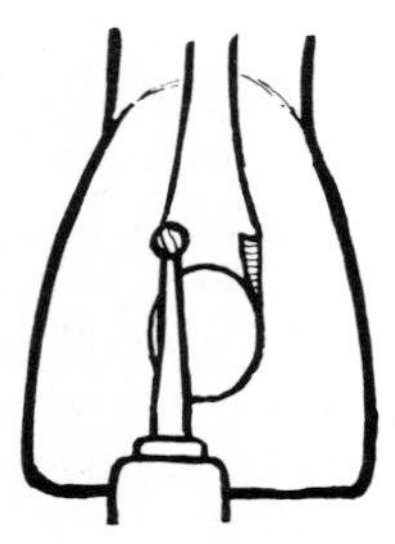

Fig. 13.8. Enlarged entry to pulp chamber.

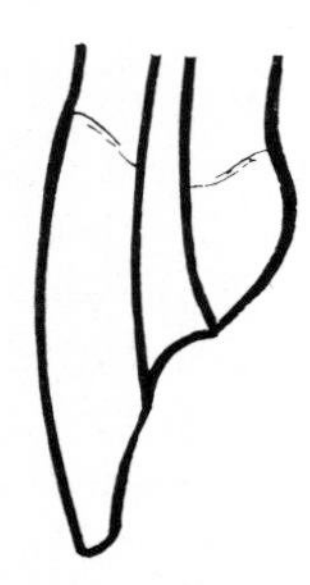

Fig. 13.9. Elimination of residual conua in maxillary central incisor.

the hand. The bur then becomes essentially side-cutting and the risk of ridge formation is reduced. Most catalogues contain details of several designs of round-ended, tapering burs used for this purpose.

When a smooth funnel-shaped entry is achieved, attention is directed to the pulp cornua, where these exist (Fig. 13.9). A No. 4 round bur is used to locate and obliterate these recesses and render them completely accessible. *The persistence of the cornua, with the debris which they contain, is a source of continued re-infection during subsequent treatment and a cause of coronal discoloration.*

Mention should be made of a common fault of approaching the pulp from the cingulum in an axis too obtuse to the pulp (Fig. 13.10). There are two possible sources of failure here. First, the chamber may be overshot, with the formation of a pit in the labial dentine. Into this pit all subsequent instruments stubbornly find their way. Second, the tip of the pulp chamber may be missed, with the ill results referred to above in connection with persistent cornua.

The approach, when completed, should show a smooth tapering entry, devoid of ridges or pits, giving good access to the axis of the canal. Into this, instruments may be inserted without hindrance.

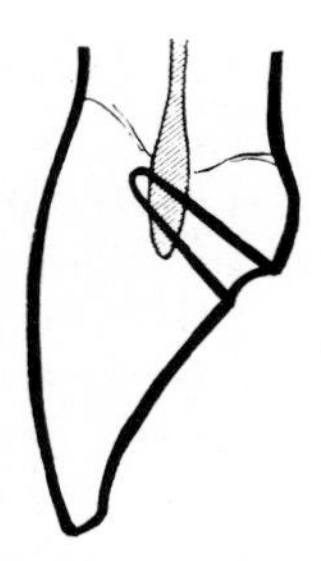

Fig. 13.10. Entry to pulp chamber through the cingulum, with risk of penetrating labial aspect.

Molars and premolars

These are approached in all cases from the occlusal surface. The initial entry is made approximately in the centre, with variations according to the relation, in shape and position, of the pulp chamber to the crown. It follows that the precise details of entry are different for each tooth in the upper and lower quadrants.

For example, the access to mandibular premolars (Fig. 13.11 A) is round and situated in the centre of the crown. The access to maxillary premolars (Fig. 13.11 B) is slightly nearer the buccal than the lingual cusp.

In the case of mandibular molars (Fig. 13.12 A) access is a triangular opening based on the mesial fossa and confined to the mesial two-thirds of the occlusal surface. In the case of maxillary molars (Fig. 13.12 B) it is a

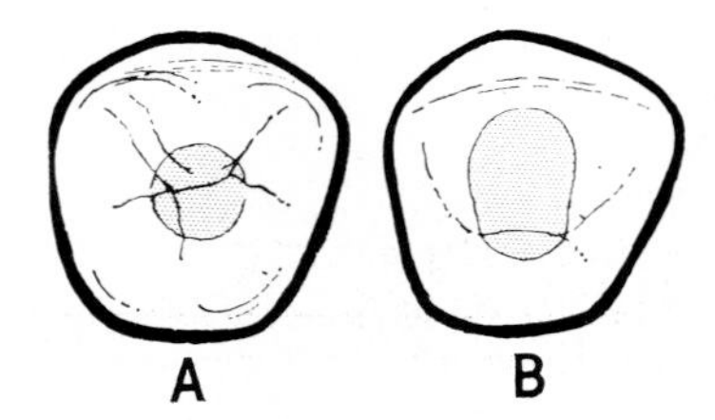

Fig. 13.11. Access to pulp chamber of (A) mandibular premolar; (B) maxillary premolar.

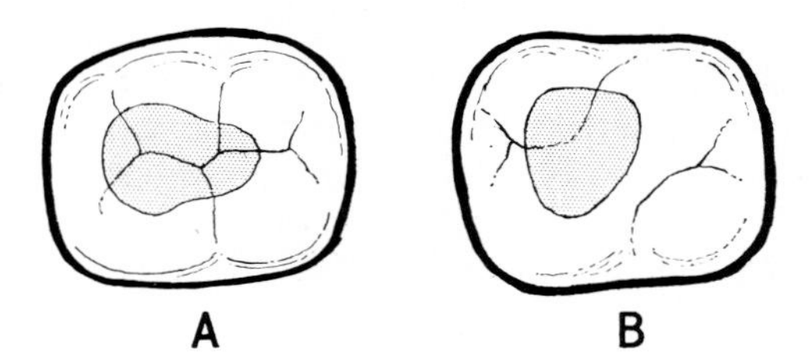

Fig. 13.12. Access to pulp chamber of (A) mandibular molar; (B) maxillary molar.

triangular opening sited in the mesial fossa and involving only the mesial aspect of the oblique ridge.

In every case the occlusal 'roof' of the pulp chamber must be removed over the whole of the area indicated so that the orifice of each canal can be defined and made accessible. If accessibility is still found difficult, then the opening must be extended until access can be gained. A study of a bitewing X-ray will give a good indication of the depth of the pulp chamber below the occlusal surface.

The walls of the opening should be smooth and the cornua of the pulp chamber should be obliterated for the reasons previously given.

In conclusion, it should be emphasized that *inadequate access is a common cause of failure* to achieve a good result in root treatment, particularly in posterior teeth. It is therefore well worth while to spend a little additional time and trouble to achieve good access. If at any time during the instrumentation of the canals difficulty is experienced, the operator should first ask himself whether access to that canal is adequate, before proceeding to locate a further cause of difficulty.

Technique of root canal procedures

The following operations have been selected as representing the commonest procedures of endodontics and those which will illustrate the application of general principles. In describing them it will be assumed that a thorough assessment of each case has been performed and the correct line of treatment decided upon.

In each operation the treatment of a single-rooted tooth will be described since this simplifies the operation and its description. The treatment of multi-rooted, posterior teeth is complicated chiefly by the difficulties of access and the complexity of the root canals. The principles remain unchanged. The skilled application of these principles to the successful treatment of posterior teeth is of greatest importance, for it is these posterior teeth which are so frequently needed as abutments for fixed and removable prostheses in the later decades of life.

It is essential that sterile instruments are available and that an aseptic working technique can be achieved by the operator. Isolation of the tooth under treatment is a prerequisite in every case unless the specific reason for not applying the dam is given.

If, by reason of the shape of the crown, rubber dam cannot be retained in position, a soft copper band should be fitted to the tooth, contoured to the gingival margin, burnished to conform to coronal contours, and cemented in place. This will in most cases allow excellent retention of a rubber-dam clamp and ease of re-application at each visit.

The existence of an extensive cavity, for example Class II, is another reason for the use of a copper ring. In these cases the ring helps to retain the temporary cement restoration in position and assures the water-tight seal which is essential.

If, in spite of everything that can be done, the application of rubber dam is impossible, the decision to undertake root treatment must be reconsidered. In such cases extraction and replacement may be the correct line of treatment.

Treatment of the vital uninfected pulp

First visit. It is only very rarely that it is necessary to extirpate a completely normal pulp and such an operation can only occasionally be justified. The commoner circumstance is the accidental exposure of a normal pulp during cavity preparation when, as is often the case, capping the pulp is deemed inexpedient. Before the decision to carry out root canal treatment can be made the tooth should be isolated with cotton-wool rolls and radiographed to confirm that no local contra-indication to such treatment exists.

When a previously uninfected pulp is so exposed it must be recognized that contamination of the exposed surface must have taken place from infection already present in the cavity. It is therefore necessary to wash out the cavity immediately, unless the preparation is already isolated by rubber dam, dry the cavity with a sterile pledget, and cauterize the surface of the pulp with pure phenol or the actual cautery. Thereafter no further contamination of the cavity can be allowed. Local anaesthesia is given if the tooth has not already been anaesthetized.

As soon as rubber dam is applied the cavity is sealed with accelerated zinc oxide-eugenol or EBA cement and, after sterilizing the crown and surrounding areas with two applications of an antiseptic solution, correct access to the pulp chamber is obtained.

A small square of sterile rubber dam about 5 × 5 mm is now pierced by a smooth broach held in a broach holder. The rubber can be adjusted anywhere along the broach to act as a marker. The broach is inserted into the root canal and gently advanced until, somewhere just short of the apex, further progress is halted. The rubber marker is then adjusted, with the broach in position, to indicate the incisal edge (Fig. 13.13). Upon withdrawal the length of the broach inserted can be measured against a stainless-steel ruler.

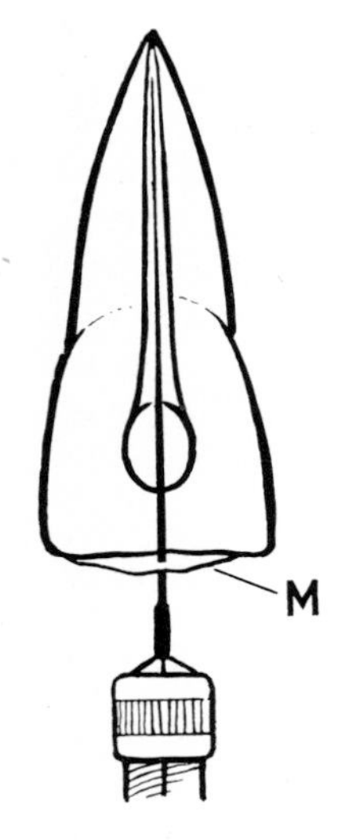

Fig. 13.13. Initial insertion of plain broach to establish patency of canal. Note rubber marker, M.

The purpose of this procedure is to establish the patency of the canal so that the barbed broach may be controlled. Because of its barbs, this broach does not usually pass quite as far as does the smooth pattern. If, with the latter, there is some obstruction before the apex is judged to have been reached, the point of the broach should be made to explore a little to seek a way through. Such an obstruction may be a small bend in the canal or a slight irregularity of the walls of the canal.

More serious obstructions in the form of calcific deposits may be present within the pulp. Sometimes warning of these can be gained by study of the radiograph. In severe cases they may render impossible any attempt to treat the canal, but possibility of this should usually have been foreseen.

A barbed broach is now inserted in a similar manner, as high up the canal as possible. It is given one full turn, to engage the barbs in the tissue of the pulp. Repeated twisting breaks up the tissue and chokes the barbs, with increased risk of incomplete removal (Fig. 13.14).

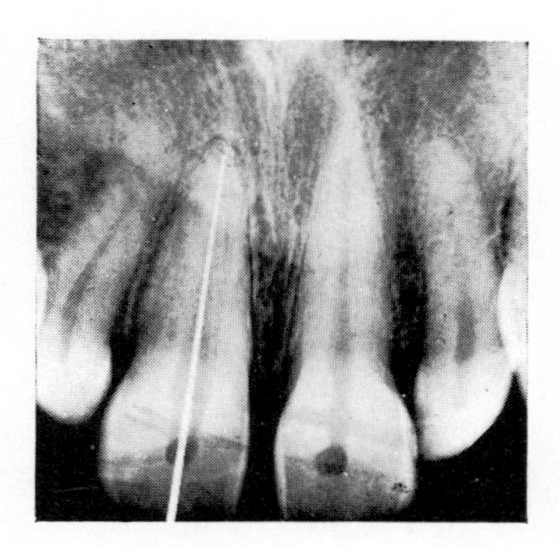

Fig. 13.14. A barbed broach inserted to the full length of the canal.

On withdrawal the whole pulp tissue up to the apical canals should be removed intact. This tissue can be teazed out with forceps and examined so that the operator may satisfy himself that the whole pulp has been removed. Should it appear that some pulp tissue remains, a new broach may be used for re-insertion to the highest point, re-engagement of the barbs in the residual pulp tissue and withdrawal.

The importance of complete removal at this stage is that any pulp remnant left until the next visit will be fully sensitive upon instrumentation and will require anaesthetization a second time for its removal.

If bleeding from the apical vessels is brisk and troublesome, which does not often happen in these cases, a paper point dipped in hydrogen peroxide solution (10-volume) and inserted high in the canal wall will soon control this.

Before reaming the canal it is necessary to measure the length as expressed by the distance from the apex to the incisal edge of the crown. For this purpose a soft-tempered smooth broach is passed up the canal until its point is judged to be at the apical foramen. Holding it firmly in this position the broach is bent at a right angle across the incisal edge, thus enabling it to be re-located (Fig. 13.15). With this broach in position a second radiograph

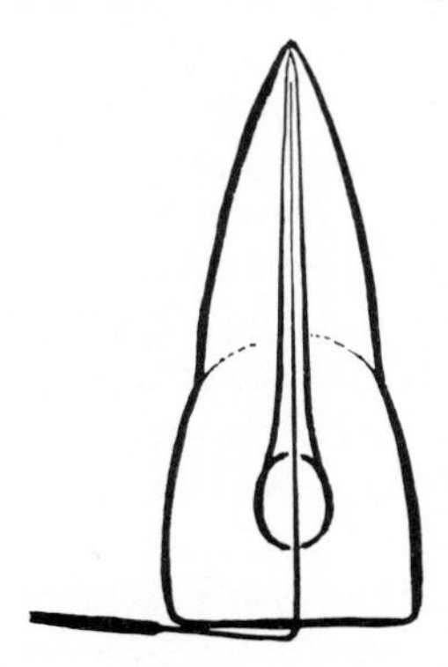

Fig. 13.15. Use of a bent plain broach for radiographic measurement of canal.

is now taken. In doing so, the position of the film and the tube angulation are of importance and need particular care because of the presence of the rubber dam and clamp. The omission of a clamp, when possible, or the use of a cervical ligature rather than a clamp can assist in these conditions.

The film must be positioned so that the canal is in the midline, with the apex just above the centre *but with the incisal edge located on the film near the lower margin.* The angulation is adjusted to give correct representation of the length, with neither elongation nor shortening. If, however, the length of the root is distorted on the film, its actual length can be calculated (see p. 265). From this radiograph and the length of the diagnostic broach in its relation to the apex, the distance from apex to incisal edge can be obtained.

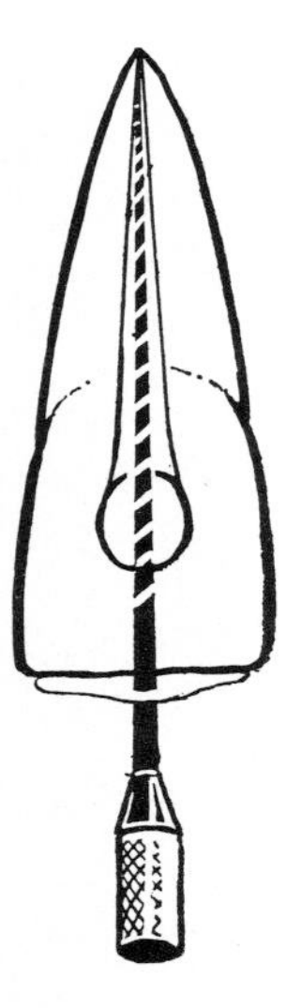

Fig. 13.16. Location of reamer in canal, with point 1 mm from apex.

The canal may now be reamed. A reamer of suitable size, say size No. 25, is selected and upon this a small rubber marker is placed a distance from the point equal to the length of the canal less 1 mm. This means that when the rubber marker is at the incisal edge the point of the reamer is 1 mm short of the apical foramen. The reamer is inserted into the canal until the point engages the sides. This usually happens at a point about 3 or 4 mm from the apex. The reamer is rotated and advanced until the marker reaches the incisal edge (Fig. 13.16). It is then withdrawn and a No. 30 reamer, similarly marked with rubber dam, is inserted and, using it in the same manner, is advanced to within 1 mm of the apex. This proceeds until reamer size 60 or 70 has been used (Fig. 13.17).

The preparation of the canal may be continued by the use of a file. This instrument, of a size slightly smaller than the last reamer used, is inserted until it engages the walls of the canal and withdrawn with a firm stroke

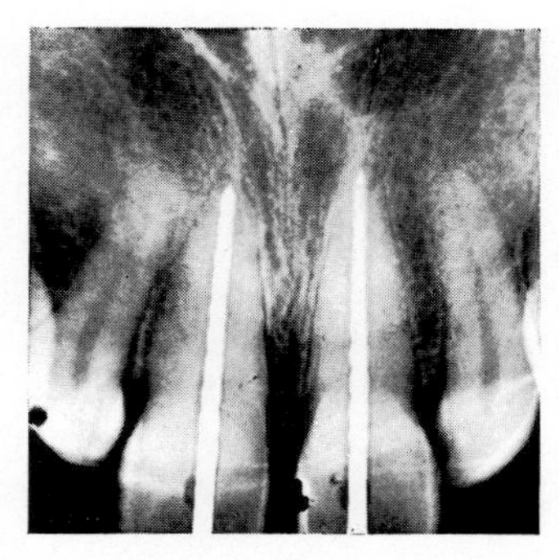

Fig. 13.17. No. 70 reamers in final position at apices.

against the walls. This is repeated on all walls of the canal until these feel smooth and regular. In so doing, considerable amount of dentine and debris is removed from the canal.

The remainder of the debris may be removed be means of alternate instillations into the canal of sodium hypochlorite solution (Milton) and hydrogen peroxide solution (20-volume). Some operators prefer the former, alone; others use saline solution alone in order to avoid any possibility of apical irritation. Irrigation can be done in either of two ways:

1. The solution, starting with hypochlorite and ending with the same, may be instilled into the canal with forceps. The points are closed, dipped into the solution, carried to the orifice of the canal, and inserted as far as possible. When the points are opened the solution runs by surface tension into the canal. Here it may be agitated by use of a plain broach avoiding a pumping action and penetration through the apex.

The hypochlorite is then absorbed on a paper point and replaced by hydrogen peroxide used in the same manner.

When no further debris is seen on the paper point after the hypochlorite solution, the canal may be assumed free of debris.

2. Using a more effective method, the canal may be irrigated with the solutions using 2-ml syringes. A large-bore needle, size 19–20, should be used and this may be bent to give easier access if required. Nylon syringes have advantages for this purpose, but they cannot be dry sterilized; they must be boiled.

The needle of the hypochlorite syringe is inserted as far as possible and *withdrawn at least 5 mm*. On no account may irrigation be commenced with the needle tip wedged in the canal. To do so would undoubtedly force the solution through the apex. The irrigation should be concluded with hypochlorite solution and not with peroxide, in order to avoid the possibility of subsequent effervescence penetrating the apical canal.

When the canal is clean, excess moisture is removed by the successive use of several paper points. It is then ready for an antiseptic dressing, which is inserted to allow a short period of periapical irritation to subside. It also allows the tissue fluid from the oedema of this irritation to drain into the canal, so reducing the mild transient apical periodontitis resulting from extirpation.

For the dressing a paper point is selected. The terminal 2 mm of the point is cut off with scissors and the blunt end also shortened so that the shortened 'point' may be accommodated entirely within the canal. The paper is moistened with parachlorphenol; no excess of the drug is allowed. This is inserted as high as possible in the canal. Since its tip is blunted, it cannot pass into or beyond the apical canal. The paper point is followed by a small pledget of cotton wool and the canal closed with quick-setting zinc oxide cement.

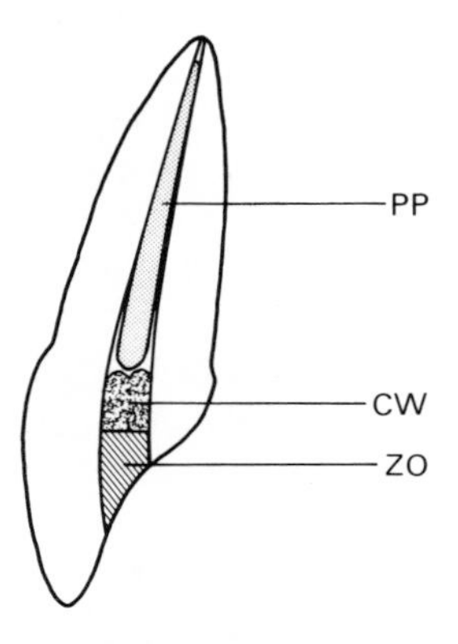

Fig. 13.18. Truncated paper point, PP, followed by a pledglet of cotton wool, CW, sealed with zinc oxide cement, ZO.

It must be understood *that gutta-percha alone is in no sense an adequate seal.* It is very susceptible to leakage and may be forced up the canal, injecting the contents of the canal through the apex.

After removal of the rubber dam, which should, as always, be inspected to see that no fragments have been retained in interdental spaces, the occlusion should be checked and relieved if required. The patient may be discharged for two days.

Summary of the first visit procedure

1. Closure of incidental cavities.
2. Access to canal.
3. Exploration of canal and complete extirpation of pulp tissue.
4. Haemostasis.
5. Radiograph with diagnostic broach.
6. Reaming.
7. Filing.
8. Irrigation.
9. Insertion of antiseptic dressing.
10. Closure of canal.

Estimation of canal length. If, in spite of attention to the precautions outlined on p. 262 the radiograph is distorted in length but shows the tooth apex and the incisal edge, the length of the canal, calculated from incisal edge to apex, can be deduced.

The length of the broach may subsequently be measured. The apparent length of the broach and the tooth can also be measured on the dry radiograph. The actual length of the canal, as defined, equals

$$\text{apparent length of canal} \times \frac{\text{actual length of broach}}{\text{apparent length of broach}}$$

This length, in the average maxillary central incisor, is about 23 mm. When this is so, the rubber-dam marker is adjusted on the reamer at a distance of 22 mm from its point.

Second visit. Upon the return of a patient after an interval following any root canal treatment, *inquiries should be made to determine the progress of the treated tooth during the interval*. This is of material consequence to the next stage of treatment and must never be omitted.

Some transient tenderness may have been experienced from the local anaesthetic. This must be differentiated from symptoms arising from the tooth itself. A sense of fullness over the apex and tenderness on biting, or an impression of the tooth being 'lifted' in its socket, all evidence of periodontitis, are of particular significance in this context.

In the case being considered these symptoms, if present at all, should only be of the mildest degree, lasting a few hours. They should certainly have subsided after two days. More severe symptoms would indicate a failure of technique. Either the apex has been penetrated, and the periapex considerably disturbed, or infection has been introduced into a previously sterile canal. The former may require a further period of rest or medication. In the latter case the canal must now be treated as infected (see p. 269).

Clinical examination follows. The tooth is lightly percussed, its mobility tested, and the sealing cement examined to confirm that it is intact. Rubber dam is applied and the crown and surrounding area sterilized.

With an excavator, zinc oxide cement and gutta-percha are removed. A barbed broach is the most convenient instrument for the removal of cotton wool and paper points.

These should be examined for the presence of pus or odour.

If the mechanical preparation of the canal was not completed at the previous visit, the final stages of reaming and filing may now be performed. This completed, the canal is irrigated to remove particles of debris produced by these operations. Drying of the canal is achieved by a series of paper points; is may be considered satisfactory when a point returns without visible wetting of its tip. Desiccation of the canal with alcohol and hot air is unnecessary and undesirable.

When the correct matched gutta-percha point has been tried in, and crimped to mark its correct length against the incisal edge, the canal sealing paste is mixed on a sterile slab. This may have been boiled or cleaned with an antiseptic such as 2 per cent benzalkonium chloride (20 per cent Roccal Concentrate) and dried with alcohol. The cement consists essentially of zinc oxide, precipitated silver, and resin; the fluid is a mixture of eugenol and Canada balsam. It is slow setting, taking four to six hours on the slab. In the canal its setting is accelerated to one hour by warmth and the presence of moisture in the dentine. No attempt should be made to clean the slab with water after use; alcohol and chloroform are effective for this purpose.

The cement is mixed to a soft consistency which strings when lifted with the spatula. A spiral root canal filler in a handpiece is lightly charged with cement from the slab and, inserted in the canal, revolved at slowest speed. This may be repeated so that the canal can be seen to be filled but the coronal portion still remains empty.

Excessive use of the filler should be avoided in order that the cement is not forced through the apex. The cement is non-irritant and a small quantity extruded into the periapical area, though better avoided, is not serious.

The distal one-third of the gutta-percha point, which is handled throughout with forceps, is dipped in the cement on the slab and carried to the canal. It is inserted slowly but firmly to the full extent required to carry it to the end of the reamed canal.

At this stage the point completely fills the apical one-third or one-half of the canal. Since the canal is usually larger in its coronal half, a means of filling this portion is required. This is done by using conical gutta-percha points. Points of suitable size are firmly inserted beside the original point and pushed well home with a root canal plugger or small flat-ended condensing instrument, No. 11. Two or three of these points may be re-

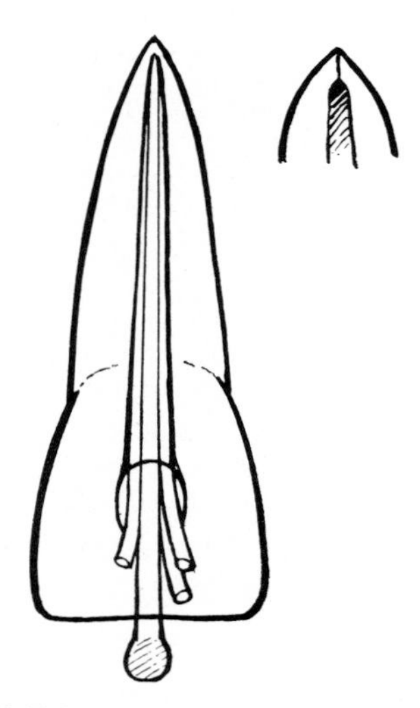

Fig. 13.19. Location of initial gutta-percha point and subsidary cones in filled canal.

quired and they may be firmly condensed to fill the canal without fear of forcing the original point too far (Fig. 13.19).

The hot blade of a small flat plastic instrument is used to cut off the excess gutta-percha as high up the canal as can be reached. The remainder is plugged home firmly and the coronal entrance to the canal is cleaned and dried with chloroform.

A temporary seal of zinc oxide and eugenol is used to close the crown and a confirmatory radiograph is taken of the apical condition (Fig. 13.20).

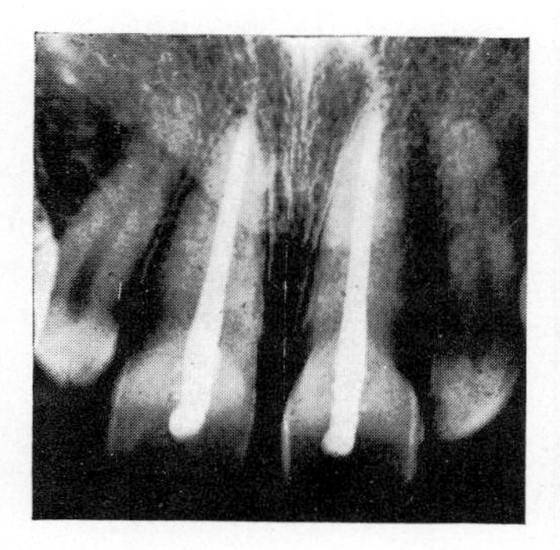

Fig. 13.20. Completed root canal fillings, using gutta-percha points and sealing paste.

At a later date the zinc oxide may be removed, the gutta-percha covered with a small application of phosphate cement, and the crown restored by composite resin (Fig. 13.21). Amalgam is a more durable restoration but in anterior teeth is liable to give rise to loss of transulence.

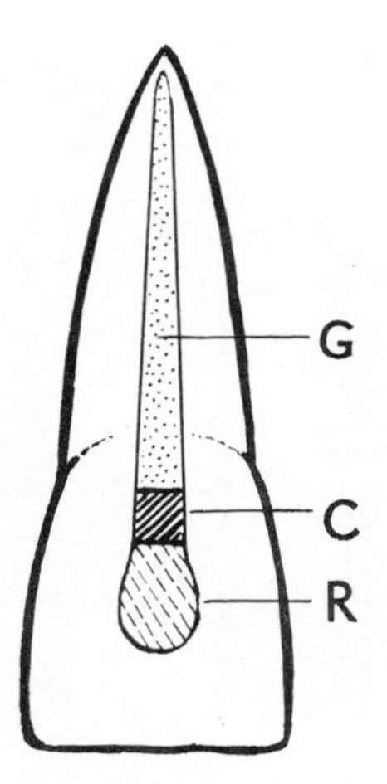

Fig. 13.21. Completed root canal filling: G: gutta-percha; C: phosphate cement; R: permanent restoration.

Summary of second visit procedure

1. Interrogation and clinical examination.
2. Rubber dam and opening of canal.
3. Reaming and filing, if required.
4. Irrigation and drying.
5. Try in matched gutta-percha point.
6. Insertion of root canal sealing cement.
7. Insertion of matched point and subsidiary cones.
8. Temporary seal and radiograph.
9. After an interval, final coronal restoration.

Notes on other forms of root canal filling

1. *Conical gutta-percha points.* When these are used for the main filling of the apical portion of the canal, the relation of the point to the root apex is more difficult to assess. A fine-pointed reamer is used. The sealing cement having been inserted, the gutta-percha point of a size comparable to the tip of the reamer is inserted until it is judged to be just entering the apical canal.

A radiograph is then taken. If the point is, say, 2 mm short of the apex, it can now be advanced with forceps to this extent. Should it protrude beyond the apex, a larger point is selected and retried in position. The insertion of additional points to fill the coronal half of the canal must be done with care to avoid the extrusion of gutta-percha through the apex.

2. *Resorbable root canal paste.* The use of resorbable root canal paste, of which the chief constituent is iodoform, is of advantage in some types of canal fillings. There are several methods of use. Probably the most effective is in conjunction with matched gutta-percha point, particularly when periapical infection is present and the apical canal has been penetrated either by intention or inadvertently.

The apical canal being patent, the radicular canal is reamed with a bevel-pointed reamer to the required size, usually slightly larger than would nor-

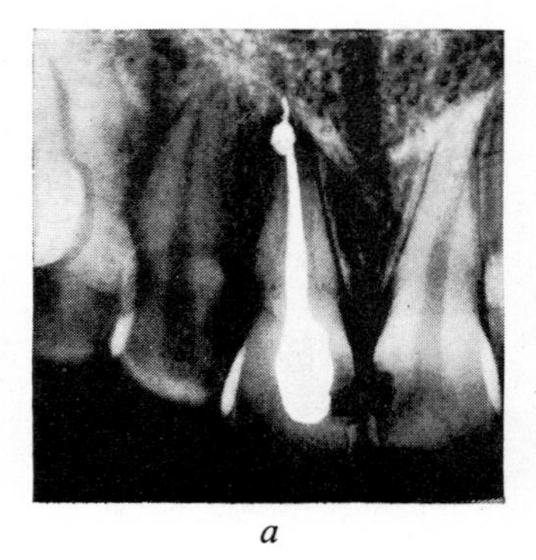

a

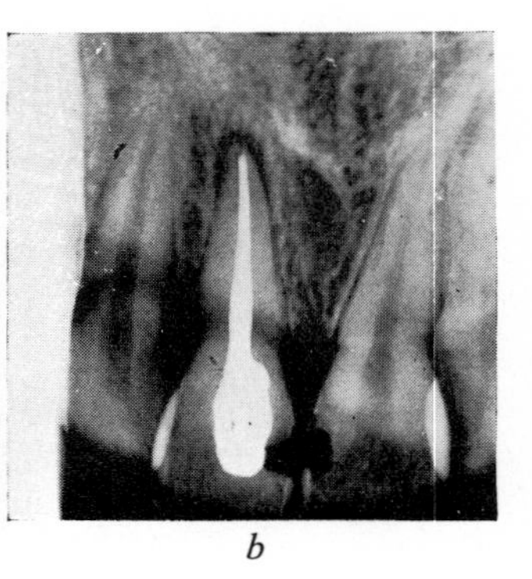

b

Fig. 13.22. Radiographs showing (a) resorbable paste extruded into the periapical area: (b) resorbed after an interval of 10 days.

mally be the case, in order to accommodate a gutta-percha point which will not protrude through the apical canal.

When ready for filling, the filler is charged at its tip with resorbable paste, inserted to the apex, and revolved for long enough to pass a small quantity of paste into and beyond the apex. The entry of the paste into the periapex is usually marked by some mild discomfort to the patient. Its presence may be checked by radiograph if required (Fig. 13.22).

Excess paste is removed from the canal with paper points, with the tips removed and finally with one or more moistened with chloroform. Canal sealing paste and the matched point are now placed in the canal as described above.

Resorbable paste of this type is irritant and is resorbed within a matter of days or weeks. It is antiseptic. The rationale of its use is that it promotes rapid regeneration of apical bone and the sealing of the apex by secondary cementum. The results of its use often appear to be very satisfactory, judged by clinical criteria.

3. *Conical silver points*. These points are of use in canals which are unlikely to be used for the insertion of a post to carry a full coronal restoration. They are stiff, yet flexible enough to conform to moderate bends in a canal. They are of particular value in posterior teeth, and in small canals where a fine gutta-percha point would buckle upon insertion.

They are used in much the same manner as conical gutta-percha points, using a root canal sealing paste. Their relation to the apex is controlled by radiography in the manner already described.

4. *Sectional silver point*. In teeth which are likely to be restored by a post crown, and this probability is high for most anterior teeth, a silver point can be used to fill the apical one-third of the canal only. The point is partly cut through at one-third of the root length, from the apex. The point is loaded with cement on its terminal third and carried into position. When the cement is set the point can be broken by twisting, leaving the apical one-third filled and the remainder empty. The canal may remain empty, though sealed at its coronal end of course, until ready for treatment with a post crown.

Treatment of the infected vital pulp

This is a condition previously described in which much of the pulp tissue is still vital, but the coronal portion is known to be heavily infected. It arises, as has been indicated, as the result of a carious exposure or following a fracture of the crown in which the pulp is exposed (Fig. 12.5, p. 244).

First visit. In the former case, the coronal cavity containing the exposure is completely excavated under local anaesthesia and sealed with a stiff mix of accelerated zinc oxide cement. Rubber dam is applied and the correct access obtained under conditions of sterility. The reason for this is that, although the pulp is infected by organisms of the patient's mouth, the introduction of extraneous organisms is to be avoided.

In the case of the fractured tooth (Fig. 12.5) it is occasionally necessary to cement a copper band on to the remainder of the crown in order to ensure the correct adaptation of the rubber dam. Entrance into the pulp canal is gained by enlarging the existing exposure until the pulp chamber is laid open.

There is an obvious risk in this type of case of transferring infection from the canal through the apex in the course of instrumentation. This, too, must be avoided by careful use of instruments and the cauterization of the exposed pulp surface with phenol or with the actual cautery or diathermy.

For this reason, the preliminary exploration with a smooth broach is omitted unless difficulty is encountered in the course of penetrating the canal. In the first place a barbed broach is inserted as far as it will pass up the canal, given a turn, and withdrawn with the pulp tissue. The complete removal of the pulp in one piece is particularly important here, but repeated penetration of the canal by the broach to the apical region must be avoided.

Haemorrhage from the apical region is often brisk because of the vascular engorgement. It must be controlled and blood clot removed from the canal with paper points.

No further preparation of the canal is carried out at this visit in order to avoid the risk of spreading infection at this stage. A culture of the infecting organisms may be of subsequent value in medication. It is performed in the following way.

Taking a culture. Two small sealed containers of sterile broth are required. The medium in common use is enriched glucose broth. A sterile paper point is inserted as far as it will pass into the canal and allowed to remain for about one minute. The cap of the first bottle is loosened without contact of the fingers on the glass neck. The neck of the bottle is flamed by holding it in the upper part of a spirit or gas flame for three to four seconds, twisting it so that the flame passes over the whole circumference of the neck. Holding the bottle now 15 cm above the flame, the cap is removed, the paper point withdrawn from the canal and inserted in the bottle without allowing it, or the forceps, to touch the neck (Fig. 13.23). The open neck of the bottle should now be flamed to kill any air contaminants and the cap replaced while the bottle is 15 cm above the flame.

If the root canal is entirely dry, it is preferable to use a moist point. To do this a tube of broth is opened, with the precautions as regards flaming described above, and a sterile point is moistened in the broth and the tube closed. The paper point is then inserted in the canal. This step provides a greater likelihood of collecting such viable organisms as exist in the canal.

A second paper point is placed in the other bottle, following the same technique. One bottle is incubated under aerobic conditions, the other under anaerobic conditions. The laboratory request should ask for identification of the organisms grown in each case.

The procedure may be carried out by the operator alone, or with assistance. In either case the technique should be performed exactly as

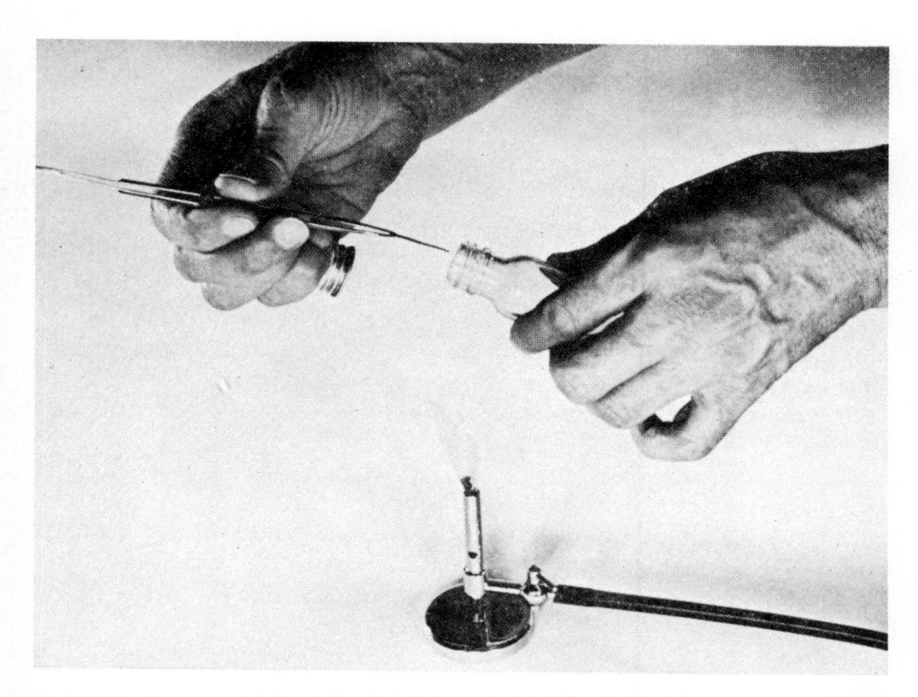

Fig. 13.23. Root canal culture. A paper point removed from the canal is placed in a container of sterile glucose broth, held about 5cm above a small flame. The screw cap is held by the little finger.

described. Close attention to detail and some practice is necessary if accurate results free from contamination are to be achieved.

The canal is now dressed with a shortened paper point moistened with parachlorphenol and sealed as described on p. 264. This patient is dismissed for three days.

Second visit. *The progress of the tooth during the interval and its present condition having been assessed*, the tooth is isolated and the canal opened.

If the progress has been uneventful, it may be assumed that the gross infection in the canal has been controlled. After removing any residual parachlorphenol with several paper points, two cultures are taken. The canal is now reamed and filed to the correct size and extent as previously detailed. Debris is removed by irrigation, the canal dried, re-dressed, and sealed as before. The patient is again dismissed for three days.

In the case of these cultures it is only necessary to know whether growth has taken place or not. The development of turbidity in the broth after twenty-four hours' incubation is an indication that viable organisms are present. If the broth remains quite clear, the canal is probably sterile. It is generally accepted, however, that two successive negative cultures are required as a criterion of sterility.

Third and subsequent visits. The patient's progress having been ascertained, and in the absence of clinical signs of periapical inflammation, the tooth is isolated and cultures again taken, taking care to remove any residual antiseptic before doing so. Irrigation of the canal may be performed and if the previous cultures were sterile, a dry paper point is sealed in the canal pending the result of the second culture.

If the evidence is that infection still exists, further medication is required and dressings must continue until all clinical signs subside and two successive negative cultures are obtained. When bacteriological facilities are not available, the clinician is guided entirely by clinical signs and experience. Apart from evidence of gross apical infection such as alveolar swelling and sinus formation, *the degree of tenderness experienced by the patient during the intervals between visits, and tenderness to percussion, are the most reliable signs of progress.*

The amount and nature of the exudate through the apex into the canal, as seen on the paper points, is also significant. A virtually dry point after three days, though in itself no guarantee of sterility, indicates that, at least, the infection is probably controlled. It will, however, be appreciated that bacteriological culture offers the best and most accurate method of assessment.

In a case such as is being considered, sterility will often be achieved in three visits and the canal may be filled on the fourth. Sometimes success is achieved in fewer visits. If infection persists beyond four visits, the operator should have first scrutinized his technique for sources of recurrent infection or contamination, and then consider alternative drugs for medication of the canal. These are referred to on p. 274. In these cases in particular, recognition of the persistent organism is of value, and the performance of sensitivity tests assists in the selection of a suitable antibiotic.

Treatment of the necrotic pulp

In these teeth the pulp has succumbed to infection (Fig. 11.24). The canal is filled with the purulent remains of the pulp and there may be an acute, subacute, or chronic abscess at the apex. This condition may arise following the death of the pulp from irritation by silicate restorations, or from a neglected carious cavity or following a blow on the tooth.

First visit. If the crown of the tooth is closed, the pus in the canal may be under pressure and a severe periodontitis exists. The tooth is extremely tender. The immediate need is drainage and the canal must be opened without delay.

No rubber dam need be applied, for the canal is to be left open to the mouth. The tooth is isolated with cotton rolls, access is gained through the correct path, using, for preference, an airotor handpiece and a small round diamond or carbide bur.

The instrument may be used with the lightest touch and very little discomfort is caused. The angulation of the bur should be watched closely since, with the reduced sense of touch characteristic of these instruments, entry into the pulp chamber is not easily perceived except by the sight or smell of pus, when present.

Should it be necessary to use medium or lower speeds, care should be taken that the handpiece has no lateral play and is running true. Vibration transmitted to the tooth may be damped by supporting it with a fingertip applied fairly firmly to the labial surface of the crown. Another method of stabilizing the tooth is the use of a small composition splint moulded to the labial surfaces of the affected crown and one crown on either side.

Sometimes, as the pulp is opened, a drop of pus exudes and the patient rapidly experiences a reduction in the severity of the pain. More often the canal is found to contain pus but not obviously under pressure and the symptoms subside more gradually. Further instrumentation of the canal is contra-indicated, and having established an opening of the approximate size of a No. 3 round bur, the canal may be left to drain freely.

The patient may be dismissed for three days with instructions to report at once if the tooth appears to become more painful. If this happens, the cause is generally the blockage of the canal with food debris. The formation of a wide funnel-shaped access at this stage tends rather to increase this risk of food impaction and does not significantly assist drainage, because the initial obstruction to drainage is more often in the apical part of the canal.

Second visit. If drainage has been adequate the patient will report the subsidence of symptoms, and on clinical examination it will be clear that the condition of apical periodontitis has markedly resolved.

Should this not have occurred to a degree where the tooth is relatively comfortable, no alternative remains but to use a barbed broach in the canal with a view to removing the existing obstruction. It must be used with all due care. In addition, if fluctuation is detected over the apex, incision of the overlying mucosa may be required to assist apical drainage. No further instrumentation of the canal should be carried out until, after a further period of drainage, symptoms have subsided and the condition has become quiescent.

If initial drainage has proved satisfactory, rubber dam is applied and access to the canal is improved to allow ease of entry.

The pulp chamber and canal are gently cleaned by the insertion of paper points, until all exudate and debris are removed. A short paper point, or a small lightly rolled pledget of cotton wool, moistened with parachlorphenol, is now inserted in the lower part of the canal, and this is sealed with zinc oxide and eugenol cement.

The patient is dismissed for two or three days, again with instructions to report if increasing discomfort should occur.

Third visit. If the progress in the interval has been satisfactory, it may now be assumed that the infection is controlled, but not eradicated.

Under rubber dam, the canal is opened, cleaned, and cultures taken.

Reamers and files may be used to render the canal patent and smooth. Care should be exercised to avoid undue disturbance near the apex. Irrigation removes debris and a medicated paper point is again used as a dressing for a period of three days.

Subsequent visits. The management of the case now proceeds as described under the headings of second and third visits on pp. 271–2. Infection is controlled bacteriologically and the canal is mechanically prepared for filling.

Alternative medication. Cases of resistant infection occur fairly frequently. If a sterile canal has not been obtained after four visits, alternative forms of medication should be considered, and a knowledge of the resistant organism is an advantage in these cases.

The available medicaments are numerous, and high amongst them stands antibiotic paste. The formula of this preparation is chosen for its bacteriostatic effect upon gram-positive and gram-negative organisms and upon *Monilia*. Its use has been advocated to the exclusion of other medicaments, but many clinicians prefer to reserve it for resistant cases, under bacteriological control.

The antibiotics penicillin and bacitracin (for the latter chloramphenicol may be substituted), together with sodium caprylate, are formed into a paste with an inert silicone liquid. Another proprietary paste, Chloromycetin Endodontic Compound, also contains Chloramphenicone as a fungicide. A third formula which has proved successful contains neomycin, bacitracin, polymyxin B and nystatin—a combination of wide spectrum antibiotics and fungicides.

The paste, supplied in a sterile container, may be carried into the canal on a blunt probe and lightly packed to fill the available space and allow room for an impermeable seal.

Alternatively, the paste may be obtained in glass cartridges and injected through a dry large-bore needle, using a cartridge syringe. The risk of forcing the canal contents through the apex should constantly be borne in mind. The needle should be inserted and withdrawn a few millimetres to avoid its impaction in the canal. Injection must proceed slowly, while the needle is gradually withdrawn as the canal fills.

Other medicaments such as clove oil, phenol, cresol, creosote, hexylresorcinol, benzalkonium chloride (Roccal, Zephiran), chloroxylenol (Dettol), chlorhexidine (Hibitane), and numerous others can be used with effect. The drug should be used with full knowledge of its bactericidal and fungicidal effect and its irritant action upon soft tissue.

Where repeated applications are required, rotation of drugs is a principle to be recommended. This implies the use at short intervals of a series of different drugs. Rotation is intended to reduce the possibility of the organisms developing a resistant condition to any one medicament.

Not infrequently the persistence or recurrence of apical periodontitis is due to over-medication with a comparatively irritant drug. When this is suspected the canal should be irrigated, dried, and sealed with a paper point moistened with glycerin of iodine in position. The dressing absorbs exudate which enters the canal. An alternative is to pack the canal loosely with a slurry of calcium hydroxide mixed as a paste with saline and inserted with a spiral root canal filler. In all cases the occlusion should be relieved. Periodontitis should subside in one or two days.

The role of apicectomy

Apicectomy is the operation of resection of the apex or apices of a tooth, the remainder of the root canal being filled with an impermeable root filling. Under local anaesthesia the soft tissue is reflected and the alveolar plate removed to expose the apex, which is then resected at the requisite level (Fig. 13.24). The canal may be filled before or after resection (Fig. 13.25), but the essential point is that *the residual part of the root canal must be closed by an impermeable seal.*

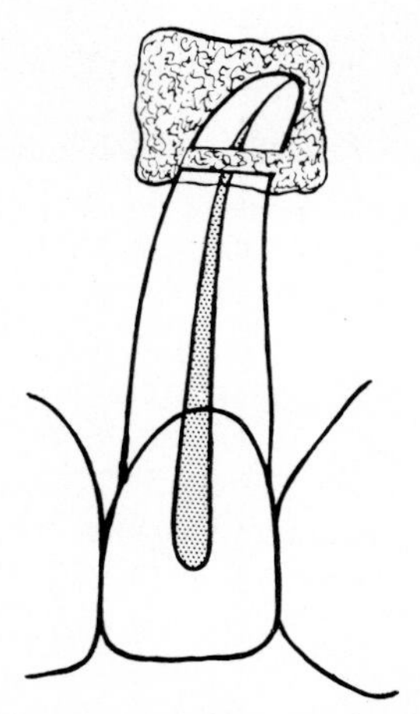

Fig. 13.24. Diagram of resection of curved root apex.

The operation is indicated in those cases where the apex is the seat of infection which cannot be controlled by medication, or where the form of the apical canals is such that they can neither be prepared nor filled. A single-rooted tooth with an acute distal curve in its apical one-third is an example of this.

It is highly desirable that conditions leading to the necessity for apicectomy be diagnosed at the outset, or at least early in the course of treatment.

On the other hand, increasing experience of root canal therapy and the constant improvement of technique which comes with practice, reduce the need to resort to this operation. It is also true that apicectomy can provide a rapid and satisfactory conclusion to cases which might otherwise prove intractable by more conservative methods.

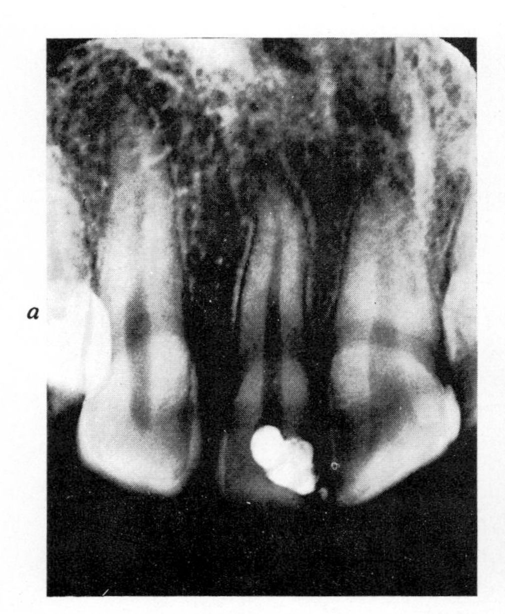

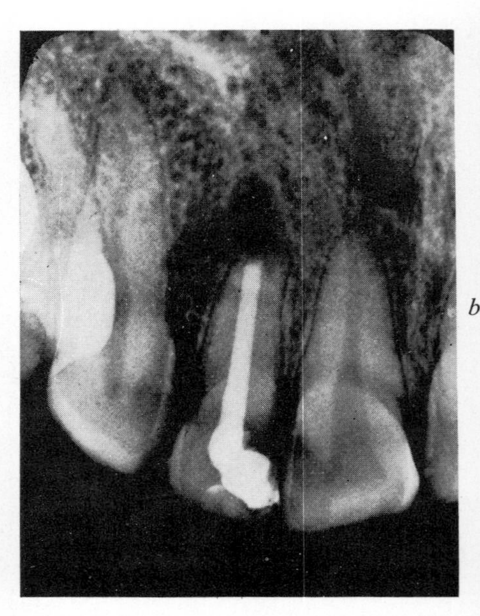

Fig. 13.25. Radiographs of maxillary lateral incisor showing (a) apical curvature and accessory apical canal; (b) condition immediately after root filling and apicectomy.

Summary

Instruments
Broaches, smooth for exploration, barbed for extirpation.
Reamers rotary cutting; files axial abrasion. Hand operated; engine driven. Rubber dam obviates handling risk. Rules for reamers, p. 252. Mechanical preparation most important factor in control of infection.

Access to root canals
Anterior teeth: never cingulum; midway between incisal and cingulum; axis of canal.
Posterior teeth: occlusal surfaces, according to anatomy of pulp chamber.
Cornua to be opened, cleaned. Never approach through Class II, V cavities. Inadequate access causes failure.

Techniques
Vital uninfected pulp. LA, dam. Extirpate, measure, prepare canal, irrigate, dress, seal. Summary, p. 264. Radiograph canal.
Second visit: assessment. Fill canal, matching point; accessory points. Complete. Summary p. 268.
Other root fillings. Conical GP. Conical silver, sectional silver, points.
Infected vital pulp. LA, dam. Cavity excavated and filled.
Access. Extirpation. Bacterial culture, aerobic, anaerobic.
Dress, seal.
Second visit: assessment. Ream, file, irrigate, dress, seal.
Third visit: assessment. Complete preparation, fill canal.
Seal, temporary restoration. Continued infection, review technique, medication, resection if relevant.

Necrotic pulp. No LA, no dam. Canal under pressure, open, drain.
Second visit, minimal interference, insert antiseptic dressing, seal. If resolution delayed, apical drainage, via canal, via mucosa. Consider resection if relevant.
Third visit: Start canal preparation. Dress antiseptic, antibiotic. Canal filled when symptomless.
Alternative medication. Antibiotics, fungicides, antiseptics. Rotation of drugs. Apical periodontitis, infective, drug-induced.

Apicectomy
Resection of apical one quarter of root.
Indications, resistant infection, inaccessible apical canal, accessible apex.
Alveolar approach. Importance impermeable seal at site of resection.

Index

Where more than one page number is given, **bold type** indicates the main reference.